Springer Geophysics

The Springer Geophysics series seeks to publish a broad portfolio of scientific books, aiming at researchers, students, and everyone interested in geophysics. The series includes peer-reviewed monographs, edited volumes, textbooks, and conference proceedings. It covers the entire research area including, but not limited to, applied geophysics, computational geophysics, electrical and electromagnetic geophysics, geodesy, geodynamics, geomagnetism, gravity, lithosphere research, paleomagnetism, planetology, tectonophysics, thermal geophysics, and seismology.

More information about this series at http://www.springer.com/series/10173

Oleg V. Petrov

The Earth's Dissipative Structures

Fundamental Wave Properties of Substance

Springer

Oleg V. Petrov
Russian Geological Research Institute
(VSEGEI)
St. Petersburg, Russia

Translated by
Thomas Alexander Beavitt
Institute of Philosophy and Law
RAS, Ural Branch
Ekaterinburg, Russia

Daria Kiseleva
Institute of Geology and Geochemistry
RAS, Ural Branch
Ekaterinburg, Russia

Maria Streletskaya
Institute of Geology and Geochemistry
RAS, Ural Branch
Ekaterinburg, Russia

Natalia Popova
Department of Foreign Languages
Institute of Philosophy and Law
RAS, Ural Branch
Ekaterinburg, Russia

ISSN 2364-9119 ISSN 2364-9127 (electronic)
Springer Geophysics
ISBN 978-3-319-93613-0 ISBN 978-3-319-93614-7 (eBook)
https://doi.org/10.1007/978-3-319-93614-7

Library of Congress Control Number: 2018961222

Translated from the Russian language edition: *Dissipative Structures of the Earth as Manifestation of Fundamental Wave Properties of Matter* by O. V. Petrov, © Federal State Budgetary Institution A.P. Karpinsky Russian Geological Research Institute, 2007. Published by FGBU VSEGEI Publishing House.

This Springer imprint is published by the registered company Springer Nature Switzerland AG
The registered company address is: Gewerbestrasse 11, 6330 Cham, Switzerland

Foreword

Today, it is becoming increasingly obvious that, following in the wake of the plate-tectonic revolution of the 1960s, geology is on the verge of another major paradigm shift, a new revolution. This revolution is maturing not only under the pressure of the accumulation of new facts—especially the achievements of seismic tomography, which first illuminated the processes going on in the mantle depths up to the boundary of the Earth's core—but also under the influence of a new direction in the natural sciences, whose founder was Ilya Prigogine. Largely thanks to the efforts of such scientists as Yu. M. Pushcharovsky and F. A. Letnikov, representations of synergetics, deterministic chaos, self-organisation, fractality, etc., are beginning to break through into our—for the most part still conservative—science.

These new trends are vividly reflected in the original published work of Oleg V. Petrov. Petrov's global geodynamic concept is based on the application of the fundamental physical principle of gravitational instability to the Earth. His internal, comparatively logical and harmonious concept of the evolution of our planet is constructed on the radical paradigm of internal standing gravity waves, to which he also applies a novel methodological approach. At the same time, his concept does not discard, but rather assimilates the abundant contemporary geological data on the structural and developmental features of continents and oceans and their interpretation in terms of plate and plume tectonics. Clearly, it is not a complete picture and some important aspects remain to be elaborated; nevertheless, a work of this nature is not to be evaluated in negative terms of its lacunae, but rather according to the avenues it opens for future science.

In comparison with mainstream geological and even geophysical literature—both in terms of content and form—Petrov's work may seem unconventional and nonstandard; consequently, it may not be easy for the average reader to grasp the significance of his argument. For this reason, it is difficult to predict the ultimate reach of the ideas expressed in this book. However, it is likely that, if not decisively (which I do not exclude), then at least to a certain extent, they will accelerate the revolutionary transition to a new paradigm. It is for primarily this reason that I commend the publication of the book as altogether timely and useful.

Ekaterinburg, Russia

Victor E. Khain
Academician of the Russian Academy of Sciences

Contents

Abstract

An independent investigation of the natural phenomenon of cellular standing internal gravity waves in hydrodynamics is presented on the example of Bénard cells—the classic example of dissipative structures in thermodynamics. In explaining these phenomena, a unified physical basis of the emerging concept of the wave nature of dissipative processes of the Earth's formation and fractal structuring is generated.

Since Newton and Fourier, it is well known that any wave movement can be represented as an infinite number of wave modes; therefore, it can be supposed that wave processes and arising structures have such noteworthy geometrical and dynamic properties as a fractal or fractional dimension being invariant with regard to the scale. The hierarchies of dissipative Earth structures including their fractal dimension are determined using a wide spectrum of morphometric, geophysical and other data. It is shown that the fractal geometry of these structures is a reflection of the wave properties of matter. On the basis of a study of the wave nature of the spontaneous organisation of density-unstable systems, essentially novel methods for the interpretation of geological and geophysical data are developed, including new three-dimensional geological mapping technologies, palaeogeodynamic reconstruction and predictive models that define the applied aspects of the wave concept. Fractal self-similarity, as a reflection of the fundamental wave properties of matter, reveals itself as the pre-eminent manifestation of structural and functional perfection: the ratio of the whole and its parts is determinative of geological and other natural processes of formation.

The book is intended for a broad audience of geologists and other experts working in the field of Earth sciences.

Reviewers
Doctor of Geological and Mineralogical Sciences I. Abramovitch
Doctor of Geological and Mineralogical Sciences B. Bluman

Introduction

While the establishment of geology as an independent scientific discipline can be attributed to the middle of the eighteenth century, it has undergone a series of fundamental transmutations during the lifetime of the present generation of geologists, the most important of which was the adoption of the mobilistic paradigm, which may justly be considered in terms of a scientific revolution. According to many specialists in the field, the primary outcome of this revolution should be seen in terms of a transformation of the predominantly descriptive geosciences into a unified theoretical discipline, capable of accurately modelling fundamental geological processes according to their interpretation as a consequence of geodynamic events associated with the motion and interaction of lithospheric plates. Despite continuing ambiguities of interpretation and a lack of precise knowledge concerning the true nature of the forces driving these plates, plate-tectonic theory has become highly influential in practically all areas of geological enquiry. The outlines of a significant breakthrough in this area began to emerge following the publication of a series of papers by the Japanese geophysicists Fukao, Maruyama, Obayashi and Inove (1994), in which a link between large tectonic ensembles and mantle heterogeneities detected by seismic tomography was demonstrated. The result was a cohesive conceptualisation, allowing processes of differentiation taking place within the Earth's interior to be linked with those phenomena traditionally studied by geology. Thus, it becomes possible to consistently coordinate mantle plume tectonics with those of the lithospheric plates and we may already refer to the emergence of a unified general geological theory of the development of our planet.

Simultaneously, studies into the physical state of the Earth's interior carried out by means of direct geophysical measurements, experiments at high temperatures and pressures, as well as emerging mathematical modelling methods, have led specialists towards an increasingly broad conclusion concerning the reality of the various manifestations of gravitational instability in the crust and mantle of the Earth. As far as we know, this phenomenon occurs when the density of any particular layer(s) is lower than that of the layers located above. Frequent anomalies in the prevailing tendency in the Earth's interior for density to increase with depth may thus be explained in terms of gravitational instability. One of the reasons for this is the differentiation of the Earth's interior, which is accompanied by the intense emission of heat. This results in a critical vertical temperature gradient and a corresponding inversion of densities. A similar effect is caused by the cooling of the Earth's surface, such that its upper layers become gravitationally unstable with respect to those underlying them (Fisher 1989).

Thus, phenomena associated with the thermogravitational instability of the Earth can be connected both with internal sources of heat and matter, as well as with its external discharges—in particular, the aforementioned cooling of the planet through its surface. It is therefore no accident that the Earth is often compared to a thermal machine (Khain 1995). In consequence, our planet may be correctly designated as a member of the class of open nonequilibrium systems—that is to say, those exchanging energy, matter and information with their surrounding environments.

In keeping with this approach, a particular emphasis is placed on the correlation of the theoretical geodynamic models and their empirical basis in accordance with the laws governing the development of open nonequilibrium systems, for which the appearance of a special kind of structural ordering is characteristic. These structures, which are generated by heat and mass transfer processes, would seem to merit their own specialised study, since diversely reflected in tectogenesis, which, in turn, as it is widely understood, control the most important geological processes, including the formation of mineral deposits. In referring to structures having this origin as dissipative, Prigogine (1986) emphasised the constructive role of dissipative processes in their formation. The list of examples of dissipative structures in natural and experimental contexts include: Bénard cells (alternation of ascending and descending convective currents in a liquid); striations in plasma; the inhomogeneous distribution of concentrations in chemical reactors; Belousov–Zhabotinsky oscillatory reactions; electromagnetic, acoustic and hydrodynamic waves; cirrus clouds; honeycomb and spiral structures of galaxies; morphogenesis in biological processes; and many other phenomena and processes in physics, chemistry, cosmology, biology, ecology and sociology.

The empirically revealed commonality in the description of dissipative structures in various fields of natural science at the beginning of the twentieth century resulted in the creation of synergetics as an interdisciplinary science that studies the paradoxical fact of the self-organisation of natural systems. This science was formed at the junction of the formal sciences and philosophy. However, today, at the beginning of the twenty-first century, a huge empirical base characterising the generality of development processes of dissimilar dissipative structures—at scales ranging from atomic to galactic and including living organisms—gives us reason to assume a fundamental role played by the wave properties of matter in determining its self-organisation. These waves often have remarkable geometric and dynamic properties, whose recognition, even on the example of well-studied objects, is by no means straightforward.

Let us consider the illustrative example of dissipative structures—Bénard cells—that was to serve as the basis for a broad generalisation, leading to an entirely new concept. This concept asserts that if there is a source of growth in entropy (e.g. heating), then the standard state of the medium consists in spatially periodic structures. These phenomena were investigated for the first time by Bénard (1901) on the example of a stationary convective motion, during which the stratification of a medium in a gravitational field generates a specific natural phenomenon—cellular standing internal gravity waves.

Historically, these waves were known before convection. The phenomenon, which was observed by oceanic seafarers in the regions of the confluence of warm and cold currents as long ago as the age of sail in the eighteenth century, represents a very dramatic occurrence where the surface of the ocean breaks up into large cellular structures.

Despite these phenomena having a common physical nature, capable of being described by the same equations (Giterman and Shteinberg 1972; Landau and Lifshits 1988), an appreciation of their explanatory significance was until recently limited by their being studied in divergent fields of natural science separated by chasms of mutual incomprehension. The internal standing gravity waves of the Earth were considered with respect to geological processes for the first time in 1992 (Petrov 1992 a, b). However, it is only today, in purposefully exploring the wave properties of the material world, that we come to understand the relevance of convective processes exemplified by Bénard cells as a special case of superposition of cellular standing internal gravity waves.

The description of convective instability phenomena on the basis of wave propagation theory or with the help of modes explains not only the reason for the spontaneous appearance of the cellular structure when the critical Rayleigh number value is reached, but also the dependence of the dimensions of the cells on the thickness of the nonequilibrium layer. According to the wave dispersion law, this dependence is determined by the dominant wavelength and amplitude increments of the dominant perturbations. As follows from the laws of fluid dynamics, the oscillation frequency of propagating cellular standing internal gravity waves cannot exceed the buoyancy frequency of the medium, which serves as the basis for the transition from gravity waves to structures of density instability—i.e. dissipative structures—or, in other words, resulting in the elimination of boundaries between them. Like all wave-based processes, dissipative structures decompose into modes and therefore possess unusual geometric and dynamic properties and a fractal (or fractional) dimension.

Having considered dissipative structures on the classical example of Bénard cells, we come to the following conclusions that determine our research topic: the spontaneous dissipative structuredness of nonequilibrium systems in terms of density ratio has a wave nature; the external appearance (prominence) of structures allows the motion of matter inside the system to be predicted; and the size of the structures varies according to the wave dispersion law. Applying these conclusions with respect to the Earth, which in the first approximation and on a geological timescale can be regarded as a density-unstable system, we are breaking new geological ground, in which wave processes are considered to be the root cause of the structuredness of the Earth's crust and mantle. In this connection, the question of the source of internal standing gravity waves, whether it be convection, plume tectonics or some other hypothetical process, loses its significance: wave laws make it possible to establish a connection between heterogeneous density structures and, in so doing, to determine the nature of the motion of matter within the Earth's interior in conformance with the external topographical structuredness of its surface. On this basis, a methodology is developed that allows for the surface forms of the geoid and its topographical relief, along with remote geophysical measurements, to be used to reconstruct the distribution patterns of the deep regions of thermogravitational instability and thus describe the motion dynamics of matter in the interior of our planet. Under these circumstances, the dissipative structures of the Earth themselves become the primary object of study.

It should be noted that the first version of the book was prepared and submitted for publication at the Nauka Publishing House in 1996. At that time, the preface was written by the RAS Academician Viktor Efimovich Khain. The author would like to take this opportunity to express his sincere gratitude to him for this, as well as for the expressed high regard to the practical significance of the concept of the Earth's internal gravity waves for understanding the fractal structure of the Earth's crust and lithosphere and the existence of a global regmatic network of faults and planetary fracturing. In the same year, the author took the decision to revise a number of sections. Unfortunately, for various reasons, this process of "revision" dragged on for more than ten years. During this time, the chapter entitled "Physical and mathematical description of the manifestation of the wave properties of matter…" was substantially updated. Moreover, an additional chapter based on a number of articles published jointly with I. B. Movchan (2000–2007) appeared under the title "Applied aspects of the theory of the dissipative structuring of a nonequilibrium geological environment". The author is sincerely grateful to I. B. Movchan for his invaluable contribution and in the general preparation of the work for publication.

Over the last 10 years since *The Earth's Dissipative Structures—Fundamental Wave Properties of Substance* had been published, the following papers were devoted to studying the wave structuring of natural systems: Shvartsev (2007), Chengzhi et al. (2008), Kopylov (2018), Nazimko and Zakharova (2017), Peters et al. (2016) and Shuman (2016). These publications revealed high significance of the investigations' results, connected with fundamental wave properties of natural systems, provided in the book.

References

Bénard H. Les tourbillons cellulaires dans une nappe liquide transportant de la chaleur par convection en regime permanent. Ann. Chem. Phys. 1901; 23:62–144.

Chengzhi C., Minyan V., Zihu Ts., Jianjie C. Structural rock hierarchy and mechanisms of its formation // Moscow University Vestnik. Series 1: Mathematics. Mecanics. 2008. No 5. pp. 31–39.

Fisher O. Physics of the Earth crust. 2nd ed. London; 1989.

Fukao Y., Maruyama S., Obayashi M., Inove H. Geologic implication of the whole mantle P-wave tomography. J. Geol. Soc. Japan. 1994; 100(1):4–23.

Giterman M.Sh., Shteinberg V.A. Izv. AN SSSR. MZhG. 1972; 2:55–61. (In Russ.).

Khain V.E. The main i of modern geology (the geology on the threshold of the 21st century). M.: Nauka; 1995. (In Russ.).

Kopylov A.L. About dissipative structures of the Earth / Concepts of fundamentals and applied studies/ Collection of papers of the International Research-to-Practice conference. - Ufa, 2018. pp. 178–181.

Landau L.D., Lifshits E.M. Theoretical physics. Fluid dynamics. M.: Nauka; 1988. (In Russ.).

Nazimko V., Zakharova L. Cluster behaviour of the ground during its irreversible movement // Acta Geodynamica et Geomaterialia. - 2017. – Vol.14. – N 4. – P. 445–462.

Peters M., Herwegh M., Paesold M.K. Boudinage and folding as an energy instability in ductile deformation // Journal of Geophysical Research-Solid Earth. - 2016. – Vol.121. – N 5. – P.3996–4013.

Petrov O.V. Nonlinear phenomena of thermogravitational instability and internal gravitational waves of the Earth. Dokl. RAN. 1992b; 326(3): 506–509. (In Russ.).

Petrov O.V. Internal gravitational waves of the Earth and nonlinear paleogeodynamic dissipative structures. Dokl. RAN. 1992a; 326(2):323–326. (In Russ.).

Poincaré H. About Science. – M.: Nauka; 1983.

Shuman V.N. Fractional dynamics and emission activity of geosystems // Geophysical Journal. 2016. Vol. 38. N 3. P. 72–83.

Shvartsev S.L. Progressive self-organizing abiogenic dissipative structures in the geological history of the Earth // The Lithosphere. 2007. 1. P. 65–89.

Chapter 1
Generality of the Wave Property Manifestation of Matter in Processes of Fractal Structuring and Generation of the Geometry of Density-Unstable Natural Formations

> Fractal structure is the universal ability of matter in nature to manifest wave properties, with the predominance of objects having a fractional dimension. The primary characteristic of the latter, as for a set of modes in wave processes, is self-similarity, i.e., the similarity of parts to the whole.

The contemporary scientific discovery of fractal hierarchies of dissipative structures—cellular standing internal gravity waves of the Earth—has, on the one hand, been seen in thermodynamic terms as thermogravitational (convective) instability phenomena, and, on the other, as the related natural phenomena of cellular standing internal gravity waves, studied independently and in parallel in terms of classic Newtonian mechanics, fluid dynamics and meteorology. In historical terms, cellular standing internal gravity waves were already being studied before convection had become a topic of scientific interest, with some of the major results of the former being obtained as far back as 1847 (Stokes 1847). From a physical standpoint, cellular standing internal gravity waves consist of fluctuations of a density-unstable medium in a gravitational field. If a volume element of such a medium is displaced from the equilibrium position, for example, as a result of a mechanical action or by means of thermogravitational instability due to heating from below, the buoyant force will cause its movement opposite to the equilibrium position. During such a movement, the stratification of the medium in the gravitational field generates a specific natural phenomenon, which in fluid dynamics is referred to as cellular standing internal gravity waves. Along with those of convective heat and mass transfer, these phenomena are connected with the fact that cold density-unstable layers descend under the impulse of the gravitational force, thus displacing upwards the lighter layers located beneath. In such cases where a medium is unstable in density terms, its entire volume, like its surface, breaks down into linear or cellular structures, i.e. standing internal gravity waves that are spatially defined by wave-like displacement lines and characterised by a certain amplitude and corresponding wavelength relative to the wave process.

Thus, it turns out that convection and cellular standing internal gravity waves have a common physical wave nature and, since gravity is a common natural

O. V. Petrov, *The Earth's Dissipative Structures*, Springer Geophysics,
https://doi.org/10.1007/978-3-319-93614-7_1

cosmic force to which all substances in motion must conform, these phenomena occur throughout nature. Gravitational equilibrium is only possible when underlying layers have a higher density than those layers overlying them. However, this condition is constantly violated, both within the earth's crust, as well as in other geospheres, the hydrosphere and the surrounding atmosphere. In this case, density inversion takes place when the underlying layer is less dense than the overlying one. Such an unstable state is referred to in terms of the so-called Rayleigh-Taylor instability, which leads to the formation of characteristic structural forms of gravitational (convective) instability or cellular standing internal gravity waves of types G, L-hexagons and Rolls. All these structural forms and typomorphic structural features have been repeatedly described in terms of natural formations, modeled in laboratories and theoretically studied by many authors, who, however, did not always refer to their common wave nature. Therefore, in this paper, special attention is paid to the following circumstance. Since all wave processes can be decomposed into modes, any structures formed on their basis must necessarily have unusual geometric and dynamic properties expressed in terms of a fractal (or fractional) dimension, which, like a set of modes, repeats itself on different scales. On the other hand, by taking this approach, due to the presence of the fractal dimension and the hierarchy of self-similar structures in natural formations serving as direct proof of the manifestation of the fundamental role of wave properties of matter in formation and structuration processes, it also becomes possible to solve the inverse problem.

With the aim of supporting the underlying rationale for this proposition, this chapter provides an overview of the relevant physical and mathematical models characterising this commonality at the level of physical and mathematical methods for describing natural phenomena. Up until now, due to these studies having developed in parallel in various fields of natural science, for example, in hydrodynamics and thermodynamics, oceanology and meteorology, geophysics and geology, chemistry and biology, astrophysics and quantum mechanics, but without making the necessary interconnections between them, knowledge of results pointing to their underlying wave nature obtained in one area of natural science have been little known to researchers working in other areas. Moreover, the question of the manifestation of the wave properties of matter turns out to be based on a very profound insight. Let us recall how scientists in the 19th century were struck by the presence of the wave properties of light, which had previously been imagined as a stream of discrete particles, and the concomitant profound revision of the fundamental concepts of physics that led in the 20th century to the development of quantum mechanics. It was with good reason that the brilliant French mathematician and physicist Henri Poincaré warned the scientific community in 1893 that the discrete and continuous properties of matter in confluent continuous media cannot be contrasted with each other due to their being inextricably linked. A lack of understanding of this fact in modern natural science led to the deep crisis between classical Newtonian dynamics and thermodynamics when describing the same natural phenomena.

Among the many structural features appearing throughout the Universe, including the Earth and its surrounding layers, the attention of contemporary scientists is drawn to the spontaneous structuring of these natural systems and the geometric commonality of the emerging structural forms, i.e. in terms of their cellular (square, hexagonal, polygonal) geometry. The structural similarity of these forms is revealed to be independent of the nature and scale of their manifestation, a factor having already been emphasised by many researchers from a philosophical perspective (V. V. Piotrovsky, M. A. Sadovsky, Yu. M. Pushcharovsky, V. E. Khain, L. I. Krasny, etc.). This has led the scientific community towards the view that the external manifestations of the spontaneous structuring and redistribution of masses in nonequilibrium density distribution in natural systems may be explained in terms of a transphenomenal physical mechanism. This perception is strengthened still further by turning to the applied natural scientific fields of fluid dynamics and thermodynamics, oceanology, meteorology and geophysics, where the same geometric commonality of spontaneous structuring can be seen as underlying otherwise apparently scattered examples. Without getting into specifics, let us consider some cases that serve to illustrate this commonality.

The natural phenomena of spontaneous structuring and density instability in the upper layers of the Earth's crust have been observed by the author in the Arctic latitudes over the course of many years. The dimensions of cellular and wave-like density-unstable structures in permafrost rocks range from tens of centimetres to tens or even hundreds of metres, typically forming the relief of a spotted or polygonal tundra. As often as not, small lakes and the first-order drainage network are confined to these structural forms. Similar structural forms of density instability are also widespread in other climatic regions and are associated, in general, with the unstable location of plastic, water-saturated less dense clay layers and denser sandy ones (Artyushkov 1964, 1965; Kostyaev 1964). The incipience of instability over a large area leads to a polygonal structuredness in the movement of sedimentary rock masses. In general, the length of the structure is several times greater than the thickness of the layers participating in the movement. The properties of the ground can vary significantly in lateral terms, gradually averaging out with an increase in the ground thickness. Therefore, larger structures have geometrically more regular forms.

Artyushkov (1964) considered the dependence of the occurrence of various forms of gravitationally unstable structures on the viscosity ratios of under- and overlying sedimentary rocks. In cases where the underlying rock has a lower viscosity, each structural form consists of a central mound and marginal downthrows (G-form). Conversely, with a higher viscosity of the lower sedimentary layer, the structural shape is represented in the centre by a downfold with culminations in the marginal parts (L-form).

Similar wave-like and cellular structures of gravitational instability in subsoils were established by us in the Norilsk region within the Kharaelakh plateau (Fig. 1.1a, b) Notably, L- and G-type formations are also among the cellular structures in this location. It is noticed that the large cells, in turn, break down into smaller ones.

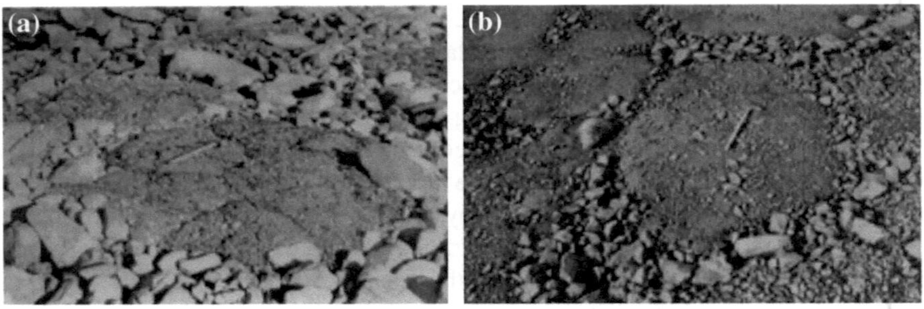

Fig. 1.1 Polygonal forms in the ground. Structures of subsidence along the edges of cells as a result of the appearance of gravitational instability in soils. Kharaelakh Plateau, River Skalistaya. Photo by the author. **a** large-scale structure, **b** fragment thereof

Fig. 1.2 Cellular morphoblocks of lava plateaus. River Talnakh valley. Materials obtained from a geological expedition to Norilisk

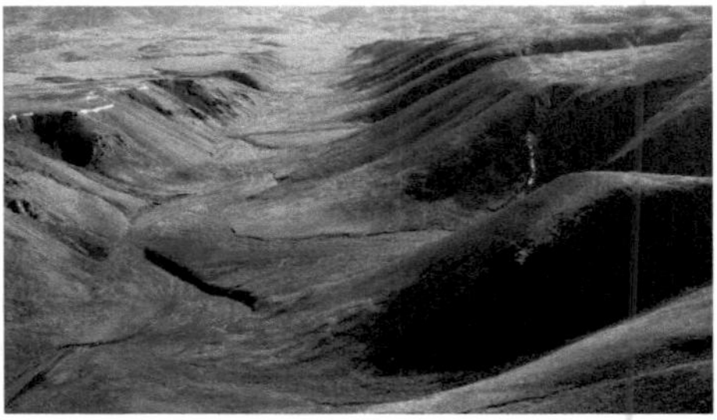

Within the Siberian trappean plateaus, including the Norilsk region, the gravitational instability of the upper layers of the Earth's crust is manifested with particular clarity (Fig. 1.2). Here, sheets of heavy basalts (with high bulk density) up to 3–5 km in thickness superpose the underlying Paleozoic rock formations, which have a significantly lower bulk density. In the composition of the latter, along with sandstones, are present argillites and aleurolites, limestones, anhydrites, marls and dolomites, in which plastic flow attributes are defined.

The area of the cellular structures of gravitational instability, as shown in Fig. 1.2, varies from 0.5 to 5 km^2. In the upper part of the earth's crust, the inversion of densities manifests itself, for example, in the formation of salt and clay diapir domes and other diapiric folds. As Belousov (1966) noted, diapir domes scattered over a wide area can be compared with rising fluid flows in Bénard cells. The difference consists in the fact that in experiments involving fluids the density decreases gradually downwards, whereas in a geological environment there are two layers having different densities and a sharp boundary between them. Such density instabilities in nature are often referred to in terms of a Rayleigh-Taylor instability.

The incipient formation of the Rayleigh-Taylor instability under laboratory conditions is considered mainly in connection with the study of convection phenomena in liquids and gases. While the states for Rayleigh-Taylor instability and stationary convection are slightly different, the motions occurring in both cases are

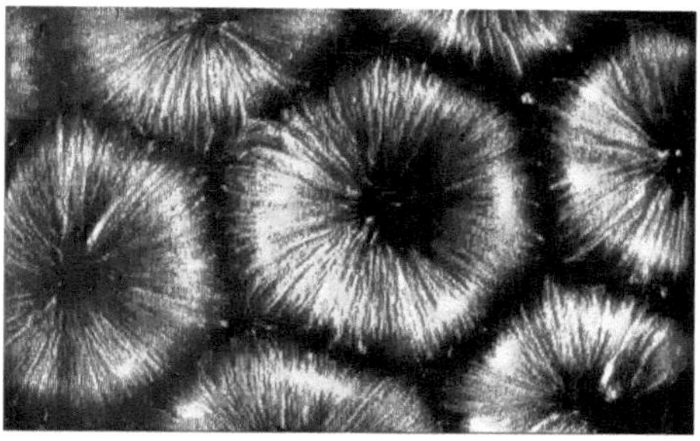

Fig. 1.3 Hexagonal convection structures. Motions of by convective heat transfer in a thin liquid layer. Photographs by Bénard (1901)

so similar in principle that convection can serve as an approximate model of what should happen under the Rayleigh-Taylor instability. It has been established experimentally that stationary convection develops in a layer of liquid heated from below once a certain critical temperature gradient is exceeded. The entire liquid breaks up into hexahedral prismatic cells, with a continuous upwards motion of the liquid taking place in the centre, while along the edges the motion proceeds in a downwards direction (Fig. 1.3a, b).

This phenomenon, as we have already noted, has a simple physical explanation. The lower layers of the liquid are heated due to thermal conductivity resulting in a decrease in their specific density. At the same time, the liquid at the surface releases its heat and its specific density increases. The cold layers descend under gravity, displacing the lower, lighter layers upwards. After a while, the lower layers heat up, become lighter and are interchanged; again, as the mathematical calculation shows, the same particle moves along a closed trajectory, repeatedly absorbing heat from the lower plate and transmitting it to the upper one. Due to the action of the viscous force, this motion occurs only at a certain critical value of the temperature, which depends on the thermal conductivity of the liquid, its viscosity, and the distance between the plates.

Since a liquid is practically incompressible, its movement in one place must cause a movement in the opposite direction within nearby zones. This interrelatedness leads to the formation of some independent structures of movement—cells that border with other independent structures of exactly the same kind. Due to its viscosity, the entire mass of liquid is involved in the motion and there are no regions where it is likely to be stationary. For this reason, the cells must fit tightly to each other, which is only possible if they are formed into trihedral, tetrahedral or hexagonal prisms. The liquid has an ascending flow at the centre of the cell and a descending one at the edges. In gases, the motion takes place in the opposite direction. The difference in the direction of motion in liquids and gases is explained by the characteristic dependence of a liquid's viscosity on its temperature (Palm 1960). As shown in Fig. 1.3, the cell boundaries are slightly blurred due to the weak inhomogeneities always present in the liquid, at least those due to the thermal motion of the molecules. Not only are the direction and spatial position of the planes arbitrary, they also depend on random inhomogeneities presented in the liquid at the beginning of the motion.

◂**Fig. 1.4** Topographic distortions of the open surface of a convective layer having a slight planar inclination. The symmetrical distortion in concaves and their framing ridges is noted (Bénard 1901)

Bénard, who used solid particles in a suspension to make the internal circulation visible, was thus able to study the topography of the exposed surface in detail, establishing a number of regularities that are important for understanding the typomorphic elements of the structure of natural formations. Thus, he noted that the concave surface is formed by the central part in each individual section of the circulation, i.e. each centre of internal ascent corresponds to the centre of descent of the exposed surface. According to Bénard, it is precisely this regularity that can be anticipated: the centripetal flow of the lower layers of the section can only be carried out by means of an effective driving pressure. It is the presence of a depression in the centre of the section that also creates an excess of pressure on the watershed lines separating the concave surfaces of the depressions from each other. From this observation follows the important conclusion that the outer topographic surface and the internal structure of the convective cell are paragenetically associated into the conformal system.

Bénard (1901) also gives a topographic description of the exposed surface in cases of a small (1°–3°) slope of the convection plane (Fig. 1.4). At the same time, he notes a defect in the symmetry, expressed in the shift of the centre of the sections to the peripheral parts of hexagonal cells. Arched (elevated) sections of watersheds (ridges), which separate one depression from the other, acquire a very characteristic arcuate shape. The concave curve of the ridge is weakly expressed; conversely, the convex frontal zone is manifested very contrastingly. As follows from Fig. 1.4, the topographic surfaces of convective structures that occur in this case acquire typomorphic features, which can also be used in the identification and study of natural formations.

In later experimental studies, hexagonal convection cells and convective rolls were obtained on the free surface of a silicone oil (Koschmieder 1967; Silverston 1958). The character of the manifestation and correlation of the roll-like and cellular convective structures has been studied. Important results for the interpretation of natural phenomena were obtained by Ramberg (1968, 1972), who was the first to use a centrifuge in modelling tectonic deformations caused by gravity. Numerous results on the modelling of salt structures and gneiss domes in laboratory were also obtained by Lebedeva (1956), Parker and McDowell (1955), Sycheva-Mikhailova (1973), Belousov et al. (1988), Dixon (1975), Talbot (1977) and Schwerdtner et al. (1978). More complex characteristic arched uplifts were obtained in similar laboratory tests (Schwerdtner et al. 1978).

Lord Rayleigh (1916) first considered the problem of convective instability of a horizontal fluid layer with free boundaries, in which there is a vertical superadiabatic temperature gradient (the Rayleigh-Bénard problem), and determined the convective stability threshold, also known as the dimensionless Rayleigh number (Ra).

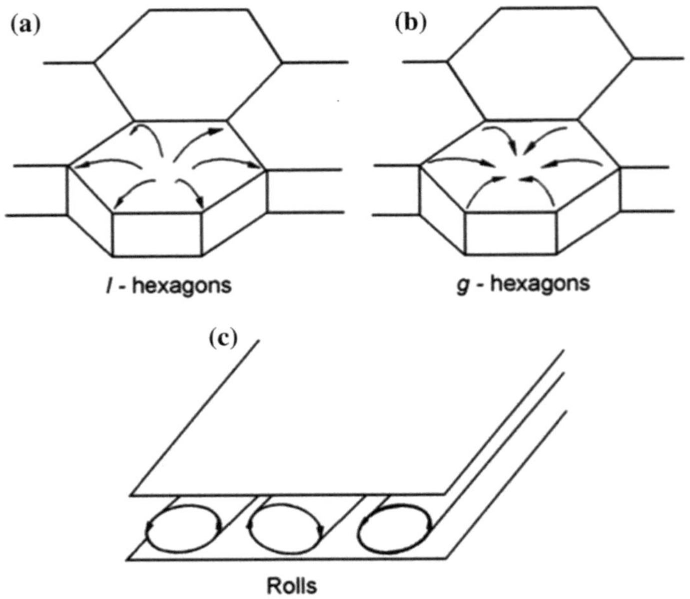

Fig. 1.5 Convective instability in the form of L- and G-type hexagonal cells (**a**, **b**) and two-dimensional rolls (**c**)

In liquids having lower viscosity at the bottom, upward motion occurs in the centre of the hexagonal cell—L-hexagons (Fig. 1.5a)—while in gases for which the viscosity increases with temperature, the convection cells show a downward movement at the centre—G-hexagons (Fig. 1.5b). An impressive demonstration of this effect was given by Tippelskirch (1956), who performed experiments with liquid sulphur, which manifests both an increasing and decreasing dependence of viscosity on temperature.

The effects of a medium's temperature dependence on viscosity is analogous to the influence of the asymmetry effect (Busse 1962, 1978). Depending on the asymmetry sign (ε), hexagonal cells either optimally manifest buoying in the centre and subduction in the edge parts according to the L-hexagon type or subduction in the centre and buoying in the edge parts, typical of the G-hexagon type (Fig. 1.6). It is worth noting that there is always an area where both of these types of hexagonal cells are stable with two-dimensional rolls. With a more detailed theoretical analysis, it has been possible to establish that the paragenetic connection between the external topographic surface and the inner depth horizons of the thermogravitational (convective) instability structures, to which Bénard pointed in his experiments, stems from the relationship between the nature of the displacements of various internal and external discontinuities and streamlines.

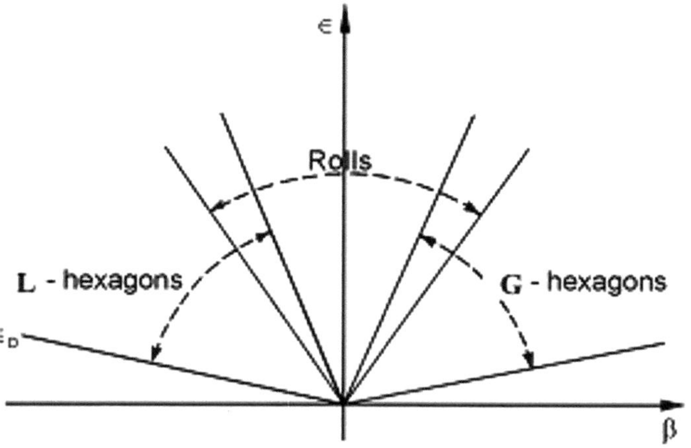

Fig. 1.6 Stable areas of convective instability in the form of rolls, L- and G-type hexagonal cells as a function of amplitude ε and asymmetry factor β

The spontaneous elevation of buoyant domes in a denser host medium under the influence of gravity has been theoretically investigated by many authors. In this regard, salt diapirism was considered by Danes (1964), Selig (1965), Biot and Ode (1965), while gneissic basement doming was analysed in the works of Ramberg (1968, 1972) and Fletcher (1972). All these authors based their calculations on models consisting of a number of horizontal layers having a given thickness, density and Newtonian viscosity. One of the boundaries between the layers separates the upper denser material from the less dense substrate. It is at this boundary that a superposition of sinusoidal perturbations of different wavelengths is designated. Each perturbation will increase with time due to the presence of density inversion; the relative increase of the amplitudes of these wavelengths can be determined by mathematical analysis. The system is dominated by perturbations that grow at maximal speed, with high-amplitude uplifts or domes developing from them; their characteristic dimensions are referred to as the dominant wavelength. Similar morphological types of cellular distribution of salt structures are unambiguously evident on the paleogeologic map showing the Peri-Caspian depression with the Upper Pliocene and Quaternary sedimentary strata removed (Fig. 1.7).

The dominant wavelength and amplitude increment of the dominant perturbations depend on the thickness, viscosity and density of the layers. The influence of all these parameters was considered in detail by Biot and Ode (1965) and Ramberg (1968). Given the characteristics of the lower layer having a reduced density, a change in the parameters of the upper layers leads to the following qualitative effects: (1) an increase in the relative thickness of the upper layer causes an increase in the dominant wavelength and the development rate of the perturbation; (2) an increase in viscosity (relative to the same characteristic of the lower layer) determines a larger dominant wavelength and a lower development rate of the

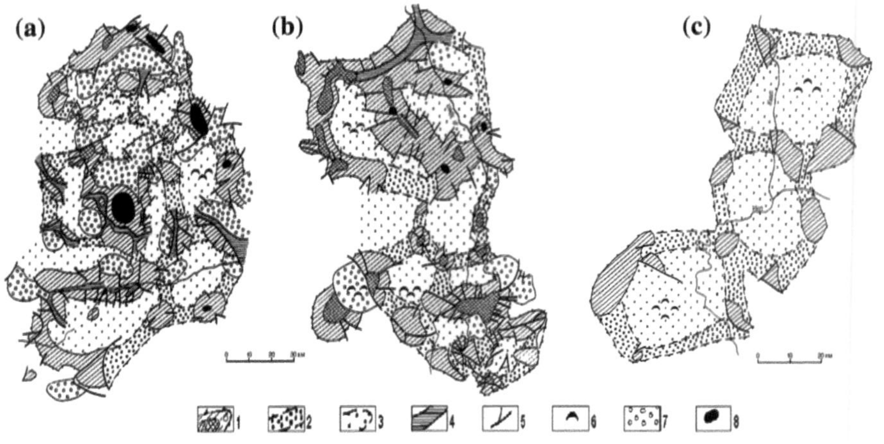

Fig. 1.7 Tectonic schemes of salt domes located on the surface of Cis-Upper-Pliocene section of the Cis-Caspian depression interior part (according to Sychyova-Mikhailova 1973). **a** Central zone, **b** Khobrin and central zones, **c** Khobrin zone. (1) salt domes (a) pierced, (b) subsurface; (2) salt tongues; (3) primary compensation moulds; (4) salt dome grabens; (5) faultings; (6) second generation domes; (7) over-arch subsidence moulds

perturbation; (3) an increase in the difference in densities hardly affects the wavelength, but leads to a significant increase in the amplitude. Ramberg (1972) applied this mathematical model to the analysis of the dome elevation of a granite basement. Figure 1.8 schematically presents a system of four layers simulating the real structure together with an indication of the corresponding dimensionless parameters. The solution obtained by Ramberg for this system yields the dimensionless amplitude factor as a function of the wavelength χ, expressed as the ratio of λ to the layer thickness h. The maximum indicated on the curve corresponds to the dominant wavelength (referring to the perturbations having the fastest development), i.e. $\lambda\Pi = 4.485$ and $h = 22.4$ km. These values are in excellent agreement with the mean intradome distance observed in many provinces of subsurface gneiss domes (Fig. 1.8).

The analysis of Ramberg's dominant wavelengths calculations (1972) carried out by John M. Dickson in his book "Structural Geology and Plate Tectonics" (Dixon 1991) draws our attention to two points. First, these results are obtained with the assumption of homogeneous horizontally located layers between whose boundaries small perturbations are experienced. Large inhomogeneities in the relief, density or viscosity can also lead to the uplift of domes in interaction with the process of selecting natural wavelengths. Secondly, the function passes through a very wide maximum: in a λ range from 3.25 to 5.9, which corresponds to λ values of 16 and 30 km in the considered example, with the χ values deviating from the maximum by no more than 10%. Considering all this, it is not surprising to observe both a significant scatter in the distances between the domes within different gneiss provinces, as well as the fact that the dominant wavelength or the preferred

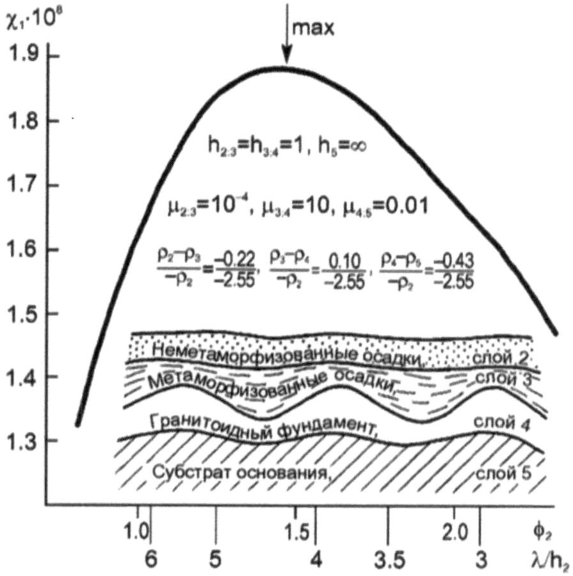

Fig. 1.8 χ function (amplitude multiplier) characterising granitic basement uplifting into denser metasediments. Dimensionless parameters and the layout of layers are indicated (Ramberg 1972)

intradome distance are not particularly sensitive functions of the physical properties characterising the system of layers.

The wave nature of the mass redistribution in the nonequilibrium distribution of densities and spontaneous structuring in the upper horizons of the Earth's crust becomes even more evident when considering the same phenomena in liquid and gaseous media such as the atmosphere, oceans and lakes characterised by stable stratification (Turner 1977; Gossard and Hook 1978). The phenomena of density instability in liquid media were observed for the first time in oceans near the confluence of warm and cold currents as long ago as the age of sail. These natural phenomena have a very impressive appearance. The surface of the ocean breaks up into large cellular structures. These phenomena, as well as those mentioned above, are connected with the descent of cold layers under the effect of gravity, which displaces the lower, lighter layers in an upwards direction. During such a movement, the stratification of the medium in the gravitational field generates a specific natural phenomenon, which, in this field of natural science, is referred to in terms of cellular standing internal gravity waves. On the surface of the sea, the accumulation of contaminants (or other processes that change reflectivity during the convergence) can create visible cells, stripes or flare spots, which indicate the presence of internal waves (Fig. 1.9).

The further development of the theory of internal waves is more due to meteorology than to the study of seas or lakes, probably because of directly observable, often very obvious effects caused by gravitational waves in the atmosphere (Fig. 1.10).

Fig. 1.9 Sea-surface slicks formed by internal waves (by La Fond 1961)

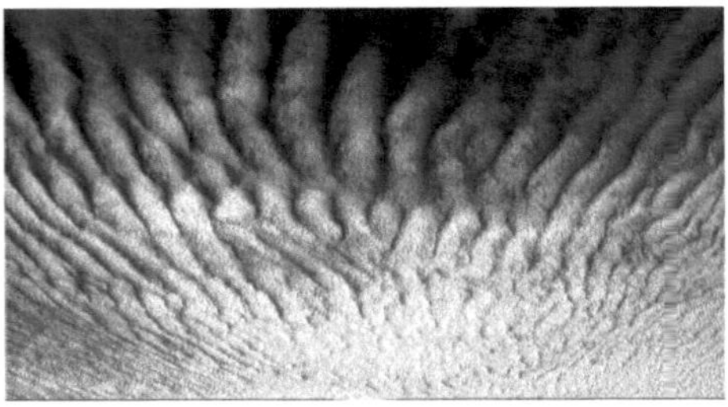

Fig. 1.10 Convective clouds in an unstable layer, elongated in lines along the direction of the velocity shift of stratified air currents (Turner 1977)

The theoretical studies of gravitational instability and convective motions in the upper mantle and Earth's deep interior became especially prominent since the late 1960s and early 1970s in connection with the problem of the mechanism of plate tectonic movement. In the works of Artyushkov (1979), Golitsyn (1980), Zharikov (1983), Myasnikov and Faddeev (1980), Pushcharovskii et al. (1989, 1990), Ramberg (1976), Seyfert (1990), Sorokhtin and Ushakov (1991), Turcotte and Schubert (1985), Fisher (1989), Gridds (1939), Hess (1962), Holmes (1928), McKenzie (1983), Kanla (1972) the development of thermal, chemical-density, gravitational, thermodiffusion and other types of convection is substantiated.

In addition to the model of mantle convection, a two-layered convection model is currently under active consideration, which occurs separately in the lower and upper mantle, as well as the multi-layered, subordinate stratification of cratons composed of Archean rocks, in which granite-gneiss domes and swells are developing. In the opinion of Belousov, granite-gneiss domes and swells are the main structural forms of the Archean eon. These structural forms were studied in the most detail in the Baltic and Canadian shields (Eskola 1949; Kranck 1954; Väyrynen 1959; Belk 1946; Buddington 1959).

Granite-gneiss domes occur in round, oval or irregular shapes. Their dimensions are very diverse, ranging from several kilometres to many tens of kilometres in diameter. The swells extend for tens of kilometres and have a width of several kilometres. Wegmann (1935) was the first to point out the similarity of granite-gneiss domes to diapirs.

In 1960, V. V. Belousov suggested (supported by many other researchers) that an important role is played in tectonic processes by the rising to the surface of crust material that happens to be less dense than the overlying rocks for one reason or another. He refers to this process as advection. In his opinion, for granitised complexes, the cause of density inversion could be the processes of granitisation itself (Belousov 1960) or metamorphic dehydration (Goncharov 1979), which in the lower parts of the crust transformed the heavier initial basic rocks into more acidic and lighter ones.

There is good reason to suppose that the process of elevation of granitised material was widely distributed during the early Precambrian (Ez 1976). In the opinion of V. V. Ez, the process of advection is evidenced by a number of geological data: the successive superposition of a different types of deformations; the connection of regional metamorphism with tectonic deformations; the emergence at the surface of deep metamorphic formations over a huge area in the absence of accumulation in the corresponding quantity of erosion products of the overlying strata; the abundance of granitisation products among metamorphic granulite and amphibolite facies now emerging at the surface; the regular change in the temperature gradient characterising the conditions of metamorphism of the same rock volumes: from the lower, appearing during descending movements, to the higher, corresponding to the ascending movements.

An unconventional approach to isolating the basic structural forms of gravitational (convective) instability, based almost entirely on the basis of geological data, was proposed by Suvorov (1978). Instead of the usual geosynclinal and platform areas, thalassocratons, etc., this author singles out tectonic pairs, by which "systems of two equal, spatially and paragenetically interrelated structural elements characterised by polar different dynamo-kinematic parameters" are referred. Each tectonic pair in this case consists of a swell and an isometric depression adjacent to it from the inside (Figs. 1.11 and 1.12). Three main kinds of tectonic pair are distinguished: continental, transitional and oceanic. The main properties of tectonic pairs are conjugate uplifts and depressions, synchronous growth, a thickened core and thin sedimentary layer on the uplifts, squeezed linear folds and shear-thrusts in uplifts with a mosaic structure of faults and expansions in depressions, predominance of dynamic metamorphism and intrusions in the case of intensified effusive activity in depressions.

Suvorov (1976, 1978) distinguishes tectonic pairs on the Precambrian platforms (the Baltic-Ukrainian uplift—Peri-Caspian depression), in the Caledonides (Kokchetav-Ulutau-Northern Tien Shan Uplift—Dzhungar-Balkhash depression—Fig. 1.11), in the Mesozoides (Verkhoyansk-Verkhnekolymsk uplift—Kolyma-Omolon depression, Fig. 1.12) and Cenozoids in the continent—ocean transition zone [island arcs, with trenches—depressions of marginal seas (Pushcharovskii 1992)]. In the volume "Tectonics and Geodynamics" [Encyclopaedic reference book "Planet Earth" (2004)], referring to the concepts of Yu. M. Pushcharovskii, he notes that similar, but larger geodynamic combinations are possible in the mantle layers of the Earth along their entire section. These concepts develop the renowned tectonic school of the Geological Institute of the

Fig. 1.11 Paleozoic regional tectonic pairs of Central Kazakhstan (according to Suvorov 1978). Kazakh-Northern Tian Shan tectonic pair (Middle Paleozoic). (1) frontal uplift; (2) back depression: Atasu-Balkhash tectonic pair (Late Paleozoic); (3) frontal uplift; (4) back depression; (5) thrusts; (6) shears; (7) buried faults; (8) faults without subdivisions. (Faults; BA—Barlykol'-Atansorsky; Ak—Akkanburluksky; Dk—Dzharkainagachsky; Zu—Zapadno-Ulutausky; K—Kaindinsky; GK—Glavny Kartausky and others {Legends on Fig. 1.11, top to bottom: Karaganda, Lake Balkhash, Lake Issyk Kul)}

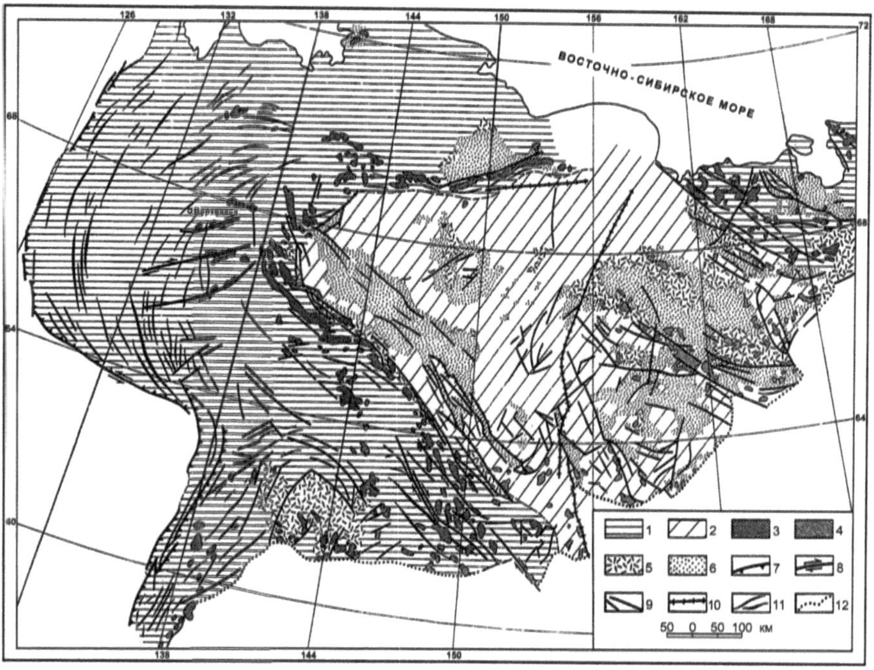

Fig. 1.12 Schema of Verkhoyano-Kolyma tectonic pair. Compiled according to the North-Eastern USSR tectonic map (V. F. Belyi, A. A. Nikolaevsky, S. M. Tilman, N. A. Shilo, 1965–1969) with amendments and facilitations by Suvorov (1978). (1) deep compression and uplift predominating zone; (2) tension and subsidence predominating zone; (3) orogenic belt granitoids; (4) depression zone granitoids; (5) volcanogenic deposits; (6) sedimentary and tufogenic-sedimentary deposits; (7) thrusts; (8) shears; (9) faults without subdivisions; (10) inferred thrust faults; (11) anticlinal axes of Verkhoyansk compex; (12) the border of Okhotsko-Chukotsky volcanic belt (Translation of Russian inscription: East Siberian Sea)

Russian Academy of Sciences and link the formation of tectonic pairs with the tectonic stratification of the lithosphere and the irregular and uneven horizontal movements of depth masses that occur in it: an outflow from under depressions, followed by the formation of the mosaic of stretching and subsidence structures in the upper horizons of the Earth's crust and the injection of depth masses into arcuate uplifts, where a system of various structures of prevailing compression is localised (Pushcharovskii et al. 1990; Pushcharovskii 2000; Suvorov 1978, 2001). According to Suvorov (1978), "compressive stresses in earthquake foci are most often oriented across the trend of regional tectonic pairs (Pacific Ring, Caucasus, Pamir, etc.). Consequently, it is evident that the arcuate uplifts in the most recent tectonic pairs are indeed horizontal compression structures." Suvorov assumes that "a relatively small depth of earthquake foci indicates the compression of the uppermost lithosphere itself, which leads to an increase in uplifts", and, moreover, that: "… the horizontal extension of the upper parts of the lithosphere (obviously, above the foci of deep-focus earthquakes that appear here) should be considered as the basis of the formation mechanism of rear depressions, which was accompanied by their subsidence, fragmentation, further stripping, and the formation, eventually, of simple mosaic-block structures."

A similar parabolic distribution of the active new faults, epicentres of earthquakes and uplifts of ridges, associated with thermogravitational instability phenomena (mantle plumes), was established by Rasskazov (1994).

As follows from the theoretical studies, despite the fact that gravitational instability occurs in various natural media in case of a decrease in the density of underlying horizontal layers with depth, this condition is not necessary for the Earth's interior. In most of the Earth's mantle, density increases with depth; nevertheless, scientists are convinced that thermogravitational (convective) motions in the Earth's mantle exist (Gridds 1939; McKenzie 1983). On the one hand, a necessary condition for the occurrence of these movements is the superadiabatic temperature gradient in the mantle (Elsasser 1972), i.e., the temperature inside the Earth must grow faster than if its growth is determined only by an increase of pressure with depth. On the other hand, the Earth's thermogravitational instability phenomena can also be caused by its nonequilibrium cooling through the surface under conditions of decreasing heat flux; as a result, the overlying formations, following cooling, become unstable with respect to those underlying them (Fisher 1989).

This last circumstance when considering the geological life of planet Earth in its entirety can be of decisive importance, since, as the Academician Khain (1994) notes, "the perennial cooling of the Earth in connection with the loss of heat stored in the era of its accretion, the decrease in the rates of deep differentiation, the gradual depletion of radioactive elements, the diminishing intensity of the lunar tides—is a proven fact."

Thus, terrestrial thermogravitational (convective) instability phenomena can be related both to internal sources of heat: (energy of decay of radioactive isotopes; gravitational energy released during the formation and growth of the Earth's core; internal energy of tidal interaction with the Moon; energy of chemical-density, crystallisational, thermodiffusional and other processes), and external heat sinks (non-equilibrium cooling of the Earth through its surface), which are typical for so-called "open" nonequilibrium systems, capable, unlike closed ones, of exchanging energy, matter or information with their environment. The appearance of these nonlinear terms (sources and sinks of heat) results in the need for a new approach for describing the entire range of chaotic terrestrial thermogravitational instability phenomena. Such an approach to the problem under consideration also leads to the emergence of a new aspect of geological research, encompassing geodynamics (Pushcharovskii et al. 1990), nonlinear metallogeny (Shcheglov 1983, 1995) and nonlinear geophysics (Kuznetsov 1981). In the field of tectonic studies this new line of research was first put forward by Pushcharovsky (1990, 2005) on the basis of data analysis about the existence in the tectonosphere of real, structural, geophysical and geodynamic heterogeneities, manifested at multiple levels and scales, as well as laterally irregular and dynamic over time. It is clear that the processes generating these mantle heterogeneities are the result of irregular, unevenly distributed (chaotic) and convective motions of varying intensity. As the Academician Pushcharovskii (1992) notes, nonlinear geodynamics studies "… the features of the disordered manifestation in time and space of individual structure-forming movements or their combinations, caused by the dynamic interaction of various interplanetary or extraterrestrial factors, or both."

Obviously, here we are confronted with one of the most widespread and at the same time mysterious natural phenomena—chaotic, irregular processes. When investigating these processes in open nonequilibrium thermodynamic systems, contemporary science has focused on the unusual geometric and dynamic properties of these processes—the Cantor—or fractal—structure repeating itself on smaller scales. The dimension of these scales turns out to be fractional.

At the beginning of the century, one of the founders of modern mathematics, Georg Cantor, constructed an example of a similar set, each point of which comprises a limit (i.e. of any set, there are points in any neighbourhood of it). In addition, this set does not contain a single interval and is uncountable (none of its points can be enumerated by natural numbers). Thus, it may be said that the set exhibits scale invariance. Let us imagine that we have a set of magnifying glasses. If viewing this set at a large scale (i.e. through a glass with a low magnification), we will see two segments. On a smaller scale, each of these segments splits into two. On an even smaller scale, each of these four segments also splits, and so on.

The very word "fractal"—derived from French fractal, from Latin fractus ("broken")—was introduced relatively recently by Benoit Mandelbrot in his 1977 book "Form, Chance and Dimension". He showed that in various areas of modern knowledge the existence of fractal sets makes it possible to explain—and, in some cases, to predict—experimental results, representing diverse complex natural phenomena. The fields in which fractal sets apply include cosmology, the theory of turbulence, chemical kinetics, the physics of polymers, the analysis of fluid percolation phenomena and many others.

One of the more unexpected examples of fractals consists in the coastline of islands. When measuring the length of a coastline L, the complex indented coastline is replaced by a broken line consisting of units having a length of at least ε. For ordinary smooth curves (circle, ellipse, part of a sinusoid) with decreasing Πε, the value L(ε) tends to a certain amount, which comprises the length of this curve. However for the coastline of Great Britain it turns out that L is dependent on ε (in the interval from 10 to 1000 km) according to the exponent law

$$L(\varepsilon) \approx \varepsilon^{1-d}, d \approx 1 \div 3.$$

An analogous dependence determines the length of some rivers, the coasts of many islands, the path of a particle under Brownian motion, or a swarm of midges, constantly changing its shape, etc. Thus, if fractals are not taken into account, then using different rulers, you can get different values of L, none of which can be considered as a characteristic of the object.

Let us consider another example. It turns out that that when a liquid with a low viscosity is displaced by another liquid with a higher viscosity, the initially flat interface converts into a surface resembling the fingers of a glove. Such structures are referred to as "viscous fingers". The gradual fragmentation of the tips of such "fingers" leads to the appearance of fractal clusters. The analysis of this phenomenon is of great practical significance, since, for example, it is observed when water is pumped into an oil reservoir to enhance oil recovery. However, as a consequence of the described effect, the water seeps considerably further than desired, and it is a mixture primarily consisting of water that is pumped to the surface.

In geoscience, the earliest ideas about a fractal dimension were obtained on the basis of seismological data analysis (Sadovskii et al. 1982, 1989). These conclusions were a consequence of the fundamental seismological law of the frequency of earthquakes. This law establishes a hierarchical, automodal character of the distribution of total energy across earthquakes of various strengths. For example, it was found that earthquakes not only exhibit a hierarchical similarity in terms of energy, but also have fractal properties in the spatial distribution. One of these properties is that, on the average, each area in the seismic region is divided commensurately into high, middle and weak seismic areas. At the same time, seismic areas with a decreased magnitude occupy an increasingly smaller aggregated area, though continuing to cover the entire seismic region. As was shown by Sadovskii (1982), this regularity, which is related to the fractal properties of self-similar sets, can in a sense be regarded as a spatial analogue of the law of repeatability: for each "seismic" area of a certain rank, on average, there is a certain fixed number of "seismic" areas of lower rank, independent of the magnitudes of the sites. Therefore, representing the fractal dimension of the geological environment, the seismic process is itself characterised by hierarchy and discreteness: the spatial distribution of earthquakes has a complex structure similar to a Cantor set.

Furthermore, it became apparent that fractality is characteristic of the geological space of oil deposits (Feder 1991) and salt-dome structures (Gramberg et al. 1994). The fracture surfaces of brittle solids have statistical fractal properties (Mandelbrot 1984; Fidlin 1992). Plastic deformations are also characterised by fractal dimensionality (Panin and Panin 1992). Seismic tomography data allowed Maruyama et al. (1994) to establish a fractal (cluster) structure of mantle plumes resembling the structure of viscous fingers. It became evident that mushroom superplumes, corresponding to the areas rising to the surface of the lower mantle, branch into several smaller plumes in the upper mantle, which in turn branch into even smaller ones at the base of the plate (Maruyama et al. 1994; Dziewonski and Woodhouse 1989).

Thus, a review of the current state of the problem shows that, in thermodynamic terms, planet Earth is a member of the class of open, nonequilibrium systems, which, unlike closed thermodynamic systems, can exchange energy, matter and/or information with the surrounding environment. Our planet's structure is characterised by the gravitational (convective) instability of its near-surface and deep layers. This instability may be experienced at local, regional, supra-regional and global levels. The nature of the gravitational (convective) instability phenomena of the near-surface layers of the crust and the deep layers of the Earth is very diverse; consequently, their manifestation both in space and in time has a clearly irregular, largely random character. This points to the need for describing the distinctly irregular or chaotic natural phenomena that, as a rule, possess remarkable geometric and dynamic properties in nonequilibrium thermodynamic systems, i.e. a fractal or fractional dimension.

On the other hand, when considering density instability and spontaneous structuring phenomena exhibited with respect to the nonequilibrium distribution of densities in the upper layers of the Earth's crust, in the oceans, seas, lakes and surrounding atmosphere, on the level of the description of natural objects accessible to our direct observation, we see that all these physical phenomena have a common wave nature. Consequently, all these phenomena may be described by the same equations (Giterman and Shteinberg 1972; Landau and Lifshits 1988; Turner 1977). The common wave nature of spontaneous mass structuring with a nonequilibrium distribution of densities explains the geometric commonality of spontaneously emerging structures, as well as their cellular (hexagonal, square and polygonal) geometry, regardless of the nature and scale of the manifestation. Behind these seemingly simple and obvious arguments there is a very important conclusion for modern natural science that convective processes (Bénard cells being their classic representation), density instability and spontaneous structuring phenomena in gaseous and liquid media, as well as in the upper horizons of the Earth's crust, can be represented as a superposition of cellular standing internal gravitational waves. When represented in terms of their wave nature and materialised in the form of expansion into modes, all these natural phenomena and processes acquire the remarkable geometric and dynamic properties of fractal (or fractional) dimensionality. In this sense, as we have been able to see, the generality of the manifestation of the wave properties of matter enables us to formulate the fundamental laws underpinning the development of the widest range of irregular chaotic natural phenomena in the language of physics and mathematics.

References

Stokes G.G. On some cases of fluid motion. Philos. Mag. 1847; 31:136–137.

Artyushkov E.V. About the physical causes of emergence of polygonal structures in soils. The issues of palaeogeography and morphogenesis in Polar countries and highlands. M.: MGU; 1964:109–130 (In Russ.).

Artyushkov E.V. Formation of convective deformations in weakly-lithified sedimentary rocks. Izv. AN SSSR. Ser. geol. 1965; 12:79–101 (In Russ.).

Kostyaev A.G. About the origin of clinoform and folded deformations of layers in quaternary sediments. The issues of palaeogeography and morphogenesis in Polar countries and highlands. M.: MGU; 1964:131–163 (In Russ.).

Belousov V.V. Earth's crust and upper mantle of the continents. M.: Nauka; 1966 (In Russ.).

Palm E. J. Fluid Mech. 1960; 8.

Bénard H. Les tourbillons cellulaires dans une nappe liquide transportant de la chaleur par convection en regime permanent. Ann. Chem. Phys. 1901; 23:62–144.

Koschmieder E.L. On convection under an air surface. J. Fluid Mech. 1967; 30(1):9–15.
Silverston P.L. Warmedurchgang in waagerechten Flussigkeitsschichten: Heft 1—2. Forschung im Ingenieurwesen. 1958; 24(1, 2):29–32, 59–69.
Ramberg H. Instability of layered systems in the field of gravity, parts 1 and 2. Phys. Earth and Planet. Inter. 1968; 1:427–474.
Ramberg H. Theoretical models of density stratification and diapirism in the Earth. J. Geophys. Res. 1972; 77:877–889.
Seyfert C.K. Convection and motion of plates. Structural geology and plate tectonics. T. 1. Mir; 1990. (in Russ.).
Lebedeva N.B. Modelling the process of diapir dome formation. Sov. geologiya. 1956; 5:163–175 (In Russ.).
Parker T.J., McDowell A.N. Model studies of saltdome tectonics. Amer. Assoc. Petroleum Geologists Bull. 1955; 39(12):2384–2470.
Sycheva E., Mikhailova A.M. The mechanism of tectonic processes in an environment of rock density inversion. M.: Nedra; 1973 (in Russ.).
Belousov V.V., Vikhert A.V., Goncharov M.A. et. al. Methods of modeling in structural geology. M.: Nedra; 1988 (In Russ.).
Dixon J.M. Finite strain and progressive deformation in models of diapiric structure. Tectonoph. 1975; 28:89–124.
Talbot C.J. Inclined and asymmetric upwardmoving gravity structures. Tectonophysics. 1977; 43:159–181.
Schwerdtner W.M., Sutcliffe R.H., Troeng B. Patterns of total strain within the crustal regions of immature diapers. Canad. J. Earth Sci. 1978; 15:1437–1447.
Rayleigh, John William Strutt (Baron). On convection currents in a horizontal layer of fluid, when the higher temperature is on the under side. Philos. Mag. 1916; 32:529–546.
Tippelskirch H. Beitr. Phys. Atmosphere. 1956; 29:37.
Busse F.H. Dissertation. Santa Monica (California). 1962.
Busse F.H. Nonlinear properties of thermal convection. Res. Prog. Phys. 1978; 41:1931–1965.
Danes Z.F. Mathematical formulation of saltdome dynamics. Geophysics. 1964; 29:414–424.
Selig F. A theoretical prediction of salt dome patterns. Geophysics. 1965; 30:633–643.
Biot M.A., Ode H. Theory of gravity instability with variable overburden and compaction. Geophysics. 1965; 30: 213–227.
Fletcher R.C. Application of a mathematical model to the emplacement of mantled gneiss domes. Am. J. Sci. 1972; 272:197–216.
Turner J. Buoyancy effects in fluids. M.: Mir; 1977 (In Russ.).
Gossard E.E., Hooke W.H. Waves in the atmosphere. Infrasound and gravitational waves in atmosphere - their origin and spreading. Ed. Golitsyn G.S. M.: Mir; 1978 (In Russ.).
Artyushkov E.V. Geodynamics. M.: Nauka; 1979 (In Russ.).
Golitsyn G.S. Convection study with geophysical applications and analogies. News and problems of science. L.: Gidrometizdat; 1980 (In Russ.).
Zharikov V.N. Internal structure of the Earth and other planetes. M.: Nauka; 1983 (In Russ.).
Myasnikov V.P., Fadeev V.E. Models of the evolution of the Earth and the terrestrial planets. M.: VINITI; 1980 (In Russ.).
Pushcharovskii Yu.M., Novikov V.L., Savel'ev A.A., Fadeev V.E. Heterogeneity and convection in the tectonosphere. Geotectonics. 1989; 5:3–13 (In Russ.).
Pushcharovskii Yu.M., Novikov V.L., Savel'ev A.A., Fadeev V.E. Heterogeneity and convection in the tectonosphere. Geotectonics. 1990; 5:3–8 (In Russ.).
Pushcharovskii Yu.M. Selected works. Tectonics of the Earth. Etudes in 2 volumes. Vol. 1: Tectonics and geodynamics. Geol. in-t. M.: Nauka; 2005 (In Russ.).
Tectonic stratification of the lithosphere and regional geological studies. Ed. Yu.M. Pushcharovsky, V.G. Trifonov. M.: Nauka; 1990 (in Russ.).
Sorokhtin O.G., Ushakov S.A. Global evolution of the Earth. M.: MSU; 1991 (in Russ.).
Turcotte D., Schubert J. Geodynamics: geological applications of the physics of continuous media. M.: Mir; 1985 (in Russ.).

Fisher O. Physics of the Earth crust. 2nd ed. London; 1989.
Gridds D.T. A theory of mountain building. Amer. J. Sci. 1939; 237:611–650.
Hess H.H. History of ocean basin. Petrologies Studies. A volume of honor of A.F. Buddington. New York. 1962; 618–620.
Holmes A. Radioactivity and Earth movements. Geol. Soc. Glasgow Trans. 1928; 28:559–606.
McKenzie D.P. The Earth's mantle. Sci. Amer. 1983; 249: 67–78.
Kanla W.M. Global gravity and mantle convection. Tectonophysics. 1972; 13:341–359.
Eskola P. The problem of mantled gneiss domes. Quart. J. geol. Soc. London. 1949; 104:461–476.
Kranck E.H. Deep structures and ultrametamorphism. Trans. N.Y. Acad. Sci. Ser. 2. 1954; 16 (5):234–241.
Belk R. Structural features of igneous rocks. M.: Gosgeoltekhizdat; 1946 (In Russ.).
Buddington A.F. Granite emplacement with special reference to North America. Geol. Soc. Amer., Bull. 1959; 70(6):671–747.
Wegmann C.E. Zur Deutung der Migmatite. Geol. Rundsch. 1935; 26:307.
La Fond E.C. Internal wave motion and its geological significance. Mahadevan volume. A collection of geological papers in commemoration of the sixtyfirst birthday of Prof. C. Mahadevan. Hyderabad (India). 1961; 61–77.
Belousov V.V. Some issues of deep tectonics. Ser. 4 geol. M.: MGU. 1960; 5:3–12 (In Russ.).
Goncharov M.A. Density inversion in the Earth's crust and folding processes. M.; 1979 (In Russ.).
Ez V.V. Tectonics of the deep zones of the continental crust. M., Nauka; 1976 (In Russ.).
Suvorov A.I. The newest global kinematics of the lithosphere (based on regional tectonic pairs. Geotectonics. 1978; 2:3–18 (In Russ.).
Pushcharovskii Yu.M. Tectonics of the oceans and nonlinear geodynamics. Dokl. RAN. 1992; 324 (1):157-161. (In Russ.).
Pushcharovskii Yu.M. The Earth's tectonosphere is a new vision. Russian Journal of Earth Sciences. 2000; 2(1):63–69 (In Russ.).
Sadovskii M.A., Pisarenko V.F. Some ideas about the seismic process. M.: IFZ; 1982 (In Russ.).
Sadovskii M.A. On the meaning and meaning of discreteness in geophysics. Discrete properties of a geophysical environment. M.: Nauka; 1989. (In Russ.).
Suvorov A.I. Regional tectonic pairs as the basis of the internal structure of geosynclinal areas and platforms. DAN SSSR. 1976; 226(6) (In Russ.).
Planet Earth. Encyclopedic reference book. Volume "Tectonics and geodynamics". Ed. L.I. Krasnyi, O.V. Petrov, B.A. Blyuman. SPb.: VSEGEI; 2004. (In Russ.).
Suvorov A.I. Tectonic stratification and tectonic movements in the continental lithosphere. Fundamental problems of general tectonics. M.: Nauchnyi mir; 2001:34–48 (In Russ.).
Rasskazov S.V. Comparison of volcanism and the newest structures of hot spots of Yellowstone and Eastern Sayan. Russian Geology and Geophysics. 1994; 35:67–75 (In Russ.).
Elsasser W.M. Viscous stratification of Earth and convection. Phys. Earth a. Planet. Inter. 1972; 6:198–204.
Khain V.E. Plate tectonics: analysis of the current state. Moscow University Geology Bulletin. 1:3–10 (In Russ.).
Shcheglov A.D. Nonlinear metallogeny. DAN SSSR. 1983; 271(6):1471–1474 (In Russ.)
Shcheglov A.D. About some questions of nonlinear geology. Regional geology and metallogeny. 1995;4:5–15 (In Russ.).
Kuznetsov O.L. Nonlinear geophysics. The issues of nonlinear gephysics. M.: ONTI VNIIYaGG; 1981:5–20 (In Russ.).
Feder J. Fractals. M.: Mir; 1991 (In Russ.).
Gramberg I.S., Goryainov I.N., Smekalov A.S., Melamud M.Kh., Goryainova G.I. Fractality of the salt-dome structures of the Gulf of Suez. Dokl. RAN. 1994; 336(1):80–83 (In Russ.).
Mandelbrot B.B., Passoja D.E., Panllay A.L. Nature. 1984; 308(4):721– 722.
Maruyama S., Kumazawa M., Kawakami S. Towards a new paradigm on the Earth's dynamics. J. Geol. Soc. Japan. 1994; 100(1):1–3.
Dziewonski A.M., Woodhouse J.H. Threedimensional Earth structure and mantle convection. Abstracts. 1989; 1(28th IGC):427–428.

Giterman M.Sh., Shteinberg V.A. Izv. AN SSSR. MZhG. 1972; 2:55–61 [Giterman M.Sh., Shteinberg V.A. Izv. AN SSSR. MZhG. 1972; 2:55–61 (In Russ.)].
Landau L.D., Lifshits E.M. Theoretical physics. Fluid dynamics. M.: Nauka; 1988. (In Russ.).
Dixon J.M. Structural geology and plate tectonics. M.: Mir; 1991. (In Russ.).
Ramberg H. Superposition of homogeneous strain and progressive deformation in rocks. Bull. Geol. Inst. Univ. Uppsala. Ser. 2.1976; 6: 35–67.
Fidlin A.Ya. The Averaging method for systems of variable order and its use in studying the dynamics of vibro-impact systems: thesis for a candidate degree in physical and mathematical sciences: 01.02.01. State Technical University. St. Petersburg; 1992.
Panin I.M., Panin M.I. Dynamic processes in the mountain range: Coursebook. M.: Publishing house of peoples' friendship university of Russia; 1992. (In Russ.).
Väyrynen H. Crystalline basement of Finland. M: IL; 1959.

Chapter 2
Physical and Mathematical Description of the Manifestation of Wave Properties of Matter in Thermodynamic and Dynamic Processes of Spontaneous Structuring of Density-Unstable Masses

The natural science of geology is in the first place a descriptive science; nevertheless, in this field a large body of accumulated empirical material is unified at the level of the fundamental laws of nature in the language of modern physics and mathematics. In this connection, we can state that with the development of various physical and mathematical methods for describing the processes of spontaneous structuring of density-unstable media, their wave nature becomes more and more evident regardless of the causes (whether thermodynamic or dynamic) of this instability.

This emerging perception of the world around us is the natural result of the developments taking place in natural science as a whole. The classical Newtonian mechanics familiar to us from our school days, which describes convection in terms of the vibrational motions of individual particles along closed trajectories, derives a given number of particles in an elementary volume (or phase point). In this case, their motion in a continuous medium is considered to be isolated from the remaining particles, without taking into account the interaction between the selected particles and the continuous medium. Conversely, since the works of Poincaré (1983), contemporary natural science inevitably leads us to the conclusion that such an approach is excessively idealistic for continuous media comprising large amounts of particles. New regularities of convective oscillatory processes are manifested from the more holistic approach taken when considering systems consisting of a large number of interacting particles as a whole. Vibrational processes in a system comprising a large number of elements are referred to in terms of waves. The states of such systems are determined primarily by the wave nature of the propagating energy, i.e. it is the wave properties of matter manifesting themselves at all levels of its organisation that acquire the determining significance in describing the fundamental laws of the development of such systems.

Summarising from these positions the disparate results of experimental and theoretical studies, we distinguish three basic approaches for describing the dynamics of the manifestation of the wave properties of matter in processes of spontaneous structuring of masses with a nonequilibrium distribution of densities:

O. V. Petrov, *The Earth's Dissipative Structures*, Springer Geophysics,
https://doi.org/10.1007/978-3-319-93614-7_2

- the classical Newtonian approach, which is based on the consideration of each individual particle and the thesis of the continuity of motion of an individual material point (or trajectory) in phase space;
- the statistical approach, in which the primary objects of description in phase space are considered not in terms of individual trajectories, but rather as probabilities in terms of ensembles of material points;
- the wave-based approach, which describes the wave dynamics of an ensemble of material points by introducing wave functions into the phase space.

2.1 Description of the Wave Properties of Matter from the Standpoint of Classical Physics in Terms of Trajectories in Phase Space

From the standpoint of classical physics, the convection phenomena have a simple physical explanation. Let us recap Bénard's experiment. Take two horizontal plates (for example, glass plates) and place between them some liquid that expands when heated. In order to detect the flow of the liquid, we mix it with fine aluminium filings, evenly distributed throughout its volume. We begin to heat the bottom plate at the same time as cooling the upper plate. Due to thermal conductivity, the lower layers of the liquid acquire heat from the hot plate, resulting in a decrease in their specific gravity. At the same time, the liquid adjacent to the cold plate releases its heat, and its specific gravity increases. Under the influence of gravity, the cold layers descend, displacing the lower, lighter layers upward. Since the temperature of the plates is kept constant, after a while the lower layers heat up, become lighter and will again be interchanged with the cooled upper layers. As a result, a continuous fluid cycle is established in which, as the mathematical calculation shows, the same particle moves along a closed trajectory, repeatedly taking heat from the lower plate and releasing it to the upper one. This is a typical approach to the description of convection processes from the standpoint of classical mechanics and thermodynamics. The main attention is given to the stream lines, which represent the continuous circulation of liquid along closed trajectories.

From the point of view of thermodynamics, it can also be said that, for small temperature gradients, heat transfer is successfully accomplished by molecular thermal conductivity. With larger temperature gradients, however, it is the convective heat transfer that becomes the more efficient mechanism.

In 1916, in considering the problem of convective instability of a horizontal fluid layer with free particles in which there is a vertical, superadiabatic temperature gradient (the Rayleigh-Bénard problem), Lord Rayleigh determined the convective stability threshold. As a result of Rayleigh's and other physicists' studies, a stability criterion was established—the dimensionless so-called Rayleigh number (Ra), which expresses the ratio of buoyancy force to the dissipation forces:

$$Ra = \frac{\beta g \Omega h^4}{\nu \chi}, \tag{2.1}$$

where $\beta = -(1/\rho)\, \partial\rho/\partial T$ expresses the coefficient of thermal expansion (ρ—density of the liquid, T—its temperature); g—the modulus of the gravitational acceleration vector; h—the thickness of the convection layer; ν—the kinematic viscosity; χ—the thermal diffusivity coefficient; Ω—the magnitude of the vertical temperature gradient.

The minimum value of the Rayleigh number, which determines the threshold for the occurrence of thermoconvective instability in the case of free boundary surfaces of the liquid layer,

$$Ra_{cr} = \frac{27}{4}\pi^4 = 657.5.$$

If the liquid is bounded by solid surfaces, then for the Rayleigh number it is possible to obtain a transcendental equation, the solution of which gives the value $Ra_{\text{cr}} = 1707.8$.

Thus, if the temperature gradient exceeds a certain critical value ($Ra > Ra_{\text{cr}}$), convection appears in the horizontal layer. For small supercriticalities, the main form of convective motion is two-dimensional waves; however, with increasing supercriticality, the stable shape of the emerging three-dimensional hexagonal structures is increasingly dependent on the physical parameters of the medium on its temperature. The temperature-dependency of viscosity is usually the dominant cause of asymmetry in the convection layers; its importance for the preference for the formation of a particular type of hexagonal cells was recognised by Graham (1933), and theoretically substantiated by Palm (1960), Segel and Stuart (1962) before Busse (1978) undertook a systemic stability analysis. The temperature dependency consists in the fact that the viscosity is minimised in the region of high tension located at the centre of the convection cell. In liquids, where the viscosity is lower in the lowermost part, motion occurs in the centre of a hexagonal cell in an upwards direction—these are referred to as L-hexagons. Conversely, in gases, for which viscosity increases with temperature, the centre of the convection cells exhibits a downward movement—these are known as G-hexagons.

To describe the Rayleigh-Bénard convection, according to Karlov and Kirichenko (2001), the Navier-Stokes equation is used:

$$\rho\left(\frac{\partial\vec{\upsilon}}{\partial t} + \left(\vec{\upsilon}, \vec{\nabla}\right)\vec{\upsilon}\right) = -\vec{\nabla}p + \eta\Delta\vec{\upsilon} + \rho\vec{g}, \tag{2.2}$$

where p—pressure; ρ and η—the density and viscosity of the liquid, respectively; $\vec{\upsilon}$—flow velocity. The last term describes the volume force $\vec{F} = \rho\vec{g}$, due to the gravitational field ($\vec{g}$—acceleration due to gravity). Here it is assumed that the liquid in question is incompressible, that is,

$$\operatorname{div}\vec{\upsilon} = 0. \tag{2.3}$$

The heat transfer is described by the thermal conductivity equation:

$$\frac{\partial T}{\partial t} + \left(\vec{\upsilon}, \vec{\nabla}\right)T = a\Delta T, \tag{2.4}$$

in which a is the coefficient of (molecular) thermal diffusivity.

Taking into account the incompressibility of the fluid, the Navier-Stokes equation is simplified by neglecting the change in the density ρ in all summands, with the exception of the force $\vec{F} = \rho\vec{g}$, i.e., we replace ρ by ρ_0. In the expression for the volume force F, we will take into account the dependency of the density on temperature:

$$\rho = \rho_0(1 - \alpha(T - T_0)), \tag{2.5}$$

where α is the volumetric thermal expansion coefficient of the liquid. The value ρ_0 represents the density of the liquid at a certain average temperature T_0. Then the Navier-Stokes equation takes the form

$$\frac{\partial \vec{\upsilon}}{\partial t} + \left(\vec{\upsilon}, \vec{\nabla}\right)\vec{\upsilon} = -\frac{1}{\rho_0}\vec{\nabla}p + \nu\Delta\vec{\upsilon} + (1 - \alpha(T - T_0))\vec{g}, \tag{2.6}$$

where $\nu = \eta/\rho_0$ is the kinematic viscosity.

In the Oberbeck-Boussinesq approximation, Eqs. (2.3), (2.4) and (2.6) are known as the thermal convection equations.

The simplest model of a nonequilibrium fluid presupposes a consideration of the vessel in which it is enclosed, not limited by the X and Y axes. In this vessel, a two-dimensional flow is realised when the velocity vector is parallel to the plane (X, Z). It follows from the incompressibility condition for the liquid that

$$\frac{\partial \upsilon_X}{\partial x} + \frac{\partial \upsilon_Z}{\partial z} = 0, \tag{2.7}$$

where the components of the velocity vector can be expressed in terms of the stream function $\Psi(x, z, t)$:

$$\upsilon_x = -\frac{\partial \Psi}{\partial z}, \; \upsilon_z = -\frac{\partial \Psi}{\partial x}, \tag{2.8}$$

when Eq. (2.7) takes the form of an identity. Taking into account (2.8), the y-component of the vector will be equal to

$$\omega \equiv (\mathrm{rot}\,\vec{\upsilon})_y = \frac{\partial \upsilon_x}{\partial z} - \frac{\partial \upsilon_z}{\partial x} = -\left(\frac{\partial^2 \Psi}{\partial x^2} + \frac{\partial^2 \Psi}{\partial z^2}\right), \tag{2.9}$$

which, according to the physical meaning of the differential operator of the rotor, is referred to as the vorticity of the velocity field.

In the language of the stream function Ψ and vorticity ω, the system of Eqs. (2.4), (2.6) can be reformulated, which requires the application of the operator rot_y to Eq. (2.6):

$$\frac{\partial \omega}{\partial t} + \upsilon_x \frac{\partial \omega}{\partial x} + \upsilon_z \frac{\partial \omega}{\partial z} = \nu \Delta \omega - \alpha g \frac{\partial T}{\partial x}. \tag{2.10}$$

In this equation, we assume that the acceleration vector of freefall has only one component $g_z = -g$. Replacing in (2.10) the components of the velocity vector according to formula (2.8), we obtain

$$\frac{\partial \omega}{\partial t} + \frac{\partial(\Psi, \omega)}{\partial(x, z)} = \nu \Delta \omega - \alpha g \frac{\partial T}{\partial x}. \tag{2.11}$$

The thermal conductivity Eq. (2.4) is rewritten in a similar way:

$$\frac{\partial T}{\partial t} + \frac{\partial(\Psi, \omega)}{\partial(x, z)} = \alpha \Delta T. \tag{2.12}$$

In Eqs. (2.8), (2.11) the vorticity ω is expressed in terms of the stream function Ψ (2.9).

In the stationary state, convection is absent, i.e. $\upsilon = 0$, which allows us to assume in this case $\Psi = 0$. Then the solution of the thermal conductivity equation has the form

$$T = T_{cr}(z) = T_1 - \beta z, \; \beta = \frac{T_1 - T_2}{h} > 0. \tag{2.13}$$

If we assume that the temperature field satisfies the equation

$$T(x, z, t) = T_{cr}(z) + \Theta(x, z, t),$$

then, as shown by the elementary calculations, the function $\Theta(x, z, t)$ satisfies the equation

$$\frac{\partial \Theta}{\partial t} + \frac{\partial(\Psi, \omega)}{\partial(x, z)} = \alpha \Delta \Theta + \beta \frac{\partial \Psi}{\partial x}, \tag{2.14}$$

which is similar in structure to (2.11), only the value of T is replaced by Θ.

Directly at the instability threshold, the stream function $\Psi(x, z)$ and the flow velocity are represented by the following formulas:

$$\begin{aligned} \Psi(x,y) &= A\sin\left(\frac{\pi x}{h\sqrt{2}}\right)\sin\left(\frac{\pi z}{h}\right), \\ \upsilon_x(x,y) &= -\frac{\partial\Psi}{\partial z} = -\frac{\pi A}{h}\sin\left(\frac{\pi x}{h\sqrt{2}}\right)\cos\left(\frac{\pi z}{h}\right), \\ \upsilon_z(x,y) &= \frac{\partial\Psi}{\partial x} = \frac{\pi A}{h\sqrt{2}}\cos\left(\frac{\pi x}{h\sqrt{2}}\right)\sin\left(\frac{\pi z}{h}\right). \end{aligned} \quad (2.15)$$

Figure 2.1 shows the dependence of the stream function on the coordinates in the vertical section of the liquid layer. Since the normal to the level line Ψ = const is determined at each point by the gradient of the function N = grad Ψ, according to relations (2.7), the direction of the velocity vector coincides with the direction of the tangent to the level line at the same point. Therefore Fig. 2.1 also illustrates the field of flow velocities. The direction of circulation for some single points is indicated by arrows.

One of the most important concepts of classical Newtonian physics—the concept of phase space and the phase plane—is closely related to such a diagrammatic representation. In this case, this concept is introduced as follows. In accordance with Fig. 2.1, the state of the system under consideration characterised according to classical Newtonian mechanics in the vertical section of the fluid layer from the standpoint of the behaviour of each individual point in the field of the lines of the stream function Ψ(x, z) and the flow velocity field grad ($\Psi(x, z)$), i.e., the state of the system being studied is represented by a point in this space at each given instant of time. A space introduced in such a way is called a phase space, and a point within this space is referred to as a phase or representative point. The use of the word "phase" in these names is due to historical reasons: the term "phase" (from the Greek phases—appearance), used in describing the state of a particular system, indicates the stage of development of a particular phenomenon or process.

Over time, the state of the system changes. The position of the representative point in phase space changes correspondingly; that is, this point describes a certain closed curve, called the phase trajectory. The field of the level lines of the stream

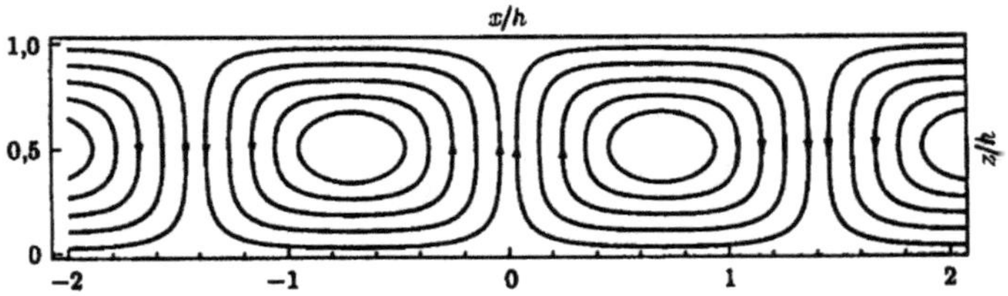

Fig. 2.1 Level lines of stream function Ψ(x, z) within the transverse liquid section. Arrows indicate the direction of convective flow circulation. After Karlov and Kirichenko (2001)

function $\Psi(x, z)$ shown in Fig. 2.1, contains all admissible phase trajectories and can be considered as the phase plane of the system under study.

The phase trajectories (or streamlines of convection currents), shown in Fig. 2.1, are "closed". This refers to the expected periodicity of motion and the possibility of harmonic oscillations occurring in the system under study. It is obvious that the oscillations of individual particles with different amplitudes in the phase plane must correspond to nested ellipses. In the central part of each circulation cell depicted in Fig. 2.1, a singular point—or centre—can be allocated. In the theory of oscillations, these singular points characterise the equilibrium position of the dynamic system. With respect to these discrete points, the overturning of overlying and underlying density-unstable layers is determined; the presence of a singular point (centre) means that the system, after being removed from the equilibrium state, performs periodic oscillations indefinitely.

In order to describe the behaviour of a system after the occurrence of an instability in a multidimensional phase space in the form of oscillatory motions of each individual point, it is necessary to construct evolution equations from the parameters under investigation. This procedure was carried out by Saltzman (1962). Saltzman expanded the functions Θ and Ψ in the double Fourier series in the variables x and z with time-dependent coefficients $\{\Psi_{mn}(t), \Theta_{mn}(t)\}$, restricting these series to a finite number of terms. Substituting these series into Eqs. (2.11), (2.12), Saltzman obtained a system of ordinary nonlinear differential equations. Numerical analysis has shown that in some cases all the variables $\{\Psi_{mn}(t), \Theta_{mn}(t)\}$ except for three, tended to zero. Using Saltzman's results, Lorenz (1963) confined himself to these three nontrivial terms from the very beginning, assuming that

$$\begin{aligned}\Psi(x,z) &= C_1\sqrt{2}X(t)\sin\left(\frac{\pi\mu x}{h}\right)\sin\left(\frac{\pi z}{h}\right),\\ \Theta(x,z) &= C_2\left\{\sqrt{2}Y(t)\cos\left(\frac{\pi\mu x}{h}\right)\sin\left(\frac{\pi z}{h}\right) - Z(t)\sin\left(\frac{2\pi z}{h}\right)\right\}.\end{aligned} \quad (2.16)$$

Here $X(t)$, $Y(t)$, $Z(t)$ are various time functions, which describe the temporal convection dynamics, while parameter $\mu = g_x h/\pi$ determines the degree of proximity to the instability threshold. In the Lorentz calculations, the constants C_1 and C_2 were assumed to be equal to

$$C_1 = a\frac{1+\mu^2}{\mu}, \quad C_2 = \frac{\Delta T}{\pi}\frac{Ra_{cr}}{Ra}, \quad (2.17)$$

where Ra_{cr} is the critical Rayleigh number, defined by the formula

$$Ra_{cr} = \frac{27}{4}\pi^4 = 657.5.$$

If the expression (2.16) is substituted into Eqs. (2.11), (2.12) and all harmonics discarded except those that are taken into account in (2.16), then we can obtain the following system of equations:

$$\begin{aligned}\dot{X} &= -\sigma X + \sigma Y,\\ \dot{Y} &= -XZ + rX - Y,\\ \dot{Z} &= XY - bZ.\end{aligned} \tag{2.18}$$

Here, the dot sign displayed above the symbols X, Y or Z denotes differentiation with respect to the dimensionless time $\tau = \pi^2(1+\gamma^2)\cdot at/h^2$, and the coefficients r, σ and b are given by formulas

$$r = \frac{Ra}{Ra_{cr}},\quad \sigma = \frac{\nu}{a},\quad b = \frac{4}{1+\gamma^2}. \tag{2.19}$$

In the ratios (2.19), the value σ represents the Prandtl number known from the heat and mass exchange theory, and r is the normalised Rayleigh number.

As the parameter *r* increases, the convective dynamics of each individual particle becomes more complicated; following a small number of bifurcations, stochastic oscillations arise, represented as a strange attractor. The qualitative form of the corresponding attractor, referred to as the strange Lorenz attractor, is given in Fig. 2.2. Lorenz investigated the chaotic behaviour of the system (2.18) with the values of the parameters σ = 10, *b* = 8/3, *r* = 28; at the indicated values of σ and *b*, the threshold value r_{cr} is 24.74.

Dynamic modes represented by such attractors correspond to the finite aperiodic motion, which, moreover, possess all the characteristics of a stochastic process similar to the random motion of a Brownian particle. The term "strange attractor" itself is explained as follows. First of all, the word "attractor", meaning something

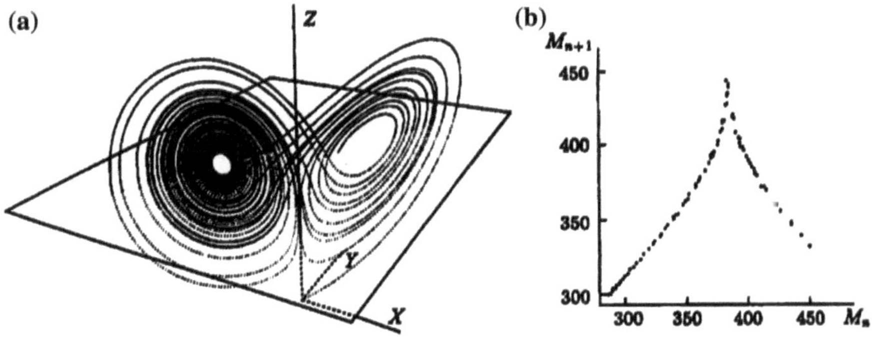

Fig. 2.2 **a** strange attractor in the Lorenz system with σ = 10, b = 10/3, r = 28 (X and Y planes correspond to Z = 27 in the figure); **b** sequential representation of the Lorenz attractor with the same parameter values (Karlov and Kirichenko 2001)

attractive, makes sense of an attractive set—in phase space, all trajectories that do not belong to this set are attracted to it without restriction. The word "strange", as noted by Karlov and Kirichenko (2001), bears in itself an emotional assessment of the unusual properties of this system, i.e. its complex, unpredictable behaviour. A characteristic feature of this attractor is that the phase trajectories periodically jump from the neighbourhood of one unstable singular point to a neighbourhood of another unstable singular point. This divagation takes place indefinitely and the phase trajectory remains open (Fig. 2.2a). However, the strangeness of the chaotic attractor lies not only in its exotic appearance, but also in its remarkable geometric and dynamic properties.

As we know from the theory of chaotic dynamics, any system with continuous time can be associated with a certain point mapping. In the case of the system (2.18), this mapping, according to Lorenz, is constructed as follows. We will fix those moments of time at which the values of the variable z reach their successive maxima M_n, n = 1, 2, … For the sequence $\{M_n\}$, the succession map $M_{n+1} = f\{M_n\}$ is constructed in the usual way. Specific values of M_n can be obtained only by numerical integration of the initial system of differential Eqs. (2.18). Figure 2.2b shows the form of the corresponding mapping. At first glance, the simple rostriform curve turns out to have a complex internal structure that manifests itself when it is viewed at a greatly enlarged scale. Initially, a splitting of the curve into two practically parallel ones is detected. A further increase in scale again reveals the same splitting of either one of these curves. If we zoom again, we will see a further splitting of one of the components, and so on. Thus, it turns out that the mapping curve in Fig. 2.2b has a Cantor structure, that is, it comprises a fractal set, which is typical for systems comprising a phase space of more than two dimensions. For the values of the parameters used in plotting the curve in Fig. 2.2b, the fractal dimension—the capacity of the set d_c—was equal to ≈2.05.

Now we will roughly intersect the phase trajectory planes of the attractor in Fig. 2.2a. Here we see a set of points lying near two continuous curves. Imagine that we have a set of magnifying glasses. If we consider this set on a larger scale, we will see that each curve splits into two; with a further increase in resolution, by another two, and so on. The strange attractor in this section is constructed in the same way as in the Cantor set. Thus, the unusual geometric object constructed by Georg Cantor at the beginning of the last century to serve as an argument in abstract mathematical reasoning turned out to be typical for most strange attractors.

Another paradoxical property of strange attractors is their sensitivity to the initial data. We choose two close points $x'(0)$ and $x''(0)$, lying on the attractor, and see how the distance $d(t) = x'(t) - x''(t)$ varies with time t. The strange attractor $d(t) \sim e^{\lambda t}$, $\lambda > 0$ has a λ value called the Lyapunov exponent, whose formal definition is as follows:

$$\lambda(\vec{x}(0), \vec{\omega}) = \lim_{t\to\infty} \lim_{d(0)\to 0} \left[\frac{1}{t}\ln\left(\frac{d(i)}{d(0)}\right)\right] \tag{2.20}$$

where $\vec{\omega}$ is the vector from the point $x'(0)$ to the point $x''(0)$ (Fig. 2.3).

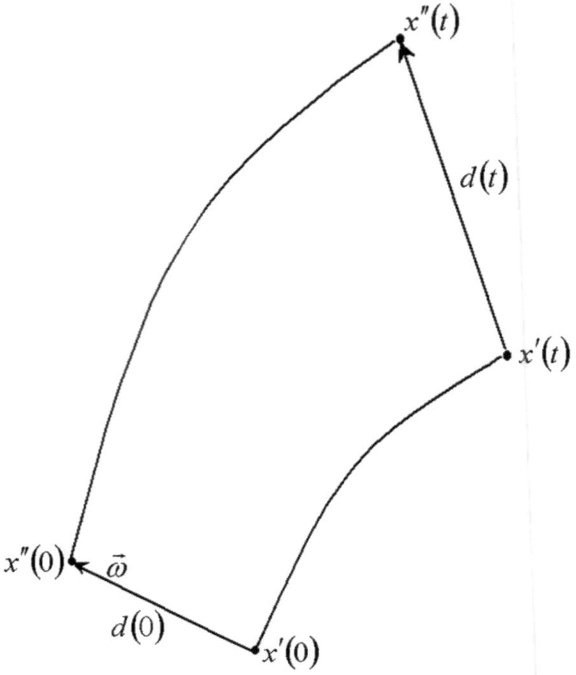

Fig. 2.3 Alteration dynamics over time of the distance between two representative points selected within the Lorenz strange attractor (Kurdyumov et al. 1989)

It can be said that the Lyapunov exponent, if it is positive, characterises the average runaway rate of two infinitesimally close trajectories. This phenomenon of sensitivity to the initial data became popularly known as the "butterfly effect". Here we are faced with an exponential runaway of trajectories, which generates non-computability and randomness.

Thus, we come to the definition of a "chaotic" regime, typical for systems with a strange attractor that has a fractal dimension. The regime is said to be chaotic if the distance $d(t)$ between any two points, initially arbitrarily small, increases exponentially with time. The runaway of trajectories is described by the function $\exp(t/\tau)$, where $1/\tau$ is by definition a positive value for chaotic systems. By analogy with λ, the value $1/\tau$ is referred to as the Lyapunov exponent and τ is the Lyapunov time (Nikolis and Prigogine 1990). Any description based on trajectories loses its meaning for times that are longer than the time horizon associated with this Lyapunov exponent.

In trying to describe the dynamics of processes of spontaneous structuring of masses in a non-equilibrium density distribution from the standpoint of classical physics in terms of trajectories in phase space, and in initially imagining the physical essence of these processes in the form of simple vibrational motions of each individual point (or elementary volume) along closed trajectories, we encountered new and unexpected dynamic and geometric properties in their behaviour. Closed trajectories corresponding to simple vibrational motions of each single point (or elementary volume) are destroyed. The distance between any two points, initially arbitrarily small, increases exponentially with time, which generates a non-computability and essential randomness in the behaviour of trajectories.

In geometric terms, the dynamics of the behaviour of these trajectories becomes a strange attractor. The trajectories controlled by such an attractor correspond to the finite aperiodic motion, which, as noted above, has all the features of a stochastic process similar to the random motion of a Brownian particle. However, these dynamic regimes, like the strange attractor itself, are not merely a disordered, amorphous structure, but are also characterised by the property of self-similarity and, in this certain sense, exhibit a high degree of regularity. An analysis of the internal ordering of the dynamic behaviour regimes of trajectories leads us to the concept of fractal sets (or fractals). Originally conceived in terms of pure mathematics, this concept today receives a wide variety of applications (Fig. 2.3).

In order to understand the fundamental nature of such remarkable phase trajectory behaviour, let us consider one of the most essential aspects of the theory of fractal sets. Examples of such sets, having a fractal or Cantor structure, were very popular in the late nineteenth and early twentieth centuries in connection with the problem of providing a rigorous foundation for mathematical analysis. As such, they attracted the attention of such outstanding mathematicians as Weierstrass, Riemann, Hermite, Cantor himself, Peano and Mandelbrot.

At the end of the 19th century, Karl Weierstrass presented one of a number of continuous functions that does not have a derivative at any point:

$$W(x) = \sum_{n=1}^{\infty} a^n \cos(b^n \pi x), \tag{2.21}$$

where $a < 1$, $b > 1$, and the product ab is sufficiently large. Weierstrass proved the non-differentiability of the function (2.21) for $ab > 1 + 3\pi/2$. Later, in 1916, G.H. Hardy showed that the non-differentiability of $W(x)$ also holds for a weaker condition when $ab > 1$. The graph of the function $W(x)$ is shown in Fig. 2.4.

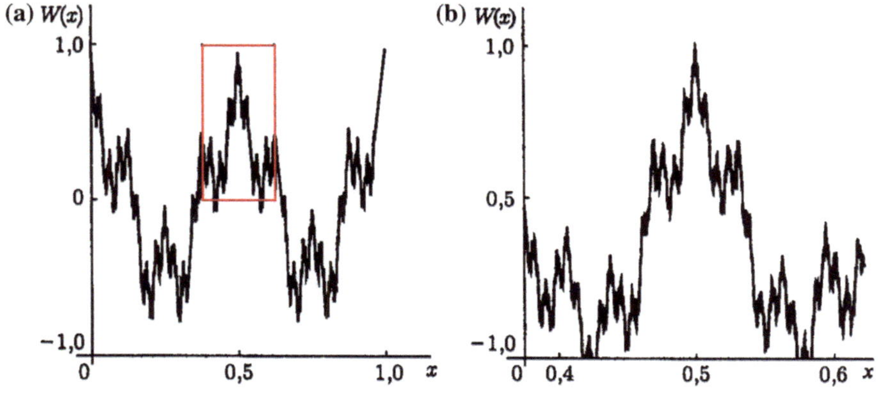

Fig. 2.4 Self-similarity manifestation in the Weierstrass function (Kurdyumov et al. 1989). **a** Weierstrass function graph; **b** curve fragment marked in Fig. 2.4a by a rectangle: a = 0.5, b = 4

It is possible to qualitatively understand the reason for its non-differentiability from the following positions. According to (2.21), a smooth wave is first used to construct W(x). Then a wavelet is applied to this wave, having a smaller amplitude and a higher frequency than $W_1(x)$. Then, an even more fine and high-frequency wavelet is added, and so on. As a result, an infinitely rugged curve appears. One of the interesting properties of this curve can be seen from the constructions performed. In Fig. 2.4a, the graph $W_1(x)$ is given for $0 \leq x \leq 1$. Consider the part of the curve in Fig. 2.4a, marked with a rectangle. This part is 1/a times smaller vertically and b times smaller horizontally than the original square. If this area is increased to the size of the original square, then we will see the original curve almost exactly (Fig. 2.4b). Repeating the construction, we can verify that the curve is reproduced at an arbitrarily small scale. In other words, there is self-similarity—the part resembles the whole.

Such an approach to the theory of fractal sets allows us to assume that the fundamental role of the wave properties of matter in the development of processes of spontaneous structuring of masses with a nonequilibrium distribution of densities is manifested even at the stage of description from the standpoint of classical physics in terms of trajectories in phase space. Here we will keep in mind that the thermodynamic parameters used in our equations were taken into account; thus, the presence of the wave properties of matter becomes crucial in the transition from classical dynamics to thermodynamics, in which, first of all, it is the nature of the interaction between particles that is being investigated.

In describing the vibrational motion of individual particles along closed trajectories, classical dynamics extracts a given number of particles in an elementary volume (or phase point) and considers their motion in a continuous medium isolated from the rest of the particles, disregarding the interactions between the selected particles and the remaining particles of the continuous medium. New regularities of convective oscillatory processes are manifested within a more holistic approach, however, when we consider the systems as consisting of a large number of particles interacting as a whole. The motion (change of state) of such systems acquires a collective character, in which the individual features of individual elements are put to one side, and the oscillatory processes in the system comprised of a large number of particles are described in terms of waves. In essence, waves describe an oscillatory process in some extended medium.

According to the above definition, a wave is a process that unfolds in time and space. Therefore, in order to describe it in the phase space, it is necessary to indicate the simultaneous state of all points differing in position in this space. In other words, the discovery of wave properties of matter in the processes of spontaneous structuring of masses in the nonequilibrium distribution of their densities requires the introduction of wave functions into the phase space and involves a transition from a consideration of a set of independent material points to that of a continuous medium.

The understanding of the inevitability of manifestations of the wave properties of matter, which is a consequence of the interaction of particles in continuous media can be traced to the work of Henri Poincaré at the end of the 19th century.

According to Poincaré, a dynamic system is characterised by the sum of the kinetic energy of its particles and the potential energy due to their interaction (Poincaré 1983). As Ilya Prigogine pointed out, free non-interacting particles could serve as a simple example: in this case, the potential energy is absent, and the calculation of trajectories is trivial. Such systems are integrable by definition. Poincaré asks: Are all systems integrable? Is it possible to select the appropriate coordinates such as to exclude the potential energy? Having showed that this is impossible in continuous media, Poincaré thus proved that dynamical systems consisting of a large number of particles are not integrable. The importance of Poincaré's discovery concerning the substantiation of the wave properties of matter is not fully acknowledged even now. Meanwhile, Poincaré not only proved the existence of nonintegrability, but also pointed out the reason for this—the existence of resonances between degrees of freedom, since each mode of the motion of interacting particles corresponds a certain frequency. If the particle interacting with other particles is removed from the equilibrium position, it begins to oscillate at a certain frequency. It is the characteristic frequencies of oscillations that lead us to the notion of resonance, which plays a major role in the Poincaré theorem. Resonances entail a nonlocal description and therefore cannot be included in the description at the level of individual trajectories associated with the Newtonian dynamics.

From the point of view of modern physics, the discovery of the wave properties of matter means a wave form of energy transfer (or pulses) in the processes of spontaneous structuring of masses with a nonequilibrium distribution of densities. The wave form of the motion of matter looks like an expansion in modes and assumes a spectral representation of energy levels and impulses, thereby destroying the basic principles of classical dynamics. As is known, these principles are based on Newton's laws—for an adequate description of the motion in terms of trajectories in the phase space, they require the coincidence of coordinates and impulses for each single point. With the introduction of wave functions into the phase space, each point (or its elementary volume) at any given time appears in the field of action of a whole spectrum of impulses. Under their influence, exponentially running trajectories arise at each end point of the phase space and the motion of each certain phase point (or trajectory) acquires a chaotic character reminiscent of the motion of a Brownian particle. With this approach, the very definition of a strange attractor and the Lyapunov exponent acquires a concrete physical meaning. Moreover, with the introduction of wave functions into the phase space, any finite point (or elementary volume) thereof, as well as any trajectory, splits into a self-similar series within its limit. Here, a fractal (or Cantor) structure of the phase space arises. All this leads to the appearance in the motion equations of new non-Newtonian terms that are incompatible with the description at the level of individual trajectories, which predetermines an attempt at a statistical probabilistic description of the processes of spontaneous structuring of masses with a nonequilibrium distribution of densities.

2.2 Statistical Description of the Manifestation of the Wave Properties of Matter in Terms of Ensembles

Trajectories are by no means the only language in which the dynamics of the spontaneous structuring of masses of density-unstable systems can be formulated: the laws of their dynamics can also be formulated at the statistical level (Prigogine 2001). This radically changes our description of the nature of the phenomena under consideration, since the primary objects of description in the statistical formulation are not trajectories, but probabilities. A significant step towards solving this problem was undertaken by the founding fathers of statistical physics—or physics of ensembles (Maxwell, Boltzmann, Gibbs, Einstein, Prigogine). This approach allows the physics of trajectories to be generalised, extending it to systems described by thermodynamics.

When studying physical processes from the point of view of thermodynamics, the state of the system must be specified, indicating not the position and velocity of its constituent parts (there are around 10^{23} molecules in a gas having a volume of 1 cm^3), as in the case of dynamics, but some set of macroscopic parameters, such as temperature, pressure, volume, etc. Consequently, the objects of thermodynamics compel us towards a distinct point of view as compared with dynamic objects. The purpose of the theory here, then, is not to predict the behaviour of the system in terms of the interaction of particles, but rather to predict the system's response to changes.

As in classical mechanics, in the theory of density-unstable ensembles, physical systems are considered in the phase space formed by coordinates and impulses. The dynamic state of a point particle (a material point) is determined by its position (a vector with three components) and the impulse (also a vector with three components). Such a state can be represented by two points (each of which belongs to its "own" three-dimensional space) or one point in the six-dimensional space of coordinates and impulses. This is the phase space. The geometric representation of the dynamic states of a single particle is generalised in the case of an arbitrary system of n particles. Instead of studying one single point (or trajectory) in the phase space, it is also possible to study the whole "population" of points (or trajectories)—an ensemble. Using this approach, J.W. Gibbs introduced the concept of population dynamics into physics (Gibbs 1902). The ensemble can be represented as a cloud of points in the phase space (Fig. 2.5).

This cloud is described by the function $\rho(q, P, t)$, which has a simple physical interpretation—it is nothing more than the probability of finding the system's representative point at time t in the vicinity of the point (q, P). The trajectory here corresponds to that particular case when the function ρ is zero everywhere, except for at the point (q_0, P_0).

The density ρ evolves in phase space as an incompressible fluid. The shape of the region can vary arbitrarily, but its volume remains constant with time for all deformations (Fig. 2.5).

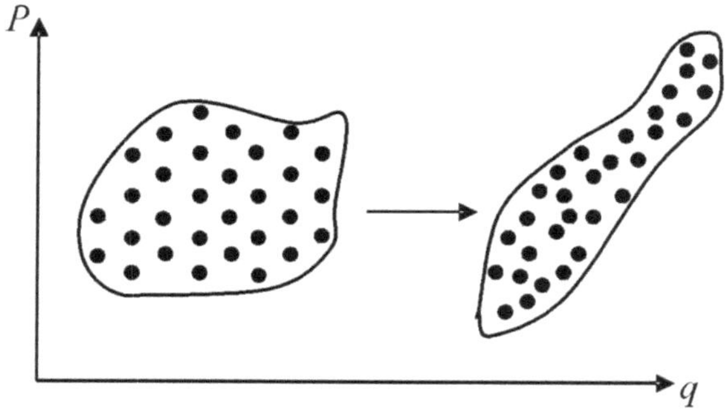

Fig. 2.5 Ensembles in the phase space. The Gibbs ensemble is represented by a cloud of particles differing in the initial conditions. The cloud's shape changes over time

The Gibbs-Einstein ensemble theory is a natural extension of the Boltzmann distribution. The density function ρ in phase space can be considered as an analogue of the velocity distribution function f, which was used by Boltzmann. However, in terms of its physical content, ρ is "richer" than f. While both ρ and f determine the velocity distribution, ρ contains additional information—in particular, as noted above, the probability of finding two particles at a certain distance from each other. The density function ρ includes everything necessary to determine correlations between particles, which are contingent upon their interaction; i.e., the probability distribution allows a complex microstructure of the phase space to be included in the framework of the dynamic description. Here it is no longer a point in the phase space that defines the initial conditions, but rather a certain region of the phase space, described at the initial instant of time $t = 0$ by the probability distribution ρ. As Poincaré showed, the reason for this is given by the existence of resonances between the degrees of freedom, that is, each mode of the ensemble's motion corresponds to a certain frequency. Thus, we arrive at a non-local description. Such systems are non-integrable by definition.

Let us now turn to the evolution of the probability distribution in time. This is determined by the so-called Liouville equation, which follows from classical dynamics. Since the derivation of the Liouville theorem is given in any textbook on statistical mechanics, we will not reproduce it here. In the operator notation, the Liouville equation has the form

$$i\frac{d\rho}{dt} = L\rho, \tag{2.22}$$

that is, the derivative of ρ in time is determined by the action on ρ of the Liouville operator L.

A particular case of the Liouville Eq. (2.22) is the Schrödinger equation

$$i\frac{h}{2\pi}\frac{\partial\psi}{\partial t} = -\frac{h^2}{2m}\nabla^2\psi + W_{pot}\psi, \tag{2.23}$$

where h is Planck's constant; $i = \sqrt{-1}$ is the imaginary unit; m is the mass of a microparticle; W_{pot}—its potential energy; $\psi = \sum_{j=1}^{n} \psi_j(x, y, z)\exp(-i\omega, t)$—the wave function; ω_j—the cyclic frequency for the *j-th* wave mode. In order to make the structural similarity between Eqs. (2.22) and (2.23) more obvious, we give another form of the Schrödinger equation:

$$i\frac{h}{2\pi}\frac{\partial\psi}{\partial t} = -\hat{H}\psi, \tag{2.24}$$

where $\hat{H} = -\frac{h^2}{2m}\nabla^2\psi + W_{pot}\psi$, referred to as the Hamiltonian operator, is a particular case of the Liouville operator L. The main formal difference here is that the Liouville operator L acts on the distribution functions ρ, and H acts on the wave functions.

The use of the operator formalism in classical statistical mechanics allows us to apply methods developed for quantum systems—the definition of eigenfunctions and eigenvalues for the Liouville operator—to density-unstable systems. This approach was introduced by Koopman (1931) and has been widely used since then in various areas of nonequilibrium statistical mechanics (Prigogine 1964). These operators now come into play whenever, for one reason or another, we have to abandon the concept of a dynamic trajectory along with a deterministic description of nature.

As in quantum mechanics, the solution of dynamic problems at the statistical level in classical statistical mechanics requires a definition of the spectral representation of the operator L. Since for a free particle the Liouville operator L is simply the differentiation operator $L = \frac{i\rho}{m}\frac{\partial}{\partial x}$ (Prigogine 2000), the eigenfunctions in this case have the form of exponentials exp (ikx), while the eigenvalues are in the form of pk/m (where $k = 2\pi/\lambda$ is the wave vector). The eigenfunctions exp(ikx) are periodic (plane waves e^{ikx}), since exp (ikx) = cos(kx) + i sin(kx). They extend throughout the space in stark contrast to the trajectory localised relative to the node point.

Further, as noted by Prigogine (2001), the classical theory of Fourier series (or integrals) proves that any function of the coordinate x, which we denote $f(x)$, can be represented as a superposition of periodic functions corresponding to the wave vectors k or, more specifically, in the form of a superposition of plane waves e^{ikx}. In this superposition, each plane wave is multiplied by the amplitude $\varphi(k)$, which depends on k. The function $\varphi(k)$ is known as the image of the function $f(x)$ under the Fourier transform.

We can briefly say that from the $f(x)$ function of the x coordinate we make a transition to the description of φ(k) in the k wave vectors. Of course, the inverse transformation is also possible. In addition, it is important to note that there is a kind of duality between $f(x)$ and 7 $\varphi(k)$. If $f(x)$ has a space interval of Δx (and is identically zero outside of it), then $\varphi(k)$ has a spectral range of $\Delta k \sim 1/\Delta x$. Any course in the theory of Fourier series indicates that when the spatial interval Δx increases, the spectral interval Δk decreases, and vice versa.

The eigenvalues pk/m of the Liouville operator correspond to the frequencies appearing in the Poincaré's resonances, which are directly measured in spectroscopy. They depend on k and p, but do not depend on coordinates. Consequently, the use of the wave vector k can be taken as a logical starting point when discussing the role of Poincaré's resonances. Plane waves allow us to describe not only trajectories, but also delocalised situations, which acquire decisive importance during the transition from classical dynamics to thermodynamics. While classical dynamics extracts a given number of particles, considering their motion in isolation from the rest of the particles, the statistical description in terms of ensembles takes into account the incessant interactions between the particles. The statistical description arises from a more holistic approach, taken when considering systems consisting of a large number of particles interacting as a whole. Thus, when the individual features of individual elements are set to one side, the motion (change of state) of such systems acquires a collective character. In describing the motion of such systems, the wave properties of matter are brought to the forefront, since oscillatory processes in a system of a large number of elements are represented by waves. In essence, waves comprise an oscillatory process in some extended medium.

Now we are able to understand from whence Poincaré's resonance effects arise in the statistical description of the dynamics of density-unstable systems. These resonances relate to the undamped wave interactions of particles in a given extended medium in much the same way as they connect harmonics in music.

Henri Poincaré showed that the total energy of such systems is given by the sum of the kinetic energy (due to the motion of particles) and the potential energy (due to their interaction). The processes of undamped wave interactions between particles that appear in our statistical description as a result of Poincaré's resonances are significant since they require the inclusion of wave functions in the phase space. In this case, the probability distribution in density-unstable systems decomposes into a sum of modes developing independently through time. Thus, the probabilities acquire an internal dynamic meaning and lead us on a statistical level to the discovery of the wave properties of matter.

At the macroscopic level, the wave properties of matter provide a coherent—that is to say, coordinated—motion of ensembles of molecules, leading to the occurrence of long-range correlations between particles whenever the thermodynamic parameters of the system exceed the convective stability threshold determined by the dimensionless Rayleigh number Ra. The solution gives the dependence of external parameters, for example, heating, determined by the temperature of hot T_h and cold T_c boundary surfaces of the layer $\Omega = (T_h - T_c)/h$, from the wave vector $k^2 = k_\perp^2 = k_z^2$ where $k_\perp^2 = k_x^2 + k_y^2$. The instability arises under the condition that the imaginary part of the frequency vanishes. As a result, the dispersion equation is

$$-k^2(k^2 - i\omega)(k^2 - i\omega \Pr) \mp Rak_{\perp}^2 = 0.$$

Here all the quantities are dimensionless. The units of length h (layer thickness) and time h^2/ν are introduced. The upper sign corresponds to heating from the top, and the lower one—from the bottom; the parameter Pr is the Prandtl number; the heating is included in the Rayleigh number.

Having satisfied the requirements of the homogeneity of the boundary conditions, we obtain the characteristic equation. It turns out that there are such boundary conditions (of free and isothermal boundaries) that, having adopted them, one can obtain the value of $k_z = \pi$, convenient for qualitative analysis of the arising effects. The dual eigenvalue problem is not solved in this case. The symmetry of the resulting cell (along the layer there is translational symmetry) is determined by minimising Ra as a function of $w = k_z^2/k_{\perp}^2$.

For ordinary convection, we have the excitation condition

$$Ra > Ra^* = \frac{27}{4}\pi^4 \approx 660 \quad for\ w = 1/2$$

Instability arises only when heating from below ($\beta > 0$). The build-up occurs aperiodically.

Table 2.1 shows the values of the critical parameters that determine $w = k_z^2/k_{\perp}^2$—the appearance of convection in a horizontal fluid layer heated from below for various boundary conditions. Here Ra_{cr} is the critical Rayleigh number; k_{cr}—the critical perturbation wavenumber characterising the periodicity of the perturbations; $\lambda_{cr} = 2\pi/k_{cr}$—the critical wavelength of the disturbance. These parameters at the macroscopic level determine long-range localised wave configurations, within which the processes of wave interactions between the particles are consistent.

The question of a set of long-range localised wave configurations, enclosed in various media, leads to a very profound set of considerations. Figures 2.6 and 2.7 demonstrate the presence of discrete jumps in the number of convection cells and, consequently, discrete changes in the critical wavelength of the perturbation with a continuous change in the thickness of cylindrical and annular liquid layers that are nonequilibrium in terms of density. In Fig. 2.7 the numbers above the curve indicate a discrete change in the number of convection modes across increasing the diameter of the cylindrical layer in experimental studies. The arrows under the curve correspond to the theoretical data. Typical forms of convection cells in

Table 2.1 Parameters determining the emergence of convection in a horizontal liquid layer

Nature of the surface boundaries	Ra_{cr}	K_{cr}	$\lambda_{cr} = 2\pi/k_{cr}$
Both free	657.11	2.22	2.828
Both rigid	1707.76	3.11	2.016
Upper free, lower rigid	1100.65	2.68	2.342

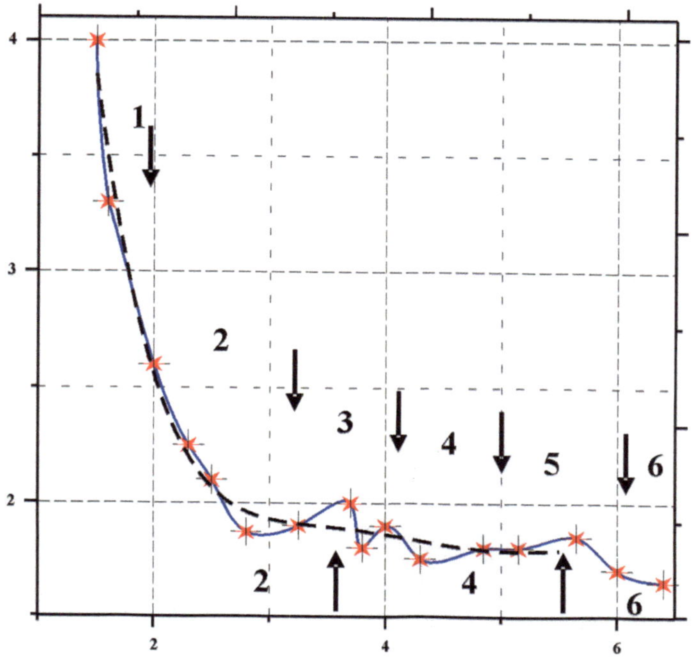

Fig. 2.6 Rayleigh number critical value as a function of cylindrical liquid layer diameter *h* (Stork and Muller 1975). The red stars and their connecting blue curve reflect the experimental data. The dashed line represents the theoretical approximation. Arrows indicate the discrete alteration of the convective mode number with an increase in the cylindrical nonequilibrium layer diameter. The theoretical data are represented below the curve; experimental—over the curve. {Legends on Fig. 2.6, left to right: Rayleigh, Ra; h (cm), diameter}

cylindrical cavities are given in Fig. 2.7. A similar change in the number of convection cells is also observed with increasing thickness of the annular layer. Analysis of the dependence of the number *Ra* on the thickness *s* of the nonequilibrium layer reveals a discrete change in the number of convection modes (Fig. 2.8). The conventional form of convection cells in annular cavities is shown in Fig. 2.9.

These experimental data clearly support the above-formulated conception at the level of statistical description that processes of spontaneous structuring of masses under conditions of a nonequilibrium distribution of densities in terms of their spatial manifestations—as well as in temporal dynamics—possess a discrete wave nature at a macroscopic level.

The establishment of these regularities opens the possibility for a transition to a description of the processes of spontaneous structuring of masses under conditions of a nonequilibrium density distribution from the positions of wave theory.

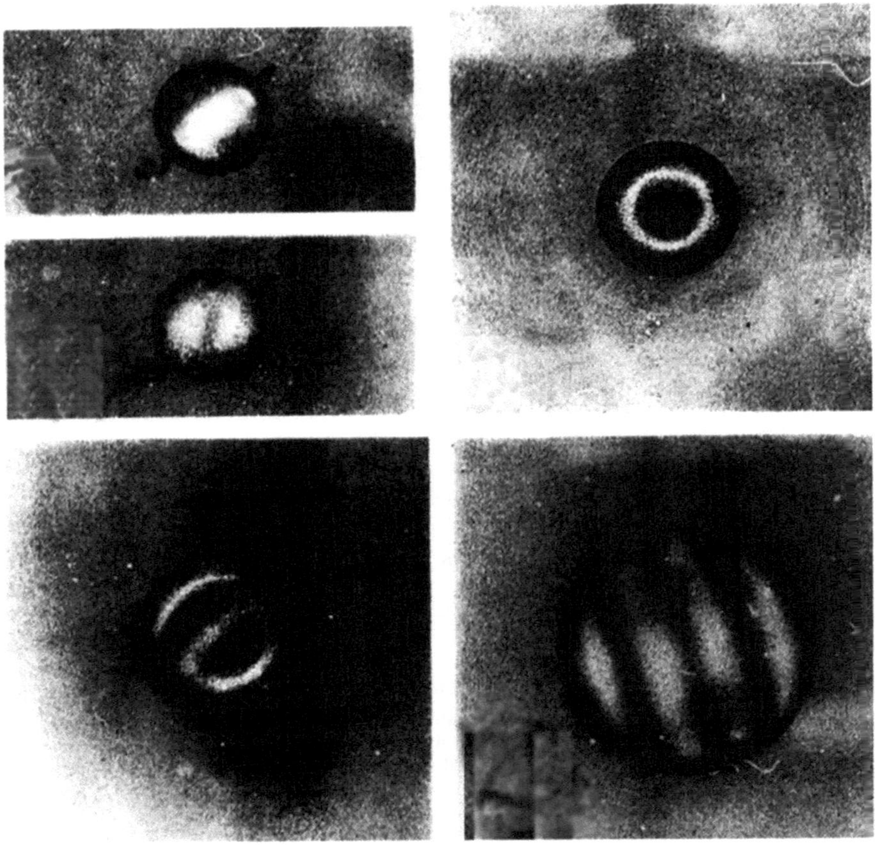

Fig. 2.7 Convective structure dynamics in the cylindrical layer (Stork and Muller 1975)

2.3 Description of the Wave Properties of Matter by Means of the Introduction of Wave Functions into Phase Space

The wave nature of linear and cellular spontaneous mass structuring with a nonequilibrium distribution of densities was first established in fluid dynamics as a result of the wide distribution of these phenomena in density-stratified media: in oceans, seas, lakes and the atmosphere.

Some elementary properties of infinitesimal wave motions in density-stratified media can be studied by considering waves at the interface between two layers of a liquid of different density (Fig. 2.10).

Due to such motions being analogous to waves on a free surface, they may already seem familiar to us. However, these familiar waves are not characteristic of

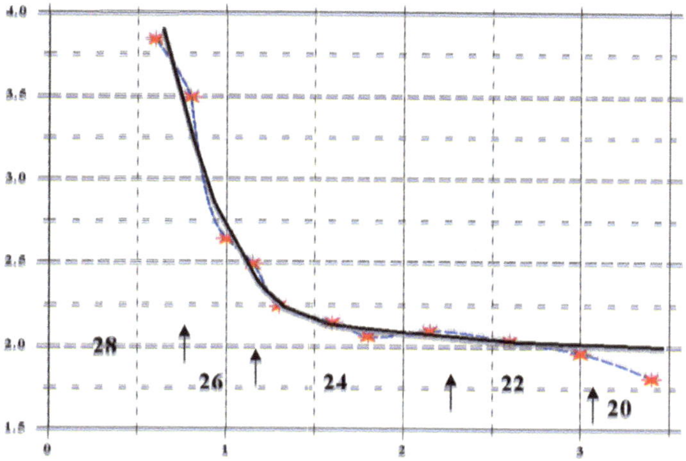

Fig. 2.8 Rayleigh number critical value for a random annular layer as a function of the ratio of inner and outer shell radii of this layer (Stork and Muller 1975) {Legends on Fig. 2.8, left to right: Rayleigh, Ra; Dimensionless thickness of cylindrical layer, s}

those wave motions occurring in continuously stratified media, since, in the latter, energy can be transmitted not only along a surface of constant density, but also at an angle to the horizon; therefore, our familiarity with the properties of surface waves will not help much here.

Internal standing waves having different modal structures in the presence of a continuous density gradient were reproduced in laboratory pools by Thorpe (1968). In Fig. 2.11, the dye stripes serve merely as labels of a continuous density gradient. They reveal a common property for density-layered media, namely, vertical oscillations of surfaces of equal density (located in one phase inside each standing wave cell).

Standing internal gravity waves refer to the fundamental properties of density-unstable media. Any perturbations in such media lead to the emergence of internal waves. However, despite the fact that convection and standing internal gravity wave phenomena have a common physical nature and are described by the same equations (Giterman and Shteinberg 1972; Landau and Lifshitz 1988; Turner 1977), the wave nature of convective processes in thermodynamics is not so obvious for the general reader and requires additional justification. To this end, experiments were carried out to study the behaviour of convective flows in a profile section. The formulation of the problem is complicated due to the necessity of taking into account the influence of the crystallisation process on the spontaneous dissipative structuring of a nonequilibrium viscous liquid layer. This is one of the refinements of Bénard's experiment, first implemented by Davis et al. (1984), as well as by Nguyen et al. (1989).

In such experiments, an alloy of lead and thallium is used. Due to the opacity of this alloy in the region close to the crystallisation front (melt–crystal phase

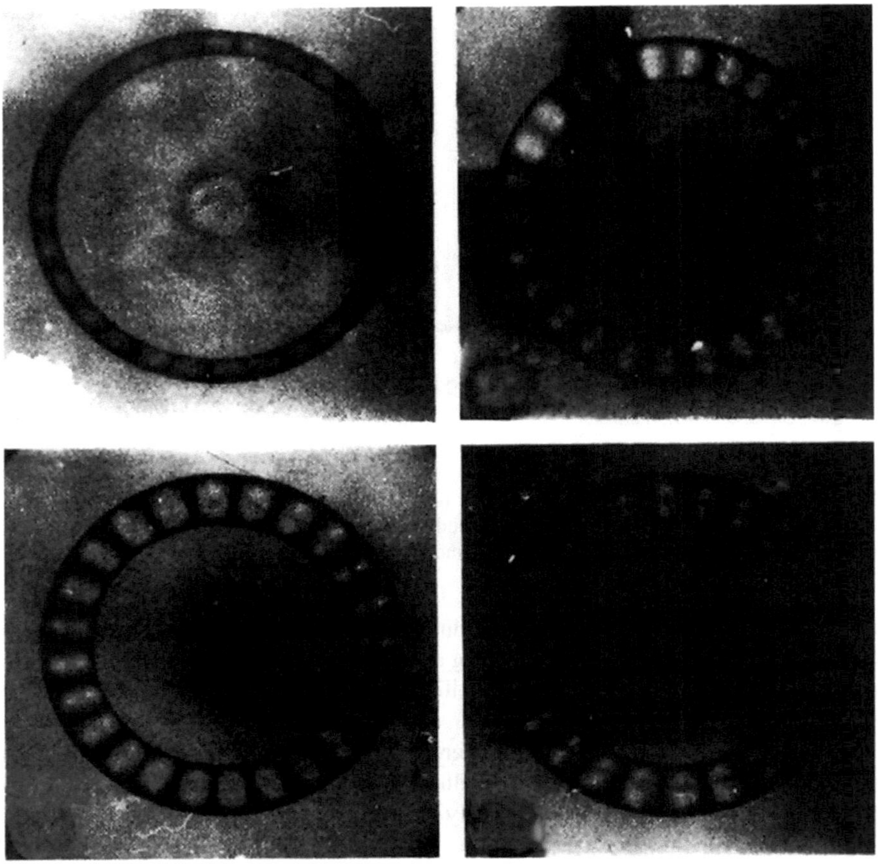

Fig. 2.9 Convective flow dynamics in the annular nonequilibrium layer (Stork and Muller 1975)

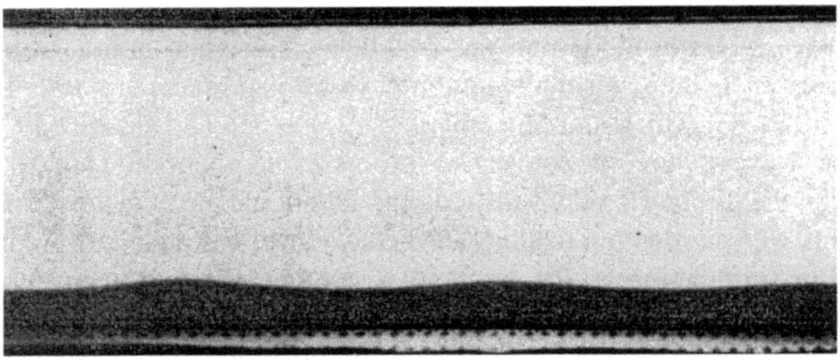

Fig. 2.10 The wave on the interface surface between two uniform liquid layers of different density and depth. The lower liquid layer is coloured (Turcotte and Schubert 1985)

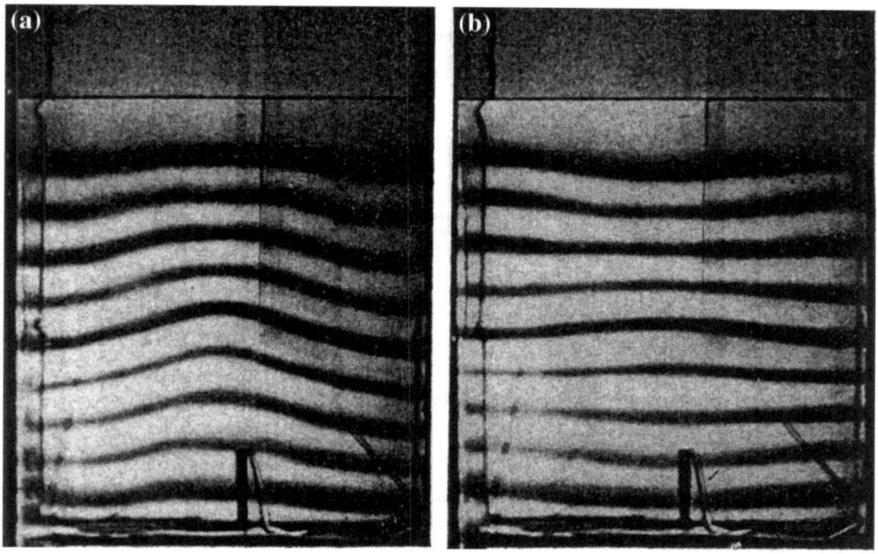

Fig. 2.11 Laboratory experiments on the studies of standing internal waves in continuously stratified liquid (Thorpe 1968): **a** modes (2,1); **b** modes (2,3). The colouring of the layers designating the constant density surfaces was managed during the filling of vessels with liquid

boundary), a laborious procedure has to be employed in order to describe the convective motion: at one-half crystallisation of the sample, the shape of the solid phase is fixed by a sharp cooling of the melt. Then a piece 0.5 cm thick, containing the areas belonging to the solid and liquid phases, is cut from the hardened part. The sequence of closely spaced transverse sections in the cut-off is quickly manifested by polishing using chemical etching, starting from the side of the liquid phase. Then, step by step, the material is erased to the boundary of the junction of the two phases. This in elaborate technology makes it possible to effectively determine the macroscopic features of the interface morphology (Fig. 2.12).

With the non-stationary growth of the solid phase, which can cause some banding of the formed material, there is no clear time dependence. The experiments are classically characterised by the convective threshold Ra/Ra_{cr}, as well as by the lateral restriction $\theta = \Phi/\lambda_{cr}$, where Φ is the inner diameter of the crucible; λ_{cr} is the wavelength at the convection threshold. At $0.52 \leq \theta \leq 1.00$, the solid phase structure has a single-oriented banding (Fig. 2.13).

As θ increases, a transition is observed from the domain-like structure of the solid phase to clearly crystalline structures (Fig. 2.14). The degree of ordering of the detected dissipative structures and their wavelength increase with an increase of θ, although this dependency is not monotonic. In all the cases considered, the section was taken at approximately the same depth level.

As the depth of the melt decreases during the formation of the solid phase, a polygonal dissipative structure appears initially on its surface; this structure is then

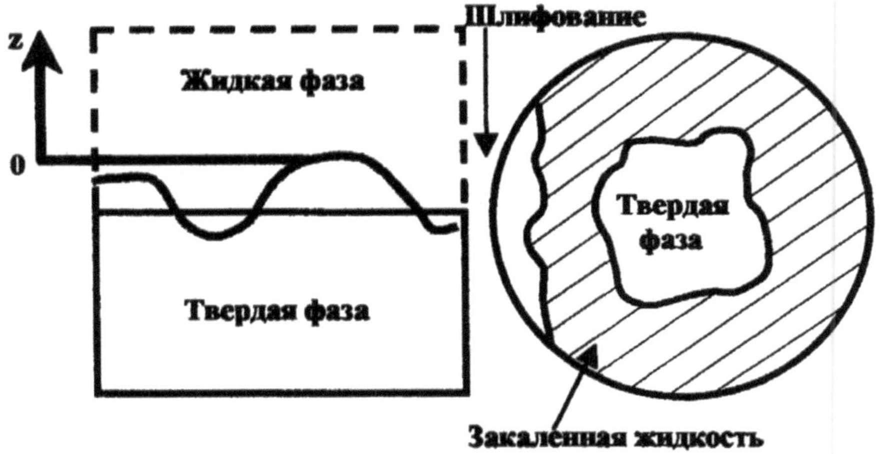

Fig. 2.12 Method of allocating the contact area between the solid and liquid phases in the crystallising melt (Thi et al. 1989) {Legends on Fig. 2.12, left to right, top to bottom: Liquid phase; Grinding; Solid phase; Solid phase; Tempered liquid}

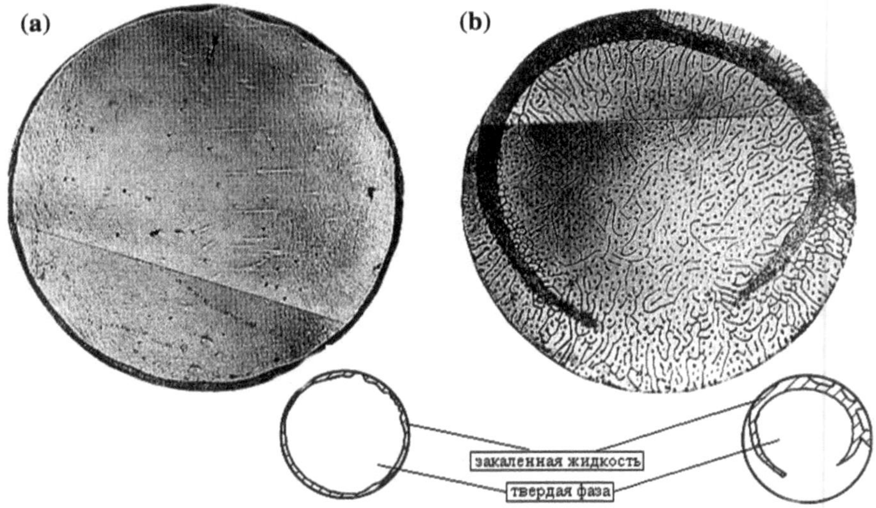

Fig. 2.13 Lead and thallium melt (Thi et al. 1989). Crystallisation speed: $v = 1.39 \times 10^{-4}$ cm/s, $\theta = 0.40$, $z = -3.8 \times 10^{-2}$ cm (**a**); $v = 1.81 \times 10^{-4}$ cm/s, $\theta = 0.52$, $z = -2.0 \times 10^{-2}$ cm (**b**). {Legends on Fig. 2.13, left to right: **a** tempered liquid; **b** solid phase

transformed into a hexagonal- and finally into a honeycomb-like- dissipative structure (Fig. 2.15).

Let us conditionally depict an unperturbed two-phase medium in terms of initially unperturbed horizontal layers (Fig. 2.16). In the case of crystallisation, a

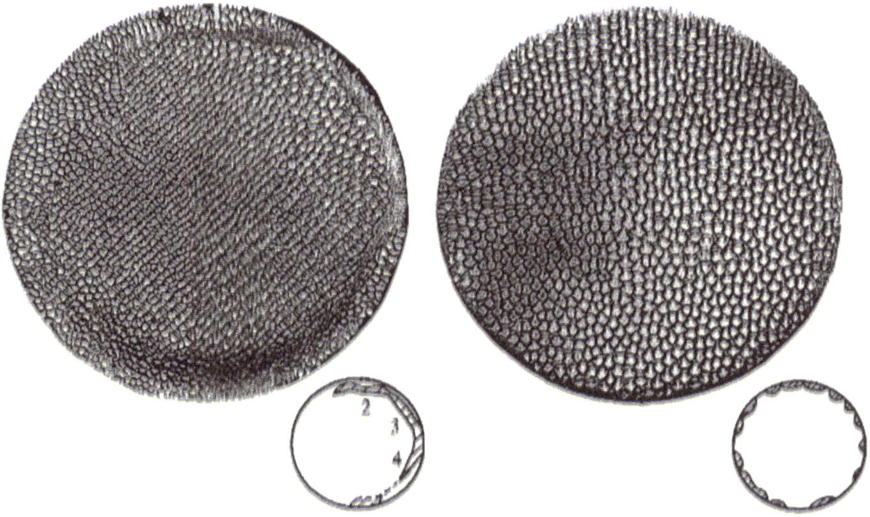

Fig. 2.14 Lead and thallium melt (Thi et al. 1989). Left: $v = 3.47 \times 10^{-4}$ cm/s; $\theta = 1.00$; $\phi = 0.95$ cm; $z = -3.1 \times 10^{-2}$ cm. Right: $v = 4.86 \times 10^{-4}$ cm/s; $\theta = 1.40$; $\phi = 0.95$ cm; $z = -3.8 \times 10^{-2}$ cm

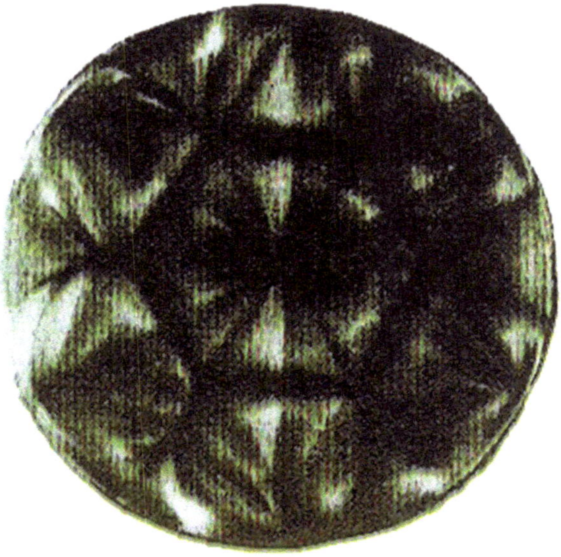

Fig. 2.15 Cell-like structure for Bénard-Marangoni convection in a cylindrical container with a diameter of 13.5 cm (Thi et al. 1989)

critical vertical temperature gradient, convection, and, as a consequence, its perturbation at the interface of the two phases, appear in the melt.

Thus, as in the case of Bénard's experiments, the convective structuring of a nonequilibrium viscous liquid layer is manifested laterally as domain-like,

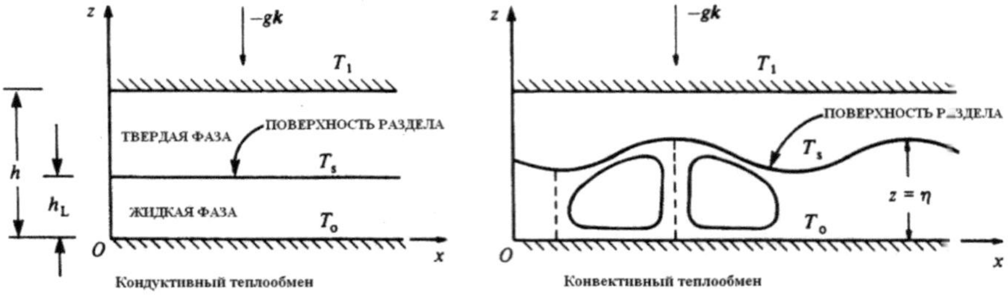

Fig. 2.16 Schematic image of the consolidation dynamics within the gravitational field of a partly crystallised melt (Thi et al. 1989). {Legends on Fig. 2.16, l to r, t to b: Solid phase; Interface; Interface; Liquid phase; Conductive heat transfer; Convective heat transfer.}

polygonal or, depending on the rate of crystallisation, hexagonal cells; both in the section and on the free surface, these are manifested as wavy shapes characterised by a certain amplitude of the wave process and the wavelength (wave number or vector). In this case, secondary effects such as the crystallisation process determine only the geometry of the dissipative structures.

The above examples of experimental studies confirm the unified physical nature of the convective heat and mass transfer phenomena as well as the generation of standing internal gravitational waves. Moreover, generalising the results of describing the wave properties of matter from the standpoint of classical physics in terms of trajectory and statistical description in phase space in terms of ensembles, we can state that in this case we are dealing not with different natural phenomena but with different ways of describing one and the same natural phenomenon—the spontaneous structuring of masses under the conditions of a nonequilibrium distribution of densities.

For continuous media and a large number of particles, such an approach is an excessive idealisation. New regularities of convective oscillatory processes are, however, manifested when taking a more holistic approach, according to which systems are considered as consisting of a large number of interacting particles as a whole. Vibrational processes in a system comprising a large number of particles lead to the generation of internal gravitational waves. The solution of this problem can be approached by two complementary methods. First we consider a method that emphasises the analogy with the problem of wave propagation in density-stratified liquids (Prandtl 1951; Turner 1977).

The statement of this problem, as well as its solution, are now classical in the theory of fluid dynamics and are described both in the fundamental work of Landau and Lifshitz, as well as in the monographs of their western colleagues Prandtl and Turner. Traditionally, the mathematical formulation of the problem begins with the basic differential equations, which in this case are represented by the equation of motion, the equation of continuity, and the law of conservation of energy. In the case of two-dimensional motions, following Turner (1977), the equations for the velocity components can be written in the form of

$$\frac{\partial u}{\partial t} + \frac{1}{\rho}\frac{\partial p'}{\partial x} = 0,$$
$$\frac{\partial w}{\partial t} + \frac{1}{\rho}\frac{\partial p'}{\partial z} + g' = 0,$$
$$\frac{\partial u}{\partial x} + \frac{\partial w}{\partial z} = 0,$$
$$\frac{\partial g'}{\partial t} - N^2 w = 0,$$

where the density distribution with depth $\rho = \rho(z)$ is assumed to be a continuous and smooth function; $N = \sqrt{-\frac{g}{\rho}\frac{\partial \rho}{\partial z}}$ is the buoyancy frequency determined by the vertical density gradient normalised to the average density; u and w are the horizontal and vertical components of the velocity vector, respectively; p' is the average pressure. According to the fourth equation in the last system of equations, the change in density at a given point of a nonequilibrium medium is simply due to the vertical displacement and has a phase shift of 90° relative to the vertical plane.

From the above system, using the successive elimination of variables, it is possible to derive various forms of equations for velocity components or stream functions. It is already clear that wave solutions of the last system of equations exist with the parameters u, w, p', g', depending on time proportionally to the factor exp ($i\omega t$). A particularly simple form of the solution is obtained for the case of an exponential density distribution, for example,

$$\rho = \rho_S \exp(-z/H).$$

which gives $N^2 = g/H$ = const, since the exponential factors can appear in all other terms. Here, following Prandtl, we write out expressions only for the velocity components:

$$u = a\omega \exp\left(\frac{z}{2H}\right) \cos(kx)\left(\frac{1}{2Hk}\cos(mz) - \frac{m}{k}\sin(mz)\right) \exp(i\omega t),$$
$$w = a\omega \exp\left(\frac{z}{2H}\right) \sin(kx)\cos(mz)\exp(-i\omega t)$$

where a is some amplitude of the oscillatory process; H—depth; k and m—wave numbers; $\exp(i\omega t) = \cos(\omega t) + I \sin(\omega t)$. Assuming that the reduced components of the velocity vector are related to the corresponding components of the displacement vector

$$u = \frac{dx}{dt}, \quad w = \frac{dz}{dt},$$

we obtain a unique geometric relationship between the stream function represented by a family of trajectories (or streamlines) at each point of which the velocity vector is tangential to the family of displacement lines. In Fig. 2.17 it can be seen that families of displacement lines are organised into ordered wave modes that are geometrically combined into spatially regular cells. These expressions describe a cellular standing wave with horizontal and vertical wavelengths, which are twice as large as the distance between the nodal lines (i.e., twice the width or height of the cells shown in Fig. 2.5, respectively). The cells are similar in size, but the amplitude in them increases with height in such a way that the energy proportional to p $(u^2 + w^2)$ remains the same in each cell. For a family of displacement lines, any motion possessing wave properties comprises a process of expansion in modes, that is, a decay into smaller cells. There is a certain frequency associated with each mode

$$\omega = N\left(\frac{k^2}{k^2 + m^2 + (1/4)H^2}\right)^{1/2},$$

with each frequency value corresponding to some discrete energy values.

Thus, for a family of displacement lines, the spectrum of waves of different frequencies makes up the total motion and determines the discrete-wave nature of the propagating energy. In this sense, the method of describing the processes of spontaneous mass structuring with the help of modes is, in principle, analogous to Newton's prism experiment, which demonstrated that a prism decomposes white light into a spectrum consisting of seven primary colours. In other words, white light is divided into simple wave components according to their frequencies. More than a century later, in 1807, the great French physicist Jean Baptiste Fourier showed that complex wave oscillations can be regarded as the result of the imposition of simple harmonic components, each of them having its own frequency, wavelength and phase.

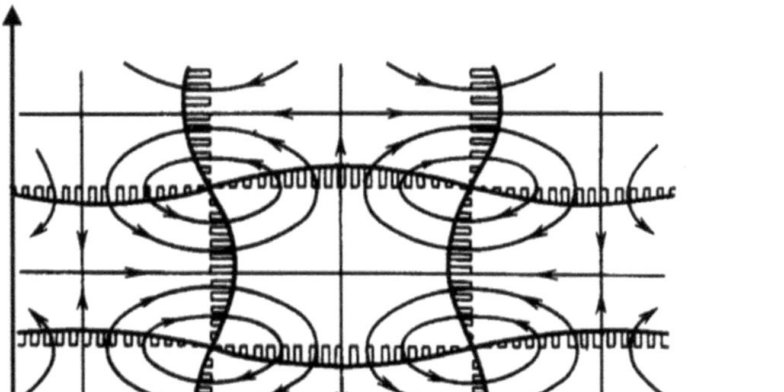

Fig. 2.17 Displacements and streamlines in cellular standing internal gravitational waves (Prandtl 1951)

The relationships between a family of displacement lines and a family of trajectories (or streamlines), as shown in Fig. 2.17, are in fact an excellent graphic illustration of the overlap of the two previously discussed ways of describing the wave properties of matter in phase space.

From the standpoint of classical Newtonian mechanics, the motion of each individual particle (or elementary volume) here occurs along individual closed trajectories (or stream lines). A space introduced in such a way can be considered as a phase space, and any of its points—as a phase or representative point. The set of phase trajectories (Fig. 2.17), corresponding to different initial conditions, forms a phase portrait of the system.

On the other hand, if we assign standing internal waves (or a family of displacement lines) to a family of streamlines, the entire volume of the nonequilibrium layer is divided into cellular sections whose boundaries represent the equilibrium levels with respect to which periodic oscillations of the ensemble of particles of a nonequilibrium medium arise in the phase space. These oscillations consist of periodic plane waves; in terms of statistical description in phase space in terms of ensembles, they correspond to the eigenfunctions of the Liouville operator. Plane waves reflect the result of the interaction of a large number of particles in continuous media and imply the transition from an individual description of the vibration of individual independent particles along phase trajectories to a description of long-range wave correlations between particles in continuous media, thereby solving the problem of introducing wave functions into the phase space.

Over time, the state of the system changes. Accordingly, from the standpoint of classical Newtonian physics, the position of the representative point in the phase space also changes; that is, this point describes a certain closed curve, referred to as the phase trajectory. For systems with one degree of freedom, the phase space is two-dimensional and is known as the phase plane; the equilibrium position is characterised by equilibrium (singular) points. All possible cases of the behaviour of phase trajectories in the phase plane near singular points are shown in Fig. 2.18.

On the basis of these special isolated structural elements, a complete classification of all structurally stable phase portraits of systems evolving in a two-dimensional phase space has now been developed in the theory of dynamical systems. Since the phase trajectories lie in one plane and do not intersect, there would seem to be only three possibilities for a dynamical system in a two-dimensional phase space:

- to perform periodic oscillations;
- to strive for equilibrium position without limit;
- to move away from the equilibrium position without limit.

Let us consider the possible connections of phase portraits with a simultaneous expansion of the phase space in terms of modes; that is, with the introduction of wave functions into the phase space under study. Figure 2.19 shows the displacement lines and streamlines (phase trajectories) in cellular standing internal

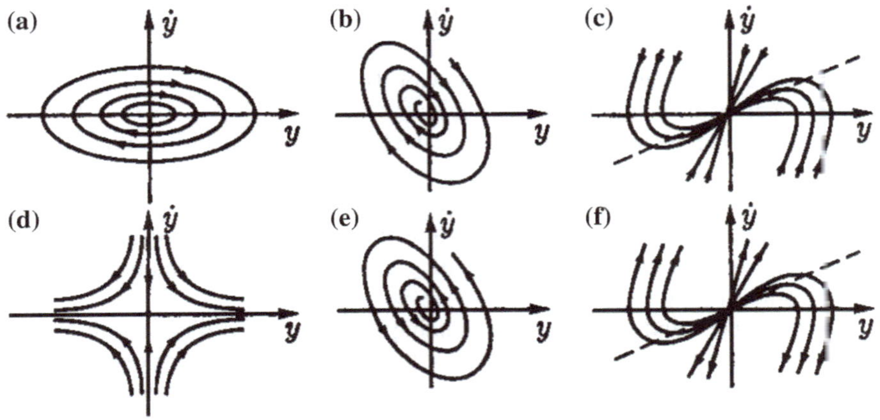

Fig. 2.18 Phase trajectory behaviour on the plane near the **a** singular points of the centre, **b** stable focus, **c** stable node, **d** saddle, **e** unstable focus and **f** unstable node

gravity waves in two layers with different density and a sharp interface between them.

This is the so-called Rayleigh-Taylor instability, which is often considered both in connection with the study of convection phenomena, as well as in describing internal standing gravitational waves in liquids and gases.

As we can see from Fig. 2.19, the behaviour of the trajectories in the phase plane is determined by the mutual arrangement of two types of equilibrium (singular) points—the centre and the saddle. The presence of a centre means that the system, after being brought out from the state of equilibrium, can perform periodic oscillations (Fig. 2.20).

The frequency and amplitude of the oscillations will depend on the mechanisms of dissipation and processes of energy and/or mass exchange with the environment. The current lines form closed trajectories around a stable centre, whose nodes are also delineated, separating the antinodes of the displacement lines of standing

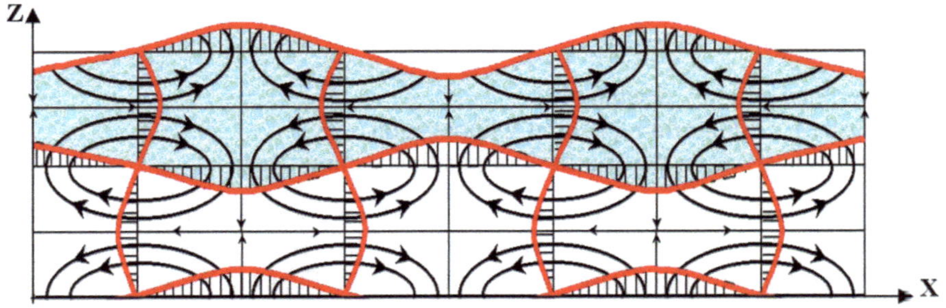

Fig. 2.19 Displacement and stream lines in the cellular standing internal gravitational waves characterising the Rayleigh-Taylor instability in two-dimensional planar phase space

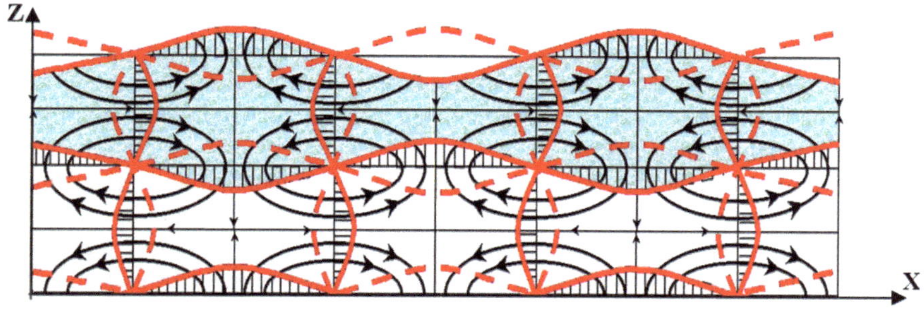

Fig. 2.20 Periodic oscillations of the displacement and stream lines in the cellular standing internal gravitational waves regarding the singular point—centre—in two-dimensional phase space

internal gravitational waves. The result of periodic oscillations with respect to an equilibrium (singular) point—centre—is the change in the direction of motion along the stream lines and displacement lines, i.e., the inversion of the structures. In this case, the displacement lines and streamlines of standing internal gravitational waves in the vicinity of the singular point acquire the dynamics of a "respiratory rhythm".

From the geometry of the sketch in Fig. 2.20 it follows that as the motion from the centre of one closed streamline to the centre of the other, the change of the radius and sign of the curvature of the intermediate streamlines is observed. As a consequence of their sign-alternating character, streamlines with an infinite radius of curvature arise. These special trajectories are called separatrices, which intersect at a singular point—the saddle (Fig. 2.18). At this singular point, a convergence of one pair of flows with an infinite radius of curvature is observed at the same time as a divergence of two other flows orthogonally to this pair, also having an infinite radius of curvature; that is, as applied to fluid dynamics, this singular point may be characterised as an unstable flow region. As can be seen from Fig. 2.18, separatrices divide the entire phase plane into four regions, whose phase trajectory behaviour differs in the nature of their asymptotics. Thus, the presence of a saddle indicates that the equilibrium point, like the entire dynamical system, is unstable. Under these developmental conditions of density instability processes, splitting of both displacement lines and streamlines occurs (Fig. 2.21). Like the phase trajectories of the strange attractor in the Lorenz system, the streamlines and the displacement lines reveal a Cantor structure; that is, they comprise a fractal set. Each displacement line, both at the interface and in the volume, splits into two, having a shorter length and smaller amplitude. If increasing the zoom, a further splitting of one of the components will be observed, and so on. With this type of splitting, the type and number of singular points on the phase plane change with each step. A singular point or centre, which characterises stable oscillatory regimes, turns into a singular unstable point—the saddle (Fig. 2.18). An alteration in the number and stability of solutions occurs; in the theory of non-linear equations, this is referred to in terms of branching—or bifurcation—of solutions. On the phase diagrams, the graphic images of these solutions carry a particular physical meaning. In losing stability, each centre of the previous level is divided into four new centres, to which the nodes of the displacement lines' antinode of the corresponding order are confined. The phase trajectories branch and close in the neighbourhood of these newly emerged singular points.

An even more complex process of discrete-wave fragmentation of displacement lines and branching of phase trajectories is revealed in the simulation of cellular hexagonal structures (Fig. 2.21). In addition to the hierarchy of closed trajectories corresponding to periodic oscillatory motions, hierarchies of trajectories with vortex and spiral components are outlined here. Complex tangles of splintering trajectories emerge. The process of such discrete-wave fragmentation of the displacement lines and branching of phase trajectories in a two-dimensional phase space can continue indefinitely. In this case, the motion of matter along the phase trajectories after one or two bifurcations acquires an irregular chaotic structure. In Fig. 2.21, any point is simultaneously located on two or three phase trajectories; thus, the direction of its motion at any given time cannot be predicted even in a two-dimensional phase space.

Such behaviour of streamlines indicates the occurrence of diffusion of discontinuities or fractions in the scheme of motions of a single point on the phase diagram. The situation with respect to the motion of each individual particle is analogous to a random walk, or Brownian motion. At each step, the future trajectory of the particle remains undetermined. The similar nature of the relation between the discrete-wave nature of the propagating energy (or the distribution of impulses) and the behaviour of a single particle in density-unstable systems poses the same problem that the originators of quantum mechanics encountered in studying the microcosmic world at the turn of the 19th and 20th centuries: the impossibility of simultaneous determination and description of the coordinate and impulse for each individual particle. The introduction of wave functions into the

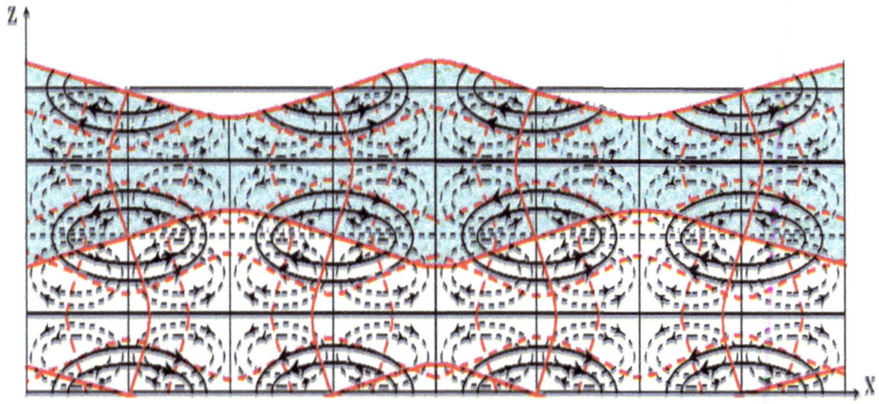

Fig. 2.21 Splitting of displacement and stream lines arising from the emergence of the density instability process in two-dimensional phase space during the formation of roll-type structures

phase space, as we have seen in the statistical description in terms of ensembles, introduces elements completely alien to the local description at the level of trajectories, which is based on the laws of classical Newtonian mechanics. The description in terms of trajectories becomes insufficient—and in some cases also impossible—since trajectories that are initially arbitrarily close, diverge exponentially with time. We have already encountered such trajectory behaviour in the previous sections when describing the wave properties of matter from the standpoint of classical physics in terms of trajectories in phase space. The typical portrayal of such behaviour in the form of a graphic image is a strange attractor.

Unlike the equations of the trajectories of classical Newtonian mechanics, which are traditionally used in describing the phenomena of spontaneous structuring of masses in the nonequilibrium distribution of densities, the introduction of wave functions into the phase space leads to wave equations. These equations are partial differential equations because, along with the time derivative, they contain coordinate derivatives, implying a transition from a set of vibrational motions of individual material points to continuous media. As we can see, this duality in the description of the wave properties of matter manifests itself whenever attempts to expand the dimension of the phase space are made, and serves as a direct reflection of the dual nature of the waves. According to the figurative expression of the famous American seismologist Bolt (1984), travelling waves inhabit the world, as observed through a time window; while standing waves entail the creation of a frequency window, going hand-in-hand with spectral representations. It is obvious that in the full description of the manifestations of the wave properties of matter, it is necessary to use both these complementary worlds of time and frequency.

From this standpoint, we cannot help but notice that the results of various ways of describing the manifestations of the wave properties of matter in processes of spontaneous structuring of masses with a nonequilibrium distribution of densities lead us to the notion that in these processes the waves themselves have an unusual dual nature, and, consequently, in this case, the wave properties must be considered from two points of view:

- firstly, in terms of characteristics that are determined mainly by the wave nature of the propagating energy;
- secondly, in terms of characteristics that are accompanied by spectral representations. The examples of wave spectra can often be found in everyday practice. We are already well acquainted with the rainbow spectrum, the first consistent explanation of which was given by the great English naturalist Isaac Newton. According to this basic principle striking spectral patterns can be constructed in the frequency region.

Interpretations of the spectra of oscillations and waves in the 20th century have acquired great importance in physical research. Many types of spectra of light, sound, seismic and other waves are used in solving inverse problems in order to obtain new data on natural objects. At the microscopic atomic level, line frequency spectra are a characteristic of vibrating atoms; for the first time, these provided an understanding of the fine structure of an atom that lies outside the limits of direct observation. The line spectra manifested following exceptionally strong earthquakes also characterise the oscillations of the Earth as a whole. The interpretation of these frequency spectra gave us much valuable information about the internal structure of the Earth. However, it is only today, in purposefully exploring the wave properties of the material world, that we come to understand the relevance of convective processes exemplified by Bénard cells as a special case of superposition of cellular standing internal gravitational waves.

Thus, having considered various methods of physical and mathematical description of the natural processes of spontaneous structuring of density-unstable masses, regardless of the dynamic or thermodynamic causes entailing this instability, we have revealed their wave nature. It appears that reflecting the main property of the wave oscillatory process, i.e. its expansion into modes, the wave properties of matter in the processes of spontaneous structuring of density-unstable masses reveal themselves in the formation of the fractal (or fractional) dimensionality of the space under study. Formation processes repeat themselves in different scales, as a result of which self-similar structures appear at different scales. With this method of perception of density (or convective) instability, we can characterise them using the concepts of amplitude, phase, wavelength, period and frequency. The change of a state of such systems acquires a collective character, in which the individual features of individual elements are set to one side. In the description of such systems, the wave properties of matter acquire a decisive importance.

Irrespective of their scale of manifestation and processes of induction, all these gravitational (convective) instability wave phenomena can manifest themselves in three types of structural form: (1) cellular hexagonal structures with emersion in the centre and immersion in the marginal parts of the L-hexagon type; (2) cellular structures with immersion in the centre and emersion in the edge parts of the G-hexagon type, and (3) linearly extended shaft-like structures of the roll-type. The results of laboratory and mathematical modelling indicate that the manifestation of hexagonal cells of the L-hexagon or G-hexagon types is generally determined by the effect of asymmetry effects and the characteristic sizes of the structures; depending on the thickness (width) and, to a lesser extent, on the viscosity and density of gravitationally unstable horizons, the amplitude increments of the dominant disturbances are determined through interaction with the process of selecting the prevailing wavelengths. As follows from the laws of hydrodynamics, the frequency of oscillation cannot exceed the buoyancy frequency of the medium; that is, in the upper layers of the Earth's crust and in the deeper structures of the Earth, it will be commensurable with the duration of geological processes. For this reason, such waves serve as the basis for the transition from waves to structures of density (convective) instability; or, more precisely, to the elimination of the boundaries between them. In addition, as a result of generalisation of empirical and theoretical data, it has been possible to establish that, since having a wave nature, all three structural gravitational (convective) instability forms exhibit a conformity between topographic surface and deep horizons, as well as possessing common typomorphic structural features that can be used for the identification and study of natural formations.

References

Graham A. Shear patterns in an unstable layer of air. Philos. Trans. Royal Soc. London. Ser. A. 1933; 232(714):285–296.
Palm E. J. Fluid Mech. 1960; 8.
Segel L.A., Stuart J.T. On the question of the preferred mode in cellular thermal convection. J. Fluid Mech. 1962; 13(2):289–306.
Busse F.H. Nonlinear properties of thermal convection. Res. Prog. Phys. 1978; 41:1931–1965.
Karpov N.V., Kirichenko N.A. Oscillations, waves, structures. M.: Fizmatlit; 2001. (In Russ.).
Saetzman B. Finite amplitude free convection as an initial value problem. Pt. 1. J. Atmosph. Scie. 1962; 19(4):329–341.
Lorenz E. Deterministic nonperiodic flows. J. of Atmosph. Sc. 1963; 20:130–141.
Nikolis G., Prigogine I. Exploring Complexity. M.: Mir; 1990. (In Russ.).
Poincaré H. About science. M.: Nauka; 1983. (In Russ.).
Prigogine I.R., Stengers I. Time, chaos, quantum. On the solution of the paradox of time. M.: Editorial URSS; 2001. (In Russ.).
Gibbs J.W. Elementary principles in statistical mechanics, developed with especial reference to the rational foundations of thermodynamics. New York: C. Scribner's sons; 1902.
Koopman B.O. Hamiltonian Systems and Transformation in Hilbert Space. Proc. Nat. Acad. Sci. USA. 1931; 17(5):315–318.
Prigogine I.R. Nonequilibrium statistical mechanics. Ed. N.D. Zubarev. M.: Mir; 1964. (In Russ.).
Prigogine I.R. The end of certainty. Time, chaos and new laws of nature. Izhevsk: NITs "Regulyarnaya i khaoticheskaya dinamika"; 2000. (In Russ.).
Thorpe S.A. On the shape of progressive internal waves. Philos. Trans. Royal Soc. London. Ser. A. 1968; 263(1145):563–614.
Giterman M.Sh., Shteinberg V.A. Izv. AN SSSR. MZhG. 1972; 2:55-61. (In Russ.).
Landau L.D., Lifshits E.M. Theoretical physics. Fluid dynamics. M.: Nauka; 1988. (In Russ.).
Turner J. Buoyancy effects in fluids. M.: Mir; 1977. (In Russ.).
Davis S.H., Muller U., Dietsche C. Pattern selection in singlecomponent systems coupling Bénard convection and solidification. J. Fluid Mech. 1984; 144:133–151.
Nguyen T.H. et al. Influence of thermosolutal convection on the solidification front during upwards solidification. J. Fluid Mech. 1989; 204:581–597.
Prandtl L. Hydroaeromechanics. 2nd ed. M.: IL.; 1951. (In Russ.).
Bolt B.A. In the depths of the Earth: what do earthquakes tell. M.: Mir; 1984. (In Russ.).
Kurdyumov S.P., Malinetskii G.G., Potapov A.B. Synergetics - new directions. M.: Znanie. 1989; 11. (In Russ.).
Stork K., Muller U. Convection in boxes: an experimental investigations in vertical cylinders and annuli. J. Fluid Mech. 1975; 71(2):231–239.
Turcotte D., Schubert J. Geodynamics: geological applications of the physics of continuous media. M.: Mir; 1985. (In Russ.).
Thi H.N., Billia B., Jamgotchian H. Influence of thermosolutal convection on the solidification front during upwards solidification. J. Fluid Mech. 1989; 204:581–597.

Chapter 3
Fractal Hierarchies of Dissipative Structures—Cellular Standing Internal Gravitational Waves in the Earth and Methods for Their Detection

The principal undertaking of the methods described in this work is to reveal multi-ordinal structures of gravitational instability—or dissipative structures of the Earth—and identify their typomorphic elements. The range of methodological approaches aimed at solving this issue is rather wide and includes: the analysis of seismic tomography data; maps of global geoid anomalies and gravitational potential; graphical and analytical methods for the decomposition of relief, figures representing the Earth's gravitational field; the deciphering of space and aerial images; morphometric, structural-geomorphological and structural-geological investigations.

A decisive role in the individualisation of the hierarchies of the Earth's gravitational instability was played by the establishment of the conformity of their deep layers with the external topographic surface. We should emphasise that this conformity refers here to the concept formulated by Gerasimov (1946), Meshcheryakov (1957), Khudyakov (1974), Khudyakov et al. (1983) concerning morphostructure as a geotectonic structure having an external geomorphological surface conformal to it. Here, the morphostructure is considered as a part of the geological space that is organised by geological processes, to which its external lithospheric relief and either structural and compositional rock complexes or the deep discontinuities of the Earth's crust, as well as the distribution of its density inhomogeneities, conform. In such an inextricable research unity of the external and internal geospheres, we are dealing with a geological body having its own surface and corresponding content uniquely integral to it.

3.1 Seismic Tomography Methods

When studying the deep zones of our planet, present-day geologists have the opportunity to draw on a wide range of geophysical data, including astronomical, gravimetric, magnetometric, magnetotelluric probing, etc. At the same time, the

O. V. Petrov, *The Earth's Dissipative Structures*, Springer Geophysics,
https://doi.org/10.1007/978-3-319-93614-7_3

results of teleseismic observations represented in the form of contemporary seismological tomography are of great importance.

In this regard, the pioneering works of American geophysicists (Aki and Lee 1976; Dziewonski et al. 1977) distinguished seismic anomalies on the basis of the first observations of longitudinal seismic waves. Coinciding with the expansion of the international network of seismic stations, in which seismographs register oscillations across a wide spectral range, providing material for seismological databanks, it became possible to diagnose the signals of a wide variety of waves. This allowed the dataset under consideration to be optimised, depending on the specific features of the problem being solved. Thus, for studying the Earth's crust, Rayleigh waves with a period of 50s and Love waves with a period of 70s turned out to be optimal; when studying the upper mantle, Rayleigh waves having a period of more than 100s were more indicative (Ritzwoller and Levshin 1998).

By now, the entire Earth is represented by global tomographic schemas, including the core and the lower mantle (Figs. 3.1 and 3.2) (Fukao et al. 1994; Su et al. 1994; Vasco et al. 1998; Resovsky and Ritzwoller 1999). Not only have variations in the seismic wave velocities within the individual layers been studied, but also the relief of all the main boundaries, including the phase transition zone at a depth of 670 km, as well as the topography of the outer Earth's core. The image of the high-velocity depth inhomogeneity is still in the process of being refined with improvements in the technology for processing the initial information. Interesting results were obtained in this connection by calculating the so-called differential

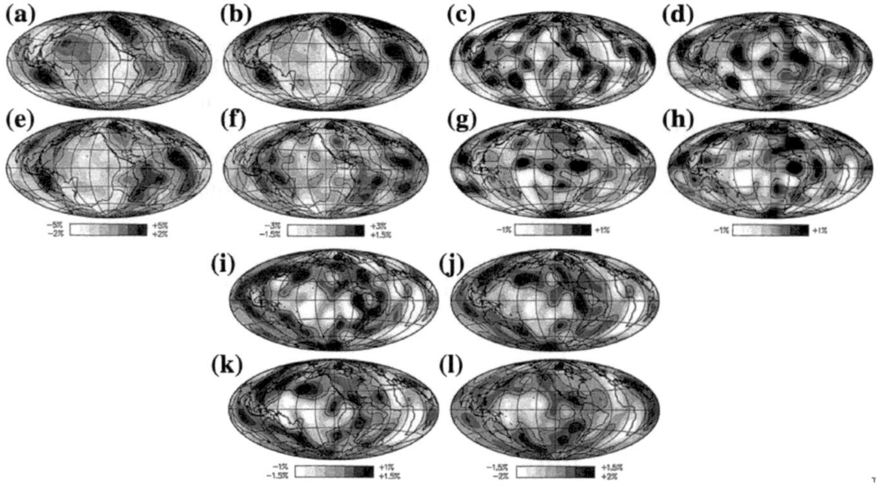

Fig. 3.1 Velocity anomaly maps for the S12 latheral seismic model (Su et al. 1994) among 12 different depth levels: **a** 50 km, **b** 200 km, **c** 400 km, **d** 650 km, **e** 900 km, **f** 1150 km, **g** 1450 km, **h** 1750 km, **i** 2350 km, **j** 2550 km, **k** 2600 km, **l** 2850 km. Double normalisation of the half tone scale (the scale of the velocity latheral component deviation from the average value expressed in percent) refers to the upper and lower maps, respectively

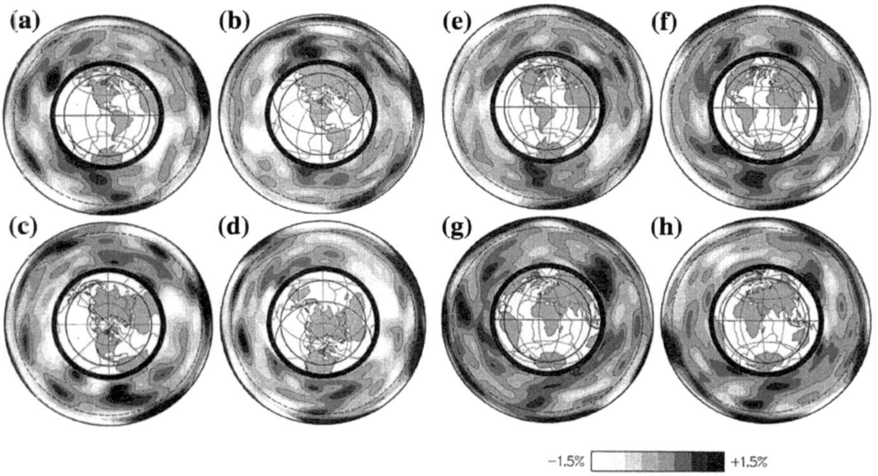

Fig. 3.2 Vertical seismic sections of the crust and upper mantle (S12 model) for the different lines of observation (denoted as a thick line on the globes in the middle of each section) (Su et al. 1994). A sharp rise in the shear wave velocity value is marked by the dashed line at a depth of approximately 670 km. Relative velocity alterations along all the sections are ±1.5%. Vertical sections: **a** polar along the 0 and 180° meridians; **b** 45° angle forming as regarded to the vertical axis (towards the North); **c** equatorial; **d** section similar to that mentioned for **b** but with a 135° angle forming towards the North; **e** polar along the 36° meridional line; **f** section similar to € but along the 72° meridian; **g** section similar to € but along the 108° meridian; **h** section similar to **e** but along the 144° meridian

velocities, which are free from the effects of local inhomogeneities in the vicinity of the source and receiver of seismic waves. The procedures for ensuring consistent exclusion of the influence of various factors (i.e. reductions due to the ellipticity of the Earth, seismic anisotropy, the age of oceanic lithospheric plates, phase transitions, station corrections, etc.) have been of great importance in understanding the nature of the detected anomalies.

The issue of the nature of high-velocity anomalies, whose discussion on the pages of scientific journals has continued over the course of the last few decades, is yet to be fully resolved. As far as is known, variations in the velocity of seismic waves can be affected by alterations in the density of rock strata associated with thermal state and phase transformations, as well as by their chemical characteristics, primarily in terms of the ratio of iron and magnesium. It is significant that an increase in the iron content of rock leads to an increase in its density, but reduces the speed of seismic waves. For this reason, the interpretation of the genesis of the seismic depth inhomogeneity can be ambiguous in some cases.

When deciphering the nature of seismic anomalies, additional information is typically adduced. Data on the nature of seismic wave absorption, velocity ratio of longitudinal and transversal waves, gravitational potential anomalies, etc., turned out to be useful in this respect (Jordan 1988; Durek et al. 1993; Cadek et al. 1994;

Su et al. 1994; Romanowicz 1995). By such means, it has been possible to indicate that the influence of the correlation of the medium's chemical and physical properties on the propagation of seismic waves varies substantially at different depths. Along with this, the lateral variations of the role of the influencing factors are found to be dependent on the type of geodynamic situation.

A clear illustration of the convergence of the elastic properties of mantle material due to changes in its composition and temperature can be found in systematic changes in the nature of the correlation between the geoid figure and tomographic data related to different depths (Cadek et al. 1995; Perrot et al. 1997). For example, thermal anomalies computed from the seismic data for the lower mantle were implausibly high, suggesting the effect of substantial iron enrichment at these depths (Yuen et al. 1993; Cadek et al. 1995). In a number of cases, however, changes in the correlation between the seismic wave velocities and the prevalence of the geoid over the ellipsoid may reflect other phenomena. In particular, it is established (Cadek et al. 1995) that the correlation between these characteristics can be reliably traced under the oceanic ridges and hotspots down to depths of ~1300 km, where it apparently disappears only to reappears at a depth of ~2000 km. The aforementioned authors tend to explain this regularity by the existence of horizons with increased viscosity in the lower mantle, which control the radial distribution of hot plume material.

In a number of attempts to understand the nature of the seismic heterogeneity of the Earth's interior, the possibility of a quantitative assessment of the influence of anomaly-forming factors within the upper mantle is of particular interest. Based on the study of the conjugated lateral variations in the velocities of longitudinal seismic waves, the gravitational potential and the electromagnetic field in the reduction, the global distribution of the generalised mantle temperature and the iron-magnesium ratio (both expressed in fractions of the standard deviation) was obtained by Yukutake and Tachinaka (1968). It was found that 58% of the fixed density inhomogeneities in the upper mantle are attributable to variations in the thermal decompaction of rocks, while the contribution of their iron-magnesium ratio is approximately 25%. At the same time, the fraction of polymorphic transformations (eclogitisation of basalt, etc.) accounts for no more than 17%. This result is in a good agreement with the spatial distribution of the Late Phanerozoic geodynamic aeon. For example, all the belts of the Cenozoic subduction zones are distinguished not only by negative temperature anomalies but also by their increased iron content, which is clearly confirmed in the known physical and chemical characteristics of submerged slabs. A similar conclusion was reached later by many geophysicists who analysed the results of seismic tomography.

Despite the remaining uncertainty regarding the nature of some seismic anomalies, the vast majority of geophysicists reasonably believes that these anomalies mainly reflect the thermally induced variations of the mantle material density. This makes it possible to use tomographic data to study the dynamics of the Earth's interior, as well as to identify the areas of gravitational instability in which dissipative structures emerge.

As an example, let us consider a three-dimensional model of the inhomogeneous distribution of transverse wave velocities in the mantle, obtained on the basis of a generalisation of 27,000 long-period seismograms and 14,000 kinematic observations (model S12-WM13 (Su et al. 1994)). In order to represent the model graphically, a spherical harmonics approximation up to the 12th order was used in order to describe the lateral change in the velocity of transverse waves as well as the expansion in Chebyshev polynomials up to the 13th order was used for describing the radial variations of this velocity. The revealed anomalies (Fig. 3.3a) extend to the depths of more than 300 km, and can be traced to the lower mantle in some cases. Some of them are located at shallow depths and underlie the basic elements of the world system of mid-oceanic ridges. There is a good spatial correlation between the main land masses and high-velocity variations recorded at depths of more than 300–400 km. The red and blue areas of the three-dimensional model correspond to a relative decrease and an increase in the seismic velocity; visualisation in this model begins from a depth of 100 km as for the discontinuity between crust and mantle. The comparison with the results of mathematical modelling indicates a very clear correlation between the three-dimensional seismic model with the isothermal surface obtained for tetragonal symmetry (Fig. 3.3b). It is obvious that low-velocity regions correspond to rising high-temperature plumes, while high-velocity regions correspond to the areas of descending (cooled) flows. It is remarkable that the coincidence is revealed precisely with that convective form, which, according to the modelling, is the most resistant to a change in the direction of the convective flow in a nonequilibrium spherical layer.

(a)

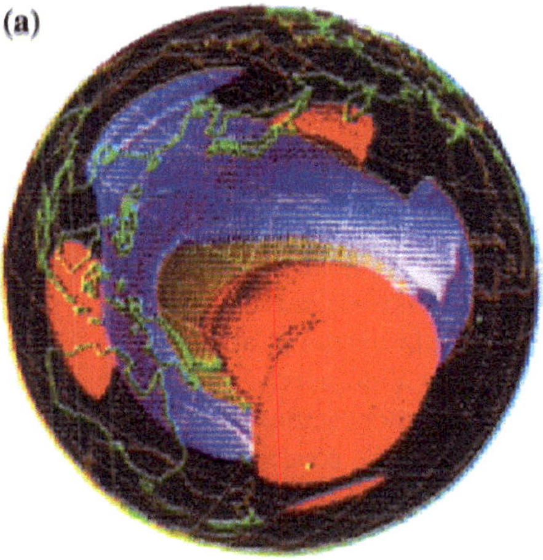

(b)

Fig. 3.3 Three-dimensional model of the distributional heterogeneity of shear seismic wave velocity (S12—WM13 (Su et al. 1994) **a** the areas of decreased velocity values are marked with red; increased velocity areas are marked with blue. The extrema (plumes) in the red and blue parts of the model correspond to the tetragonal shape; **b** comparison with the mathematical modelling of the free convection in the spherical shell

The authors of recent tomographic generalisations almost unanimously note the comparability of the results obtained by different researchers. In particular, this applies to long-period anomalies. In a number of cases, the configuration and distribution of short-period anomalies need to be refined.

Tomography has made significant additions to our understanding of the spherically zoned structure of the Earth's interior, resulting in the manifestation of additional boundaries to those previously known. Thus, the boundary at 220 km is considered to correspond to the maximum depth of penetration of the continental lithosphere, while a change in the character of the seismic anomalies is identified at the 920 km boundary (Kawakatsu and Niu 1994; Cadek et al. 1995). The hypothesis that a stagnation of the cold slab material, which, at a certain moment, as a result of gravitational collapse, begins to rapidly submerge into the lower mantle at the boundary of the endothermic phase transition (about 670 km) during the period from 100 to 400 Ma was convincingly confirmed by Fukao et al. (1994).

With a layer-by-layer comparison of the high-velocity anomaly spectra, an undisputed tendency towards "redshift" takes place during the transition to greater depths. The dimensions of the anomalies at the core-mantle boundary and in the lower areas of the lower mantle are larger than those in the upper mantle (Kennet and Bowman 1990; Su et al. 1994). Nevertheless, the change in the anomalous spectrum is not uniform and it has been possible to identify the layers where this trend is violated (Fig. 3.3).

Obviously, in and of itself, seismic tomography is only capable of revealing some aggregated picture of the long evolution of the Earth's interior. However, the resultant pattern of seismic heterogeneity can also be used for empirical control of other theoretical models of processes that occur in the core and mantle of the Earth. At the same time, the tomographic data allow a "time-base sweep" to be performed, i.e., when the results of paleogeodynamic reconstructions are involved in an analysis of the dynamics of the deep layers of the Earth. Thus, for example, the restoration of the geographic location of the subduction paleozones in the Late Phanerozoic made it possible to restore the dynamics of the immersion of paleoslabs, which are characterised by increased velocities of seismic waves (Cadek et al. 1994). Similarly, it is possible to trace the evolution of mantle upwelling. The superplumes nascent at the core-mantle boundary eventually result in a split of the continents, determining the main parameters of the kinematics of lithospheric plates over long time periods (Fukuyama et al. 1994; Maruyama et al. 1994).

As one of the results of the analysis of seismic tomography, taking into account the spatial-temporal variations in the density inhomogeneities of the mantle noted above, it was concluded that it is necessary to revise the concepts of convection as a steady-state process. Here, episodic convection models become the most adequate where the role of a particular trigger is played by large blocks of material of former slabs approaching core-mantle boundary and accumulating in layers of the upper and lower mantle (Maruyama et al. 1994). Mantle upwelling, globally controlled by superplumes with the rising of heated material, leads to the appearance of a multilayered heating system, accompanied by the generation of secondary, relatively shallow plumes. Let us note here that the possibility of the hotspots function originating from the upper mantle was hypothesised earlier when accounting for general theoretical considerations (Anderson 1980).

From this perspective, the general pattern of the dynamics of the Earth's interior appears as an "order born from chaos", with all the attendant phenomena: dissipative structures, bifurcations and attractors. Examples of the latter, in particular, are those structures that determine the stability of spreading axes and subduction zones. It can be imagined that passive rifting initially emerged as a reaction to the relative movements of lithospheric plates, generating the ascent of the asthenospheric substance, which in turn begins to play the role of a thermal attractor. The latter localises stable mantle upwelling, so that rifts acquire the features of typical active structures. Similarly, the downwelling attractor consists in the flows caused by submerging slabs.

The question of the extent to which the dynamics of the crust and mantle are reflected in the relief of the Earth's surface is of particular interest in terms of methodology. In recent times, there has been an accumulation of compelling evidence that modern topography is capable of demonstrating multiscale and disparate density inhomogeneities of the Earth's crust and mantle. In order to compare these external topographic surfaces of the identified natural objects with their internal deep structure, data was derived from seismic tomography, deep seismic and magnetotelluric probing, small-, medium- and large-scale geological mapping, structural geological and geotectonic studies. From a methodological point of view, the presence of this connection makes it possible to confidently isolate all three morphotypes of the Earth's gravitational instability structures on the basis of a complex combination of morphometric, structural-geomorphological, geological and geophysical data.

Direct visual analysis of the maps of global geoid anomalies and gravitational potential obtained from satellite data, as well as data from the relief of the Earth's core according to seismic tomography, is used to distinguish the global structures of the Earth's gravitational instability of G-hexagon type, with immersion in the centre and emersion in the marginal parts of the deep mantle substrate in the eastern hemisphere, and of L-hexagon type, with emersion in the centre and immersion in the marginal parts of the deep mantle substrate in the western hemisphere. The central parts of both structures are characteristically arc-shaped, demonstrating the conformal nature of the ratios of the core relief and the outer surface of the geoid (Figs. 3.4 and 3.5).

The direct correspondence between the anomalies of the geoid and the global gravitational anomalies is shown in Fig. 3.4. Positive anomalies of the geoid correspond to positive anomalies of gravity (Turcotte and Schubert 1985). In addition to gravimetric inhomogeneities, the direct connection of these anomalies with different levels of density inhomogeneities of the Earth is confirmed by seismic

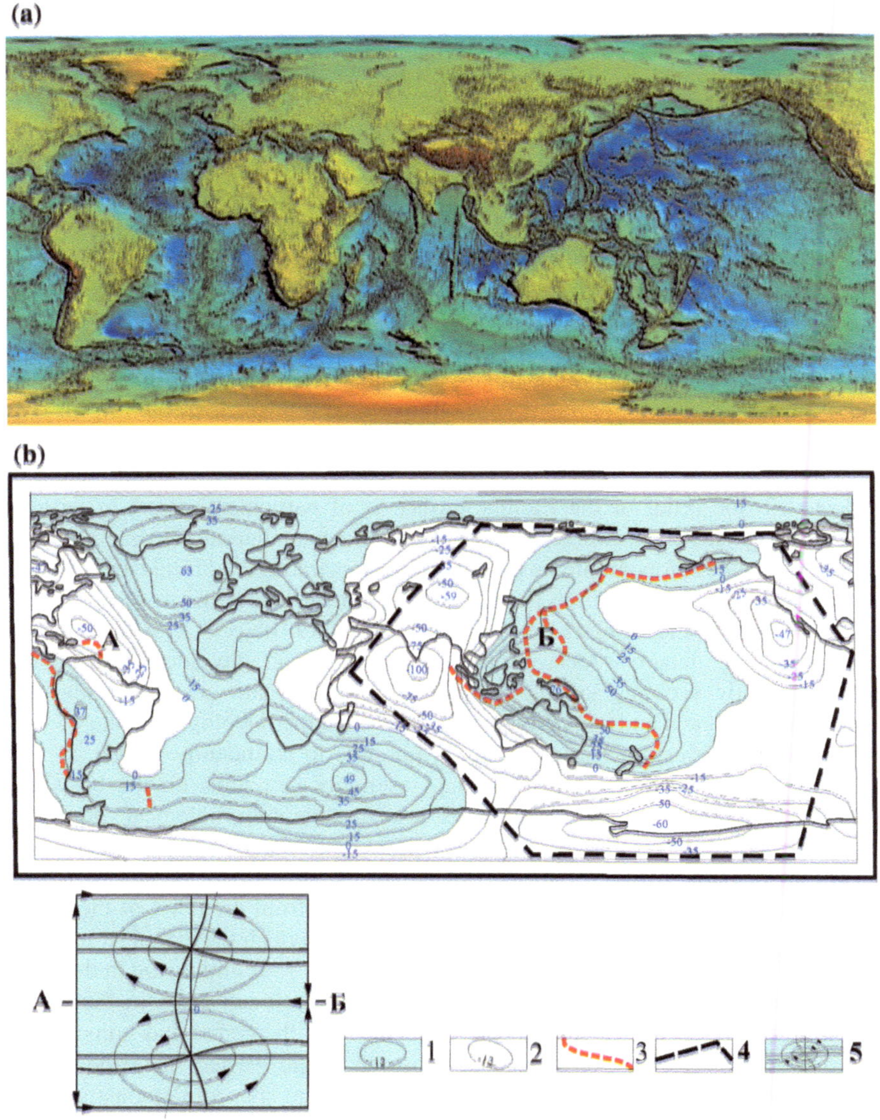

Fig. 3.4 Global structures of the Earth's gravitational instability of the first order in the latest normal cylindrical projection **a** surface relief map (based on the digital model in the ER Mapper 6.0 shell); **b** global geoid anomalies according to satellite data, GEM-9 model. Goddard (NASA) Earth model 9. Positive (1) and negative (2) excesses over the geoid's surface (isoline section 10 m); 3—subduction zones of the Earth's crust; 4—structural boundaries: A—cell with emersion in the centre and immersion in the marginal parts, B cell with immersion in the centre and emersion in the marginal parts; 5—displacements and streamlines in the structures

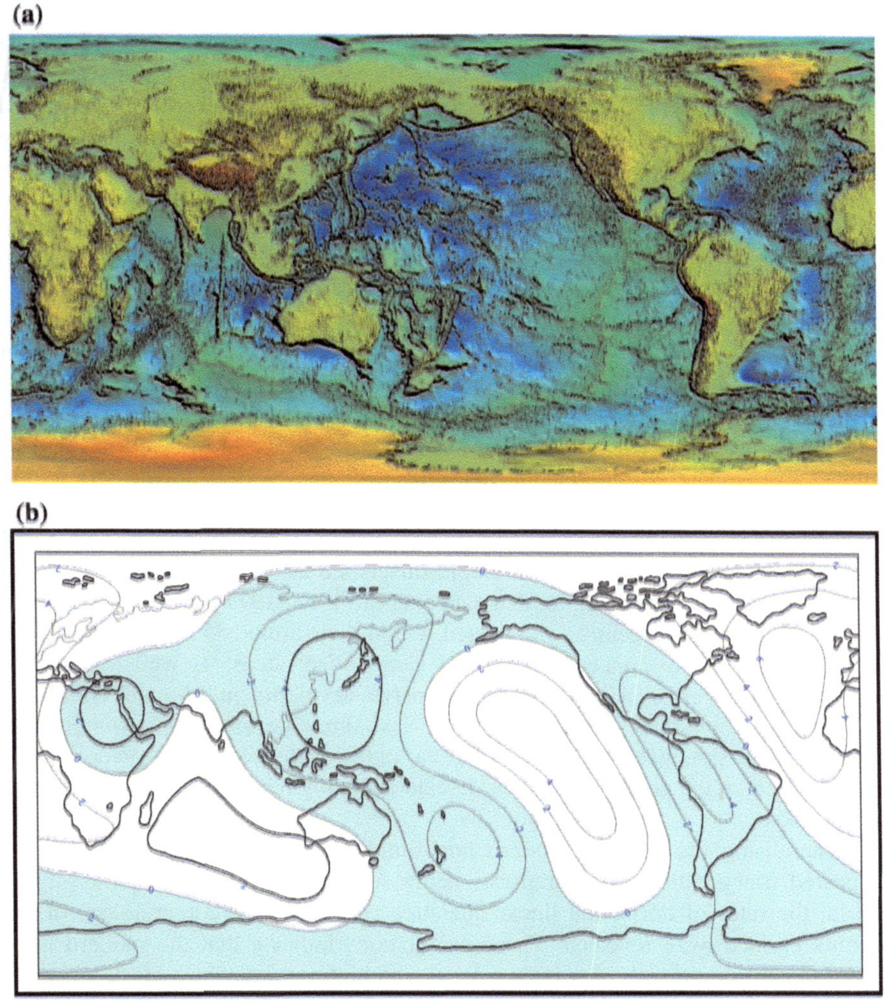

Fig. 3.5 Relief of the Earth's core according to seismic tomography data (isolines are drawn over 2 km). After Morelli and Dziewonski (1987) **a** surface relief map (based on the digital model in the ER Mapper 6.0 shell); **b** geoid global anomalies according to satellite data, GEM-9 model (Goddard (NASA) Earth model 9)

velocity data obtained by seismic tomography methods (Dziewonski and Woodhouse 1989). Seismic studies and their current version—seismic tomography—convince us that the large cells of G-hexagon type in the eastern hemisphere and the L-hexagons in the western hemisphere, identified on the basis of analysis of global anomalies of the geoid and gravity anomalies, are reflected at the core-mantle boundary, where the peaks and depressions with a total amplitude of 8–12 km are detected (Fig. 3.5).

In the eastern hemisphere, the central part of the G-hexagon type structure, with its immersion in the centre and emersion in the marginal parts of the deep mantle substrate, is fixed by large depressions along the relief of the core (Fig. 3.5). On the other hand, an arc-shaped positive anomaly of the geoid is confined to the central part of this grand cellular structure, including the western and central parts of the Pacific Ocean, most of Australia and the marginal part of the Asian continent. The marginal parts of this structure are fixed by a large negative cellular anomaly that occupies the northeastern edge of the Pacific Ocean, stretches up to the Indian Ocean, with a maximum formed by the south of India, and is further traced through Central Asia up to the north of Siberia (Fig. 3.4). In general, the outlines of this cellular geoid anomaly are conformal to the uplifting parts of the core relief.

Nikolaev (1988) reports on interesting calculations and comparisons carried out by E.S. Shtengelov. Comparing the shape of the geoid with the features of the manifestation of modern geotectonic activity, E.S. Shtengelov contrasts the data on epicentres of earthquakes and active volcanoes of the world, as well as seismological arcs and active continental margins, with that relating to the offsets and depressions of the geoid. It appeared that all seismic volcanic island arcs (except the Caribbean) and all active continental margins are located in the areas of geoid excess over the geodesic ellipsoid. Geoid offsets are also characterised by a more intense seismicity. According to M-6 data over the 1963–1977 period, 82.7% of the world's earthquake epicentres are located within their limits. 85.5% of the world's active volcanoes are concentrated at the geoid offset areas, as well as the vast majority of the areas with the most intensive development of mud volcanism (Azerbaijan, Taman and Kerch Peninsulas, Sakhalin, Oceania, South America). In addition, E.S. Shtengelov points out that the uplift of the sea shores predominates on the offsets of the geoid, while their descent is mainly observed in the depressions of the geoid.

Generalising the data on modern horizontal movements of the Earth's crust (repeated triangulations, trilaterations, stadia surveys) in different regions of the world, the rates of horizontal linear dilatation (expansion or compression) of the Earth's surface were revealed. These calculations indicate that the modern horizontal expansion prevails over compression, and higher expansion rates dominate at the offsets of the geoid. Almost all areas of modern horizontal compression of the surface are confined to the depressions of the geoid. The greatest rates of modern horizontal expansion are at the boundaries between the offsets and depressions of the geoid.

Within the Pacific tectonic segment, the central area of the immersion of the Earth's gravitational instability structure, such as the G-hexagon type, is emphasised by the arc-shaped curved world system of islands in the west of the Pacific and East of the Indian Oceans. The latter, as is known, is a complex of closely related tectonic formations, which include a chain of volcanic islands, a deep-sea trench adjacent to it, an obliquely falling seismic activity zone, as well as marginal basins and other structures of lesser magnitude. All these formations play a key role in the concept of plate tectonics, since they are confined to places of volcanic activity associated with the subduction of the oceanic crust.

The marginal parts of the global structure of the gravitational instability of the Earth, such as G-hexagons, are characterised by the extensive development of inland and oceanic rift systems over the zones of the emersion of the deep mantle substrate (Fig. 3.6). A single system of mid-oceanic ridges and uplifts extends from the top of the Gulf of California in the eastern Pacific to the Gulf of Aden in the northwestern part of the Indian Ocean.

In the eastern part of the Pacific Ocean, the emersion of a deep mantle substrate is fixed by the colossal East Pacific Rise divided by a large transverse fracture (shear) Eltanin system into two morphologically distinct rises: the South (South Pacific) and the East (East Pacific) itself. In the cross-section, the East Pacific Rise is a broad and very gently sloping elevation extending over a distance of about 8000 km. The East Pacific Rise, as well as the vast, deep-sea floor spaces adjacent to it on both sides, are divided into segments by giant latitudinal faults. The morphology and structural features of the uplift on these segments vary to some degree. A central rift valley along the crest of the ridge is mostly absent; this is one of the structural features that distinguishes the East Pacific Rise from the Mid-Atlantic ridge. According to the morphological features and specific spreading features, the East Pacific Rise is also contrasted with the Mid-Atlantic ridge, representing a different type of the structural formations of this category. The analysis of the structural development of the East Pacific rift belt generally shows the non-simultaneous incipiency of its individual parts, as well as the reorientation of spreading zones in time, accompanied by the dying and destruction of some systems and the emergence of new ones. The South Pacific Rise extends sub-latitudinally in the form of a slightly convex arc to the southeast of the Macquarie Fault at about 160° E. as far as the Eltanin Faults, having an extent at 4 km isobat of about 6000 and a width of 800–1200 km (Zhivago 1979).

This consists in a gently convex swelling of the Earth's crust with a large block relief and peaks of about 2.5 km. A rift valley is absent on the western and eastern flanks of the ridge, but is projected in its middle part. In the west, the South Pacific Rise stretches into the Indian Ocean, being separated from the middle Australian-Antarctic Rise to the south of Australia by a large fault system with submeridional shifts. The length of the Australian-Antarctic Rise is about 5500 km, having a width in the eastern and central parts of 800–1000 km; in the western part —up to 1500 km. This rise is also characterised by the absence of an axial rift valley.

Deeper within the Indian Ocean, the marginal part of the global Earth's gravitational instability structure of G-hexagon type is fixed by the Central Indian, Arabian-Indian and Aden ridges, in the axial part of which, as a rule, a well-defined rift valley is established. These ridges are intersected by transverse transform faults running along its entire length. In addition, a large number of relatively young intracontinental rift zones in North America, Antarctica, West Africa, Eastern and Western Siberia are associated with the marginal part of this global structure. These rift zones cross the continents and their margins and cause the formation of large crustal depressions (sedimentary basins or young platforms) both within the continents themselves and under the seas of the Antarctic and the Arctic.

(a)

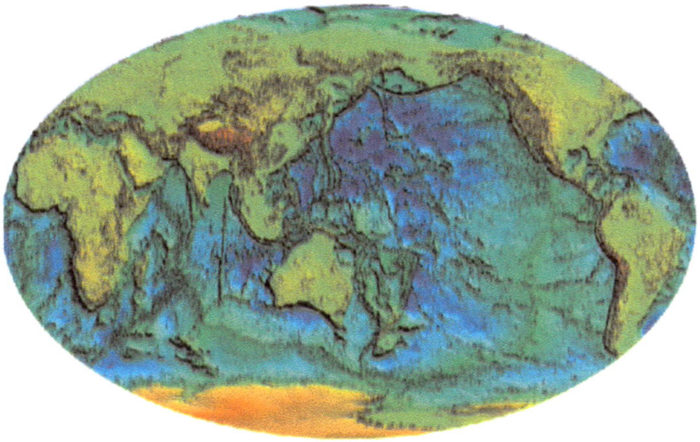

(b)

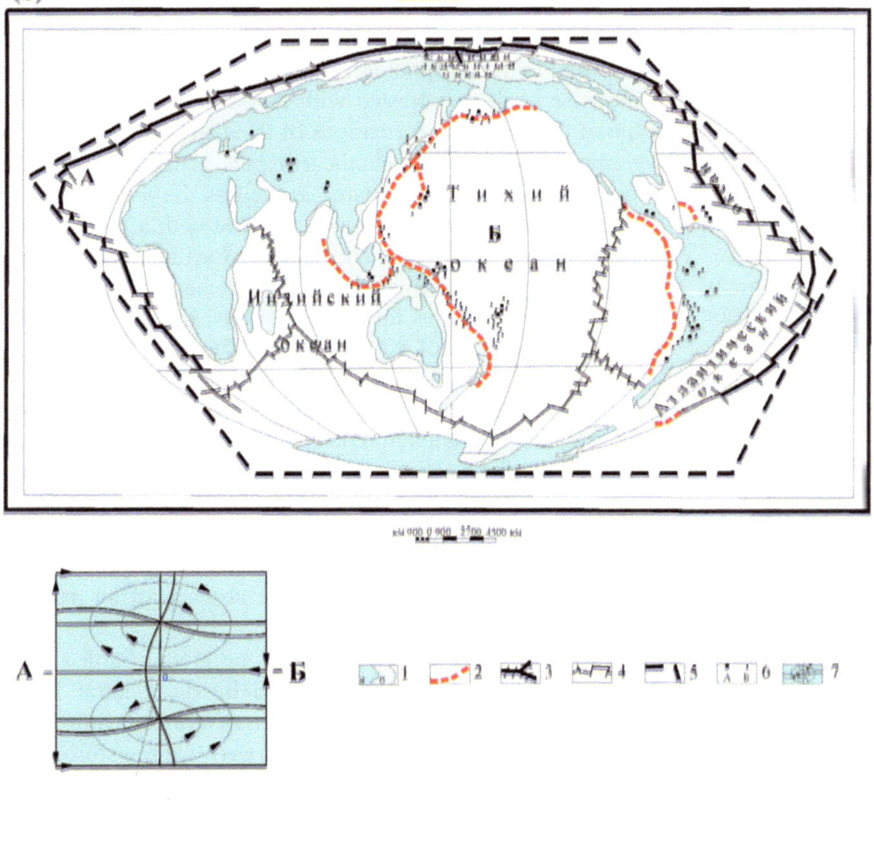

◂**Fig. 3.6** The global structure of the Earth's gravitational instability of the first order in the latest modified pseudocylindrical elliptical equal-area Molveide projection **a** surface relief schema (based on the digital model in the ER Mapper 6.0 shell); **b** the image of hexagonal Earth's structuring on a planetary scale. 1—continental Earth's crust (blue) and shelf area (light-blue); 2—subduction zones; 3—boundaries of the central zones and the direction of spreading of the Earth's crust in the marginal parts of the spherical cell of the first-order thermogravitational instability of the Earth; 4—the same, for lower orders; 5—boundaries of the structure of the thermogravitational instability of the Earth of the first order; 6—epicentre distribution of the strongest intermediate and deep earthquakes; 7—displacements and stream lines in the cellular structures of the Earth's gravitational instability of the first order

Thus, in considering the tectonic plan of the Earth's global gravitational instability structure of G-hexagon type, it is possible to elucidate for the first time both the confinedness of the modern island arcs as well as the reason for their asymmetric distribution on the Earth's surface, and the close contingence of the inter-continental and oceanic rift systems, at the same time as outlining a particular structural position regarding the boundaries of the aforementioned global structure of the North America and the West Siberian lowland. In the heterogeneous basement of the latter, according to Aplonov (1989), residual depressions having preserved the lithosphere of the Paleozoic oceans can be identified, as well as numerous Permian-Triassic rift zones that continue developing until the spreading of the incipient Early Mesozoic ocean.

According to seismic tomography, the central part of the structure of the Earth's gravitational instability of the L-hexagon type is manifested in the western hemisphere, with its characteristic pattern emerging in the centre and immersing in the marginal parts of the deep mantle substrate. This also has a characteristic arc shape which embraces both the Atlantic and Indian oceans, uniting them in the Indo-Atlantic geodynamic system (Fig. 3.5). There is no morphological threshold between these two oceans; the mid-oceanic ridge acts like the African-Antarctic element of the mid-oceanic ridge of the Indian Ocean, stretching uninterruptedly into the Atlantic Ocean. In the vicinity of Bouvet island and the Captain Spiess bank, the African-Antarctic Ridge connects with the American-Antarctic branch and then stretches to the north forming the Mid-Atlantic Ridge. The morphological and geological-geophysical parameters of the Mid-Atlantic Ridge undergo significant variations. Its axial zone is heavily intersected by a system of transform faults; the underlying idea of the spreading nature of the ridge structure is violated through the manifestations of heterogeneity in both the longitudinal and transverse directions.

The peculiarity of the morphostructural plan of the united Indo-Atlantic geodynamic system is emphasised by the distribution of the continental and oceanic crust. In the terms of the tectonics, the central part of the global structure of the Earth's gravitational instability of the L-hexagon type is represented by the spatially conjugated oceanic rift systems of the Atlantic and Indian oceans, while its marginal parts are fixed by subduction zones at the eastern and western boundaries of the Pacific Ocean (Figs. 3.4, 3.5 and 3.6). A wide annular belt with wave velocities that increase by 0.3% is indicated by seismic tomography in the lower mantle of the

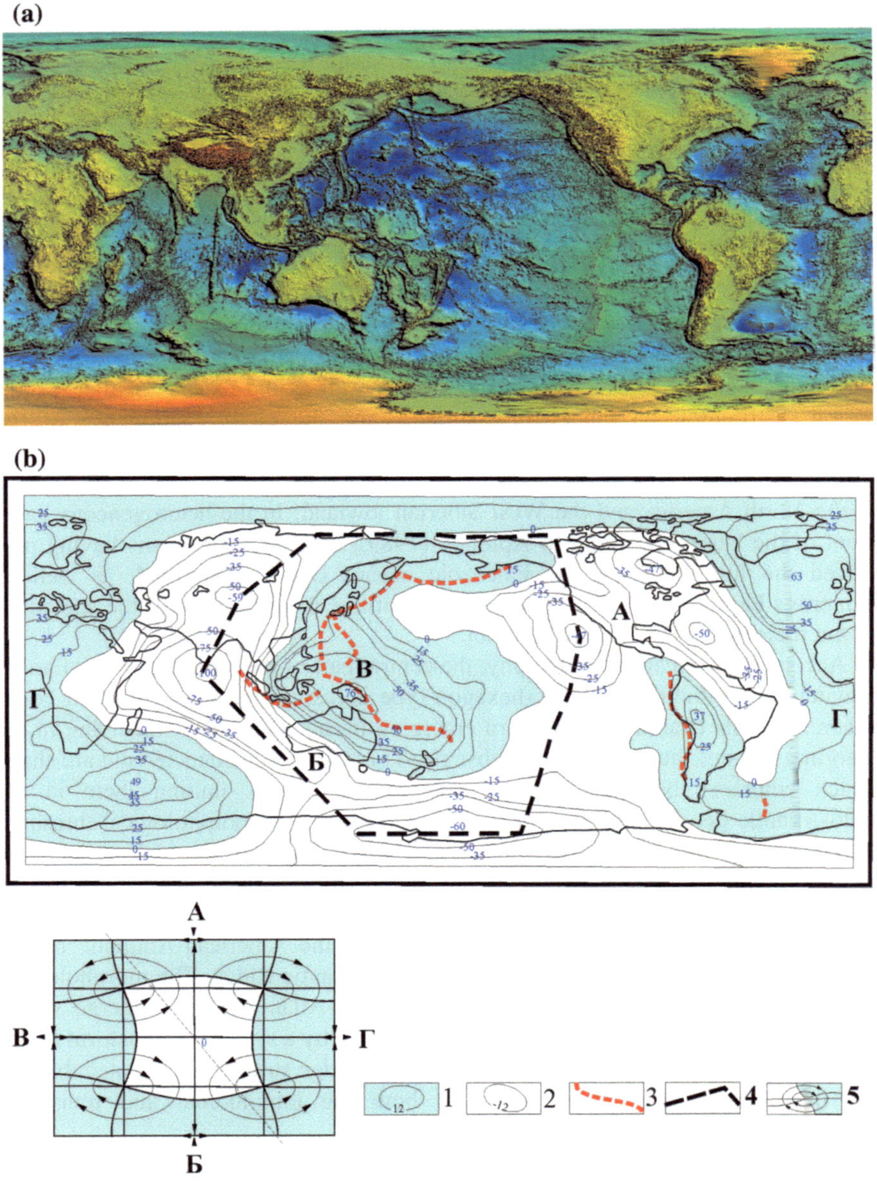
(a)
(b)
А
Б
В
Г
1
2
3
4
5

◀**Fig. 3.7** The global structures of the Earth's gravitational instability of the second order in the latest normal cylindrical projection **a** surface relief map (based on the digital model in the ER Mapper 6.0 shell); **b** global geoid anomalies according to satellite data, GEM-9 model (Goddard (NASA) Earth model 9). Positive (1) and negative (2) excesses over the geoid's surface (isoline Sect. 10 m); (3) Earth's crust subduction zones; (4) structure boundaries: A and B—structures of gravitational instability with emersion in the centre and immersion in the marginal parts, C and D—structures of gravitational instability with immersion in the centre and emersion in the marginal parts; (5) displacements and streamlines in the cellular structures of the Earth's gravitational instability of the second order

Pacific region at a depth from 1300 to 2300 km. The middle line of this ring passes under the subduction zones of the Pacific mobile belt and East Antarctica (Kropotkin and Efremov 1994). Dziewonski and Woodhouse (1989), as well as Maruyama et al. (1994) interpret the ring zone of increased densities in the lower mantle as being associated with the immersion in it of the material of lithospheric plates that have undergone subduction.

The relief of the core does not coincide completely with the outer surface of the geoid. A large arc-shaped negative anomaly of the geoid encompasses the central and southern parts of the Atlantic Ocean, interlocking through the North America in a form of a negative anomaly in the northeastern Pacific (Fig. 3.4). However, in both cases, the marginal part of this cellular structure in the eastern hemisphere is represented by the arc-shaped central region of the G-hexagon type considered above; that is, the analysis of the Earth's shape and the relief of the Earth's core indicates that these structures are interconjugated and interpenetrating formations. This characteristic of the correlation between the Earth's global G-hexagon and L-hexagon types of gravitational instability structures is illustrated by the schemes of displacements and streamlines indicated in Figs. 3.4, 3.5 and 3.6. In such a structure, each point comprises the limit (i.e., there are set points in any of its vicinities). In this regard, these structures consist in interconjugated and interpenetrating formations and it is not always possible to draw the boundary of the structure in the form of a dividing line.

At the same time, as was confirmed by our studies, such structures are scale invariant—that is, they repeat themselves in various scales and, in this regard, have a complex hierarchical structure (Figs. 3.7 and 3.8, Table 3.1). At the global level, the large-scale hierarchical cellularity of the planet's general structure is clearly revealed during the analytical and graphical deconvolution of the global geoid and gravitational potential anomalies, as well as during the visual analysis of multi-depth maps based on global seismic tomography data (Dziewonski and Woodhouse 1989; Maruyama et al. 1994). The polar asymmetry of the Earth is associated with global structures of gravitational instabilities of the third-order; this asymmetry is known to be manifested the following way: above the northern polar zone, the geoid prevails over the ellipsoid by 20–30 m; above the southern, on the contrary, it lies below the ellipsoid by about the same amount (Fig. 3.8). This gives rise to observations about the pear-shaped (or onion-like) form of the Earth and the appearance of the term "onioid", meaning onion-like, by analogy with the geoid.

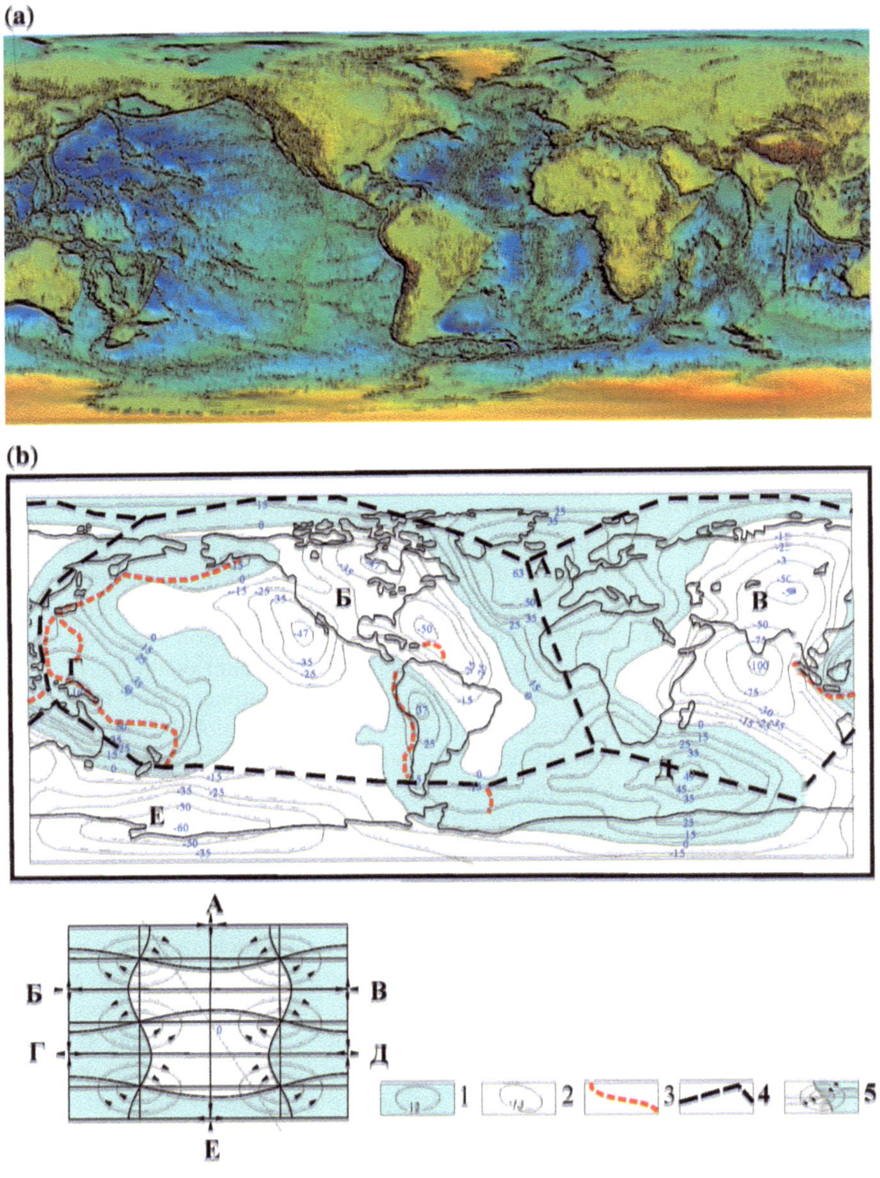

Fig. 3.8 The global structures of the Earth's gravitational instability of the third order in the latest normal cylindrical projection **a** surface relief map (based on the digital model in the ER Mapper 6.0 shell); **b** global geoid anomalies according to satellite data, GEM-9 model (Goddard (NASA) Earth model 9). Positive (1) and negative (2) excesses over the geoid's surface (isoline section 10 m); (3) subduction zones of the Earth's crust; (4) gravitational instability structures with immersion in the centre and emersion in the marginal parts or vice versa; (5) displacements and streamlines in the cellular structures of the Earth's gravitational instability of the third order

Table 3.1 Different-order swell-like and cellular neostructures of the earth's gravitational instability

Order of the structure of gravitational instability of the Earth	Diameter, km	Area, km^2	Geophysical and morphometric parameters	Tectonic structural elements	Structure type of gravitational instability of the earth
First	3.4×10^4	5.1×10^8	Geoid global anomalies of the first and the second order (up to 100 m). According to seismic tomography data, structures are recorded up to the depth of 2885 km	Distribution of inland and oceanic crust. Global rift induced and marginal plate systems. Offsets and depressions c with an amplitude up to 12 km in the lower part at the core discontinuity	Fractal cellular structure of L- and G-hexagon type (Figs. 3.4 and 3.6)
Second	2.4×10^4	1.7×10^8	Geoid global anomalies of the third order (20–30 m). Seismic tomography (depth 1500–1700 m)	Global rift induced and marginal plate systems. Rift induced and intraplate systems. Offsets and depressions c with an amplitude up to first kilometers of the lower mantle	Cellular structure of L- and G-hexagon type (Figs. 3.7 and 3.8)
Third	1.4×10^4	8.5×10^7			
Fourth	8.5×10^3	6– 7.5×10^7	Global gravimetric anomalies of the fourth order. Seismic tomography up to the depth of 1500 km	Plates at the Earth's surface. Global rift induced and marginal plate systems. Intraplate rift induced systems	Figures 3.10, 3.16 and 3.17
Fifth	4– 6×10^3	1.25– 2×10^7	Gravitational anomalies of the fifth order according to satellite data, obtained during the potential expanding by spherical functions	Large segments of rift induced and marginal plate planetary and supra-regional discontinuities	Figure 3.11

(continued)

Table 3.1 (continued)

Order of the structure of gravitational instability of the Earth	Diameter, km	Area, km^2	Geophysical and morphometric parameters	Tectonic structural elements	Structure type of gravitational instability of the earth
Sixth	2– 2.5×10^3	3– 41×10^6	Supra-regional relief forms of the sixth order and corresponding gravimetric anomalies and isothicknesses of the Earth's crust	Structures of the platform and subplatform formation. Large oceanic depressions (geoblocks, according to L. I. Krasniy). Rift induced and marginal plate supra-regional discontinuities. Supra-regional fault zones. Volcanic horsebacks, bordering abyssal depressions	Figures 3.12, 3.13 and 3.14
Seventh	1.25– 1.5×10^3	1– 1.5×10^6	Morphostructural relief forms of the seventh order. Basal surfaces and goniobasites of the second order. Scale 1: 10,000,000	Structures of the shield range, median masses. Abyssal depths. Systems of supra-regional depth faults. Volcanic horsebacks, bordering abyssal depressions	Figure 3.18
Eighth	4– 5×10^2	1– 1.5×10^5	Morphostructural relief forms of the eighth order. Basal surfaces and goniobasites of the third order. Scale 1: 5,000,000. According to gravimetric and seismic data, structures are recorded up to the depth of 80–50 km	Geoblocks, bordered with systems of supra-regional depth faults. Fixed with depressions and uplifts of the discontinuities of the Earth's crust and upper mantle up to depths of 80–50 km	Figure 3.19

(continued)

Table 3.1 (continued)

Order of the structure of gravitational instability of the Earth	Diameter, km	Area, km^2	Geophysical and morphometric parameters	Tectonic structural elements	Structure type of gravitational instability of the earth
Ninth	1.75–2×10^2	2–3×10^4	The same, of the ninth order. Basal surfaces and goniobasites of the fourth order. Scale 1: 2,500,000. According to gravimetric and seismic data, ECWM and DSS, structures are recorded up to the depth of 50–30 km	Geoblocks, bordered with systems of regional depth faults. Depressions and uplifts of the discontinuities of the Earth's crust and upper mantle up to depths of 30–50 km	Figures 3.20 and 3.21
Tenth	1–1.5×10^2	0.5–1, 5×10^4	The same, of the tenth order. Basal surfaces and goniobasites of the fifth order. Scale 1: 1,000,000		Figures 3.22 and 3.24
Eleventh	5.5–6.5×10	2–3×10^3	The same, of the eleventh order. Basal surfaces and goniobasites of the sixth order. Scale 1: 500,000		Figure 3.25
Twelfth	3.5–5×10	1–1.75×10^3	The same, of the twelfth order. Basal surfaces and goniobasites of the seventh order. Scales 1: 200,000 and 1: 100 000	Geoblocks bordered with systems of local faults. Depressions and uplifts of the Earth's crust up to depths of 10–30 km	Figures 3.26, 3.27 and 3.28
Thirteenth	2–2.5×10	2–3×10^2	The same, of the thirteenth order. Basal surfaces and goniobasites of the eighth order. Scale 1: 50,000. According to gravimetric and seismic data, structures are recorded up to the depth of 10–30 km		Figures 3.29 and 3.30

(continued)

Table 3.1 (continued)

Order of the structure of gravitational instability of the Earth	Diameter, km	Area, km^2	Geophysical and morphometric parameters	Tectonic structural elements	Structure type of gravitational instability of the earth
Fourteenth	1–1.5 × 10	5–10 × 10	The same, of the fourteenth order. Basal surfaces and goniobasites of the ninth order. Scale 1: 25,000. Depths 5—10 km	Geoblocks bordered with systems of local faults. Depressions and uplifts of the Earth's crust and sedimentary cover. Depths 2.5—10 km	Figure 3.31
Fifteenth	3–5	1.5–2 × 10	The same, of the fifteenth order. Basal surfaces and goniobasites of the tenth order. Scale 1: 10,000. Depths 2.5–5 km		Figure 3.32

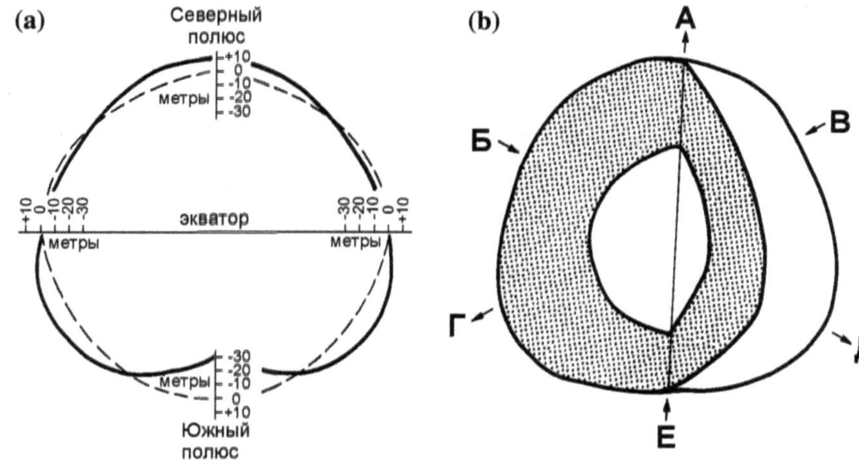

Fig. 3.9 Height of geoid (continuous line) relative to the spheroid with 1/298.25 compression (dashed line) assuming that the Earth is symmetrical relative to the polar axis (**a**) Schematic image of onioid (**b**) (not scaled). For explanation refer to Fig. 3.8. {Legends top to bottom, clockwise North Pole, metres, South Pole, metres, metres, equator

The figure of the onioid is shown in Fig. 3.9 (of course, here one should keep in mind the conventionality of the scale of the drawing).

The global structures of Earth's gravitational instability of the fourth- and fifth-order are revealed both during the direct analysis of the multi-depth maps of seismic tomography, the global anomalies of the geoid and gravitational potential, and during the analytic expansion of these anomalies to spherical functions (Figs. 3.10 and 3.11).

The supra-regional cellular structures of the Earth's gravitational instability have a diameter from 2000 to 2500 km, an area of 3–4 million km^2 and are revealed using both geophysical and essentially geological and morphographic methods.

The boundaries of the G-hexagon type structures, with immersion in the centre and emersion in the marginal parts of the deep mantle substrate, are determined by a system of intercontinental and oceanic rifts within which, according to deep seismic probing, the uplift of the underlying mantle horizons is established, while the continental crust here is heavily thinned or completely absent. The immersion of the bottom of the Earth's crust and the increase in its thickness are distinguished in the central parts of these geoblocks. The most recent supra-regional geoblock structures of the Earth's northern polar region type under consideration (Siberian, East European and Greenlandic) coincide with the ancient platforms of the same names in the terms of tectonics. The Arctic platform is a metaplatform formation, which includes a framework of separate stable ancient blocks (Fig. 3.12).

The L-hexagon gravitational instability structures, in contradistinction to those described above, are established within the Earth's northern polar region at the supra-regional level, showing an emersion in the centre and immersion in the

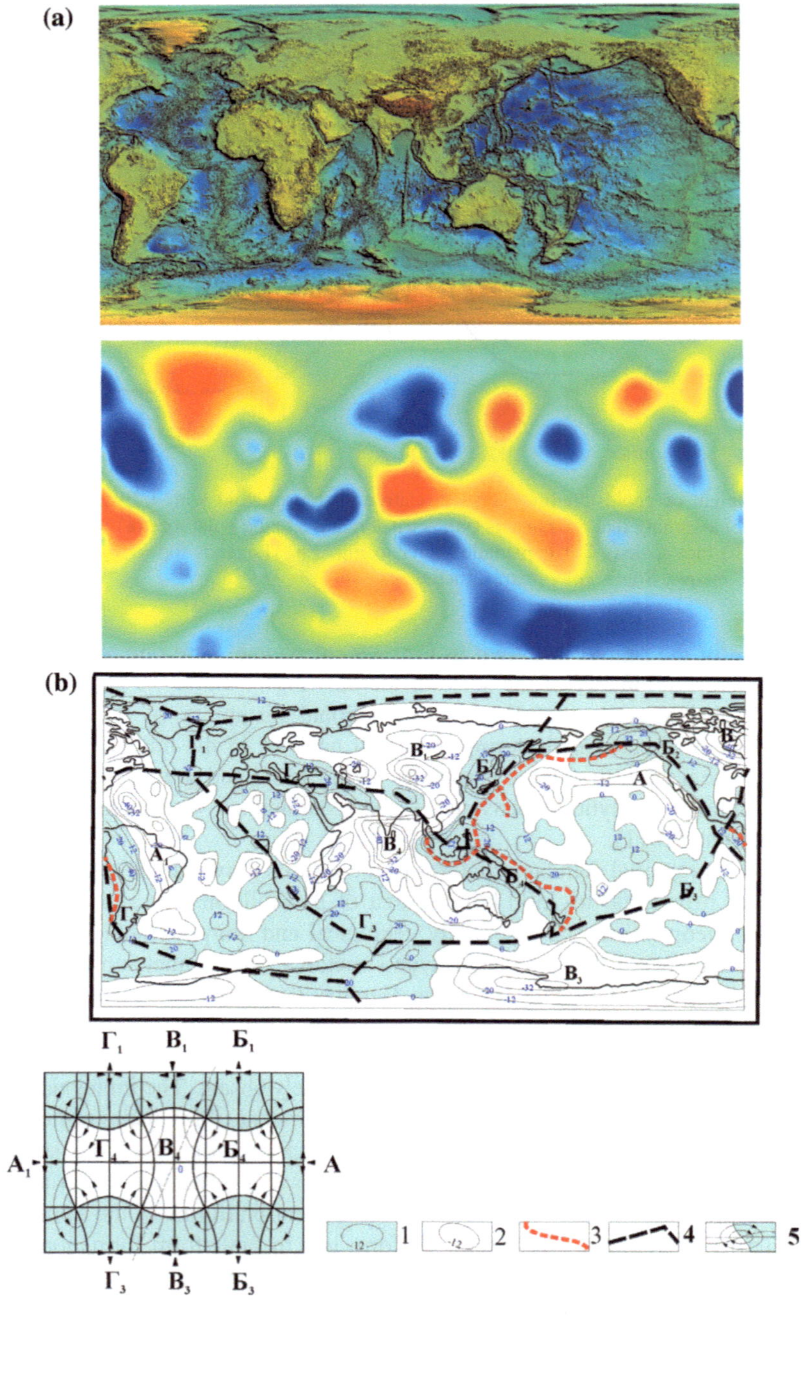
(a)
(b)
Г₁
В₁
Б₁
А₁
Г₄
В₄
Б₄
А
Г₃
В₃
Б₃
1
2
3
4
5

◂**Fig. 3.10** Most recent global structures of the Earth's gravitational instability of the fourth order in normal cylindrical projection **a** three-dimensional model of the Earth's day relief surface; **b** the frequency filtration of Δg global anomalies with the geoid anomaly localisation of the fourth order. 1, 2—Δg global anomalies, GEM-9 model (Goddard (NASA) Earth model 9); 3—the Earth's crust subduction zones; 4—structures of gravitational instability with emersion in the centre and immersion in the marginal parts and with immersion in the centre and emersion in the marginal parts; 5—displacements and stream lines in the structures of the Earth's gravitational instability of the fourth order

marginal parts of the deep mantle substrate. External morphographic typomorphic elements of their structure are determined by large isometric depression, confined to the central part of the morphostructure. The large watersheds framing the depression are the orographic boundaries of this geoblock type. The watersheds of the supra-regional level by themselves are often formed by echelon conjugated mountain ranges. The conformity of the structural deep and surface elements as well as their reversed character are retained for this type of morphostructures. In deep layers, the central regions of geoblocks are characterised by isometric uplift of various deep horizons, and their marginal parts—by the immersion of the corresponding crust-mantle discontinuities. With this kind of correlations at the supra-regional level, the thickness of the crust increases significantly in the marginal parts of the morphological blocks, and in the central regions the continental crust is strongly thinned or completely absent. The structural boundaries of supra-regional geoblocks of this type are often emphasised by volcanogenic belts and subduction zones within the Earth's northern polar region (Fig. 3.12). A similar tectonic frame is typical for the uni-ordinal geoblocks within the Caribbean and the Mediterranean and most of the coastal zones of the Pacific Ocean. However, this geoblock type is not always so exposed in terms of tectonics at the supra-regional level. Within the African continent, as well as within the majority of the entire southern hemisphere, where this type of gravitational instability is predominant, the uni-ordinal geoblocks are not so tectonically evident (Fig. 3.13).

The boundaries of the Earth's gravitational instability structures and geoblocks, established by the morphometric and geomorphological methods (Holmes 1944, 1965; King 1967), fully coincide with the boundaries of the supra-regional blocks of the Earth, identified by Krasnyi (1984) on a tectonic basis both within the continental and oceanic lithosphere. Carey (1991) emphasises the conservation of the dimension and shape of these polygonal formations without any noticeable difference between continents and oceans (Fig. 3.14). In the terms of tectonics, the boundaries of these structures at the oceanic crust very often represent the volcanic ridges that border the abyssal basins.

At the regional and local levels, alongside geophysical and geological methods, quantitative methods of relief analysis, in particular, the morphometric methods developed by Filosofov (1963), play even more important role in identifying the hierarchy of the Earth's gravitational instability structures as well as the typomorphic elements of their constitution. The majority of cellular structures detected by these methods can be isolated directly both during the analysis of river drainage

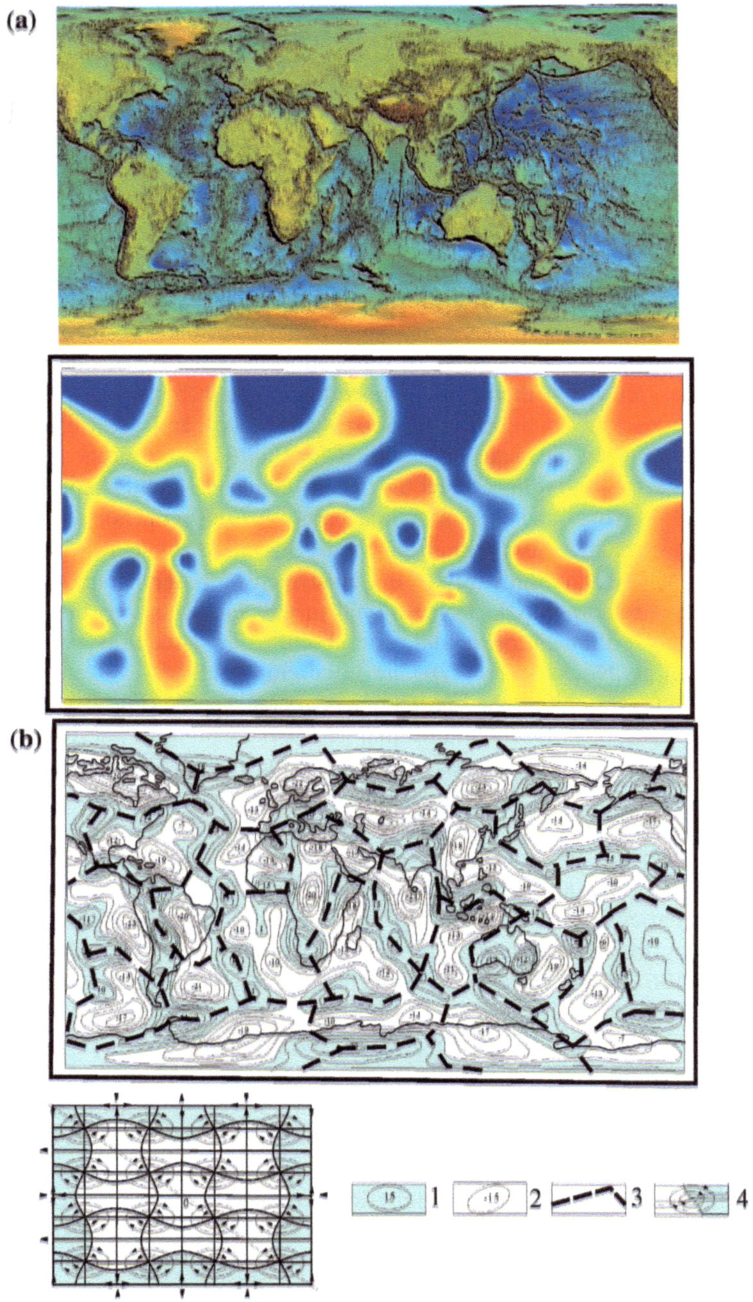
(a)
(b)
1
2
3
4

◂**Fig. 3.11** Most recent global structures of the Earth's gravitational instability of the fifth order in normal cylindrical projection **a** three-dimensional model of the Earth's day relief surface; **b** the frequency filtration of Δg global anomalies with the geoid anomaly localisation of the fifth order. 1, 2—gravitational anomalies derived from satellite data obtained by the potential expansion into spherical functions (isoanomaly lines are drawn via 5–10 mg/s); 3—boundaries contouring the structures of gravitational instability with emersion in the centre and immersion in the marginal parts; 4—displacements and stream lines in cellular structures

systems and the disjuncture of the relief based on the use of different-scale topographic maps, aerial and satellite images, as well as by geological and geophysical methods (Fig. 3.15). However, the methods developed by V.P. Filosofov allow this to be done purposefully and most consistently. Essentially, the morphometric method is a method of graphic decomposing of the relief into series, depending on the order of the valleys and watershed lines and then drawing up special maps of the base surfaces and goniobase lines on the basis of this series.

3.2 Morphometric Methods

Filosofov's morphometric approach, which presupposes quantitative estimates, combines the fields of geomorphology, geotectonics and neotectonics. He considers the gravitational force to be the unifying element, in the field of which arise both the forms of topographical relief and the structure of the Earth's crust. As such, the gravitational force comprises the common regular cosmic factor. All mass transfer processes, which determine both endogenous and exogenous processes, do so in conformity with it. The contemporary theory of the Earth's surface relief formation takes into account the potential gravity force. This means that the height of the relief, and consequently the degree of its dissection, is influenced not only (and not so much) by the surface processes and lithology of the relief-forming rocks, but rather by the deep rock bedding of various densities creating the main constituent of the gravitational field.

The observations of Filosofov and many other authors testify that the difference in the anomalies of the gravitational field of the order of 0.5 mGal may be reflected in the relief on maps having a scale of 1: 50,000. Each tectonic multi-ordinal structure and each structural floor of the Earth's crust creates gravitational fields that are mutually superimposed. In order to understand the complex gravitational field of the Earth, it can be expanded into series by harmonic analysis methods. Relief, like the gravitational field, can also be expanded into series. Moreover, along with the analytical methods, good results are obtained in many cases by methods in which the relief is graphically decomposed into series on the basis of the valley and watershed lines, followed by the compilation of special maps based on this series. The order of the valleys is determined according to the method proposed by Horton (1948) and Panov (1948) and refined by Filosofov (1959a). A valley of the first order is a valley that does not receive any inflow; in other words, it is unbranched.

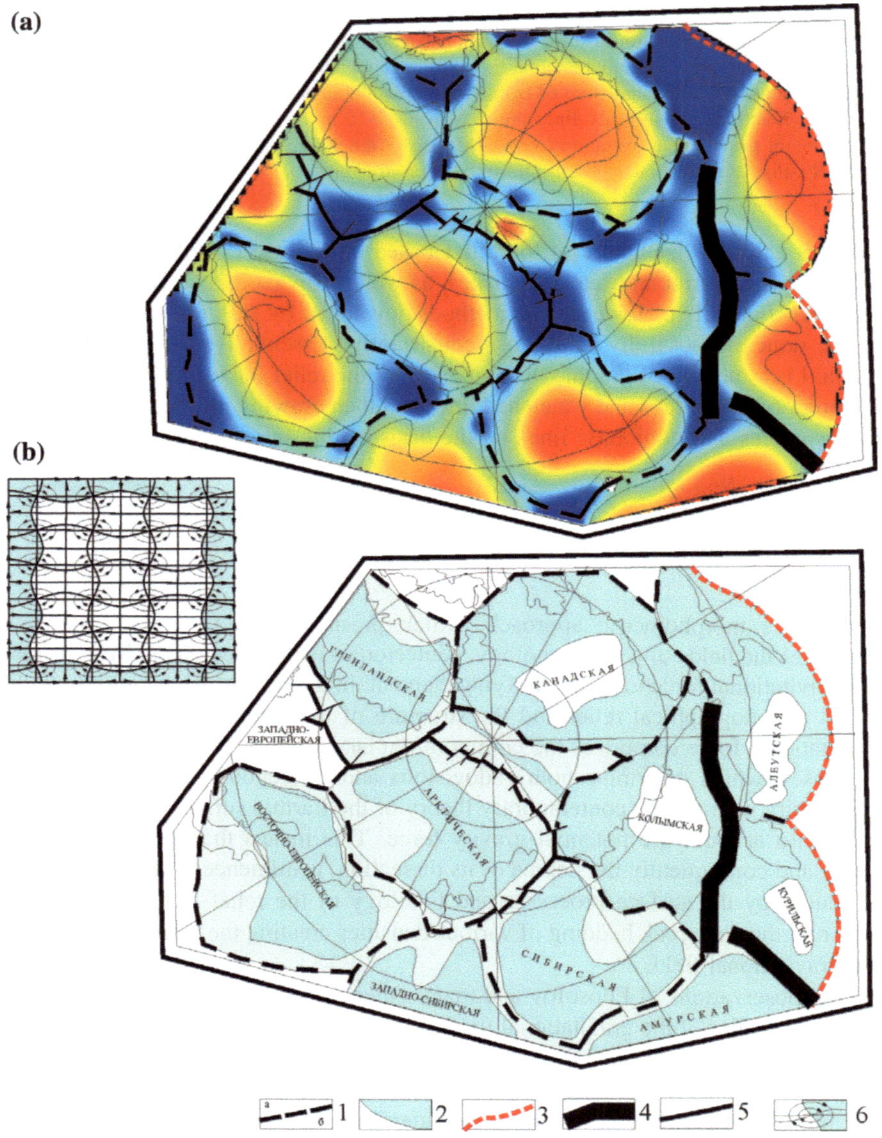

Fig. 3.12 Most recent supra-regional structures of the Earth's gravitational instability of the sixth order of the Earth's Northern Polar region in the latest normal azimuthal equiangular (stereographic) projection. Scale 1:20,000,000. **a**, **b** geoblock divisibility revealed on the basis of orographic and tectonic elements. 1—boundaries of the central zones of Earth's crust spreading (**a**) and the zones of manifested to a different degree continental rifting (**b**); 2—the Earth's crust subduction zones; 3—volcanogenic belts; 4—parts of the gravitational instability structures

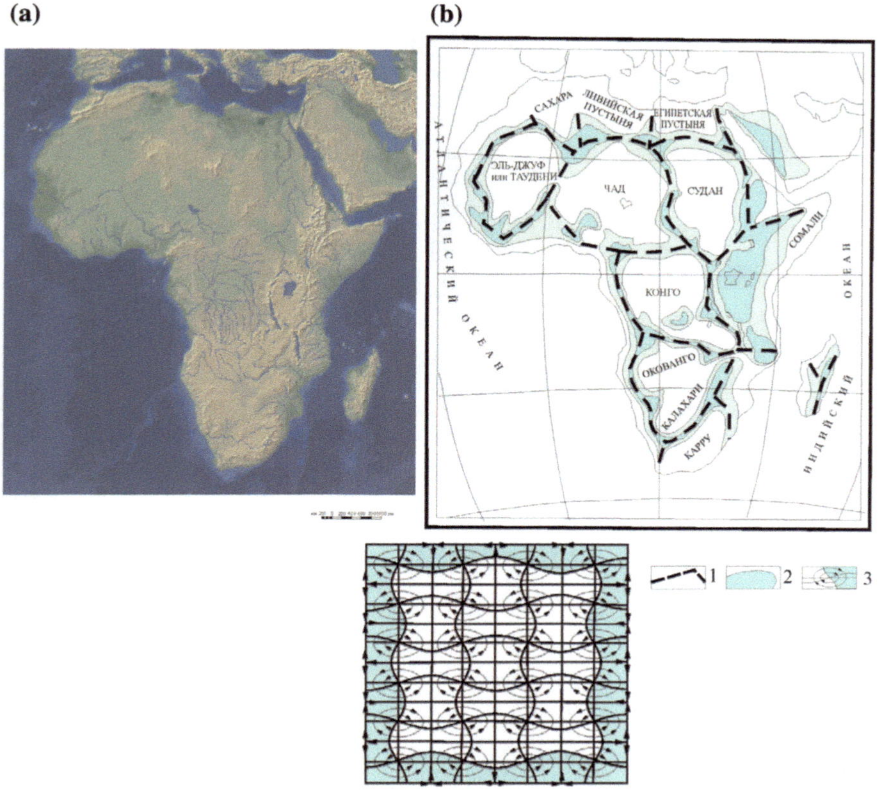

Fig. 3.13 The most recent supra-regional structures of the Earth's gravitational instability of the sixth order of the African continent. **a** Hypsometric scheme; **b** African troughs and bending folds —tectogenic relief forms (according to Holmes (1944, 1965). 1—boundaries of the gravitational instability structures with emersion in the centre and immersion in the marginal parts; 2— immersing (white) and emerging (blue) areas of gravitational instability structures; 3— displacements and stream lines in cellular structures. {Legends on Fig. 3.13, top to bottom, right to left: Sahara, Libyan desert, Egyptian desert, El Djouf or Taoudeni, Chad, Suydan, Somalia, Congo, Okovanggo, Atlantic Ocean, Kalahari Desert, Karoo, Indian Ocean}

A valley of the second order results from the merging of two valleys of the first order. A third order valley is formed from the confluence of two valleys of the second order, and so on. It should be noted here that receiving the inflow of a valley of a lower order does not increase the order of the main valley (Horton 1948).

According to Panov (1948) and Filosofov (1980), the order of a particular river (more precisely, its valley at the same section) varies depending on the scale of the maps used to determine them. Our observations show that within the Siberian platform, river valleys of the first order presented on maps of a scale of 1:1,000,000 correspond to third-order valleys on a 1:100,000 scale map. At the same time, first-order valleys on 1:100,000 maps usually correspond to third-order valleys

(a)

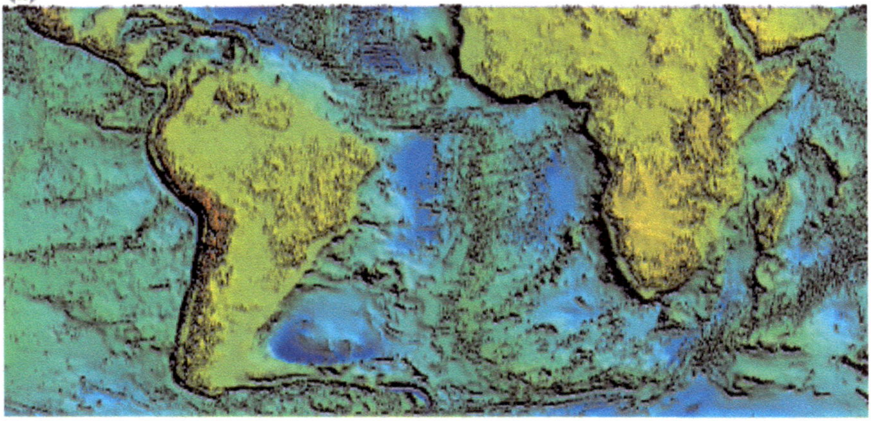

(b)

Fig. 3.14 Distribution of the troughs and uplifts of the sixth order over the continental and oceanic crust (Carey 1991) **a** hypsometric scheme; **b** cellular-block composition {Legends on Fig. 3.13 b, top to bottom, right to left: Arabian, El Djouf, Chad, Cocos Plate, Guaina, Somalian, Congo, Angolan, Peruvian Nazca, Brazilian, Cubango, Mozambique, Kalahari, Karoo, Chilean, Argentinean, Cape, Madagascar, Agulhas, Bellingshausen, Antarctic}

depicted on 1:10,000 scale maps. For the map scales 1:1,000,000, 1:500,000, 1:100,000, 1:50,000 and 1:10,000, the order of the valley was calculated. Within the scales of the maps 1:1,000,000, 1:500,000, 1:100,000, 1:50,000 and 1:10,000, an increase in the order of the same section of the valley per unit is established. The order of the valleys on the maps of scale 1:300,000 correlates with the order of valleys on maps of scale 1:1,000,000, and, in the same way, 1:200,000 corresponds to 1:500,000. The order of valleys on the maps scaled at 1:50,000 corresponds to the order of valleys on maps of scale 1:25,000.

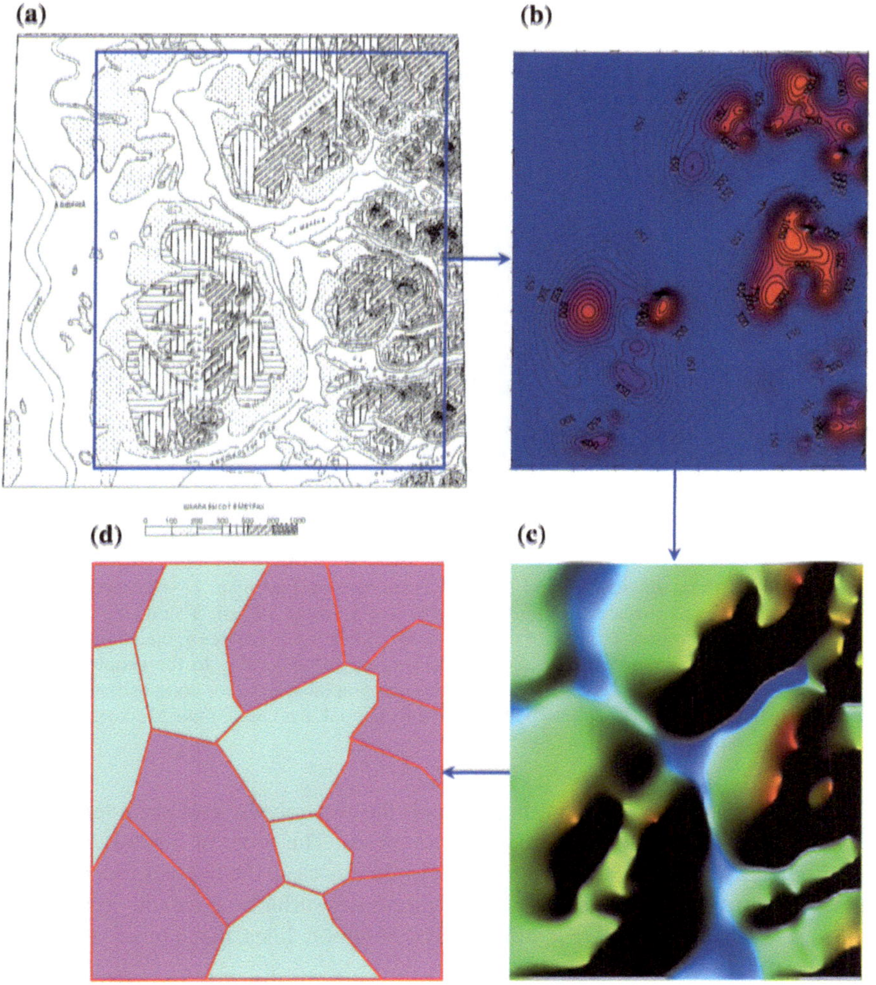

Fig. 3.15 Reconstruction of cellular structuring **a** orographic scheme of the Norilsk district; **b** isobase line map; **c** structural feature allocation by lighting modelling; **d** reconstruction of cellular forms

The principle for compiling the morphometric maps, as proposed by Filosofov (1963), is the base surface method. The base surface is a complex curve enveloping the surface through the thalwegs of the valleys, which comprise their skeleton (Fig. 3.16). Filosofov and Chernyaev (1963) developed a method of goniobase lines, based on the connection between the valleys and linear discontinuities, in order to identify faultings.

The order of the base surface map compilation is described in detail by Filosofov (1959b). The intersection points of the horizontals and valley thalwegs are

(a) (b)

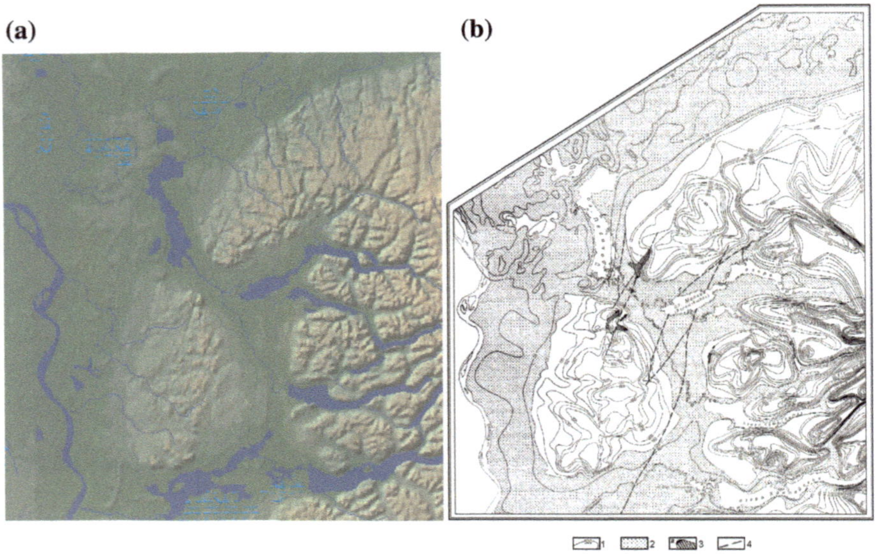

Fig. 3.16 **a** Hypsometric scheme and **b** fourth order base surface map. 1—isobase lines of the fourth order; 2—basis surfaces having the isobase line value of 120 m; 3—outcrops to the day surface and sub-Quaternary deposits of the industrial-ore-bearing intrusives **a**, their contours on the deep horizons **b**; 4—separate zones of the faultings mapped according to the geophysical data. 1:500,000 scale topographic maps are used as the topographical base for composition of isobase lines. {Legends on Fig. 3.16b, top to bottom: Lake Pyasino, Lake Lama, Lake Melkoe, Lake Keta, Lake Khantayskoye.}

connected by smooth lines passing through the watersheds, and are referred to as isobase lines, i.e., the lines of equal erosion bases. The order of the base surface is determined by the initial order of the valleys. In this case, the base surfaces of lower orders are as if they were embedded in higher-order surfaces. It should be noted that the base surfaces of different orders have common lines that coincide with the valleys of high orders.

When compiling the goniobase line map, the points of intersection of the horizontals with the valley thalwegs are connected by straight lines. These lines cannot be mixed with isobase lines. As it is known, the isobase lines have the same properties as the horizontals and cannot intersect or tangent each other. On the contrary, the goniobase lines can be tangential to and even intersect each other. The isobase lines have the appearance of smooth curved lines, while the goniobase lines represent broken lines. The isobase lines cannot be drawn above the horizontals of the same name. On the contrary, the goniobase lines can cross the horizontals that outline the lower areas. Goniobase maps are compiled according to a consideration of the valley order, as well as the base surface maps.

For the territory under consideration, the base surface maps and goniobase lines for scales of 1:10,000,000, 1:5,000,000, 1:2,500,000, 1:1,000,000, 1:500,000, 1:300,000, 1:200,000, 1:100,000, 1:50,000, 1:25,000 and 1:10,000 were compiled

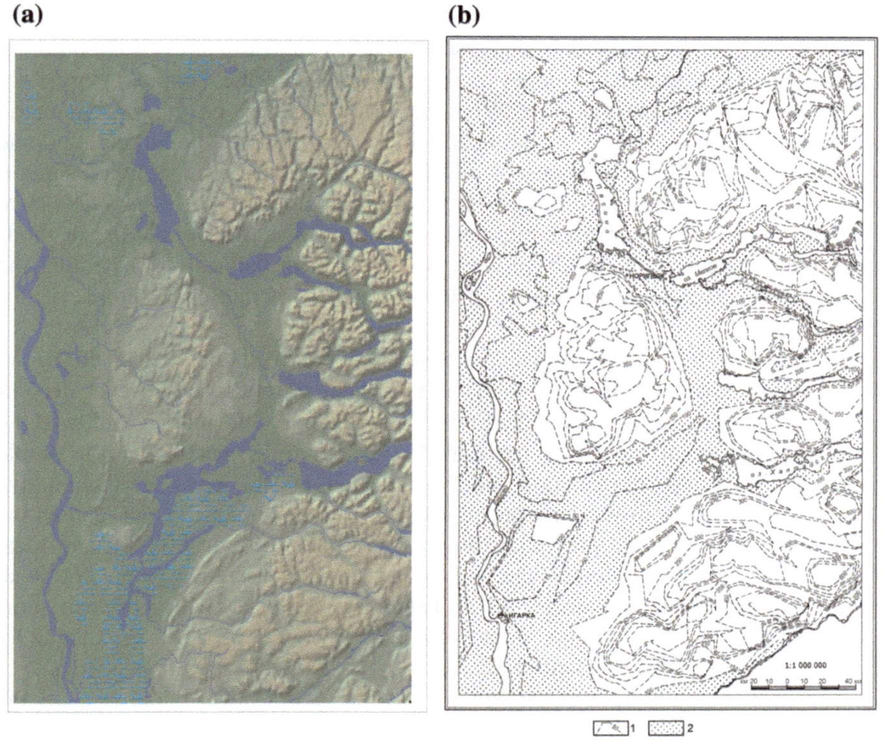

Fig. 3.17 **a** Hypsometric scheme and **b** fourth order goniobase line map 1—goniobase lines and their values; 2—base surfaces having goniobase line values of 100 m. 1:500,000 scale topographic maps are used as the topographical base for goniobase line composition. {Legends on Fig. 3.17b, top to bottom: Lake Pyasino, Lake Melkoe}

based on the morphometric methods. Taking into account the degree of geological and geophysical study, the main part of large-scale maps was compiled within the Norilsk ore region, including its central part and individual ore fields, which are well attested by drilling activity.

Figure 3.17 shows the hypsometric diagram of the central part of the Norilsk ore region as well as the map of base surfaces and fourth-order goniobase lines, where these lines (i.e. lines of equal erosion bases) clearly outline isometric—and, rarely, linearly extended—morphostructural blocks.

When comparing uni-ordinal and single-scale maps of base surfaces and goniobase lines of the territory under consideration, it is found that the latter not only allow the boundaries of multi-ordinal morphostructural blocks to be revealed, but also their tectonically conditioned structure features to be reliably established. In this case, the basic morphological elements of the uni-ordinal morphostructures, expressed in relief, retain their configuration both on the base surface maps and on the goniobase maps.

Different-scale maps of base surfaces and goniobase lines were compiled for both the northern and southern hemispheres. The morphostructures of the northern hemisphere were optimally studied using the example of the central sector of the Arctic and the Siberian platform. When carrying out work on the morphometric analysis, Filosofov (1963) recommends the following additional cartographic resources: maps of the asymmetry of valleys and interfluves, residual relief, summit plane, goniobase lines, drift, base surface residuals. In this case, these maps have not been compiled within the entire territory under consideration, but the methods underlying their compilation have been widely used in delineating, revealing the history of formation and the nature of the correlation of morphostructures of different orders.

In terms of their final results, the morphometric methods developed by V. P. Filosofov, appeared to be close to the analytical methods of the relief decomposition, anomalies of geoid and the Earth's gravitational field and have allowed the natural hierarchy of blocks of the Earth's crust to be revealed using a simple algorithm. The morphostructural blocks allocated on the basis of quantitative decomposition methods can be grouped according to the dimension and scale of the maps on which they are presented: regional (1:10,000,000–1:500,000) and local (1:200,000–1:10,000). The average widths and areas of discrete regional and local morphological blocks are presented in Table 3.1; the peculiarities of their morphography are indicated in Figs. 3.18, 3.19, 3.20, 3.21, 3.22, 3.23, 3.24, 3.25, 3.26, 3.27, 3.28, 3.29, 3.30, 3.31 and 3.32.

The analysis of morphometric, geological and geophysical parameters of the regional and local morphological blocks allowed the conformal nature of the correlations between surface and deep structures to be established and three morphotypes to be identified among them, with each being characterised by its typomorphic structural elements and preferential spatial distribution on the Earth's surface. At the same time, the regional and local morphostructural blocks of the northern hemisphere were considered on the example of the central sector of the Arctic, while the blocks of the southern one—on the example of South Africa. In total, 12 morphostructural blocks of twelve orders were allocated within the central sector of the Arctic, based on the quantitative relief decomposition methods, having an area of 1–1.5 million km^2 to 5–7 km^2 both at the regional and local levels (Figs. 3.18, 3.19, 3.20, 3.21, 3.22, 3.23, 3.24, 3.25, 3.26, 3.27, 3.28, 3.29, 3.30 and 3.31). The shape of these morphoblocks is isometric (cellular), less often linearly extended (swell-like). Their boundaries are confined either to the valleys of modern rivers of the appropriate order or to the large relief depressions. The central parts of isometric morphostructural blocks are morphologically expressed in the relief by an arc-shaped curved rise in the plan (Figs. 3.18, 3.19, 3.20, {3.21, cited later}, 3.22, 3.24, 3.25, 3.26, 3.27, 3.28, 3.29, 3.30 and 3.31).

As a rule, the concave part of the rise is weakly expressed in terms of morphology and tectonics. Conversely, the convex frontal part is very clearly manifested both morphographically and tectonically. It should be emphasised that the revealed morphographic aspects are preserved within all multi-ordinal

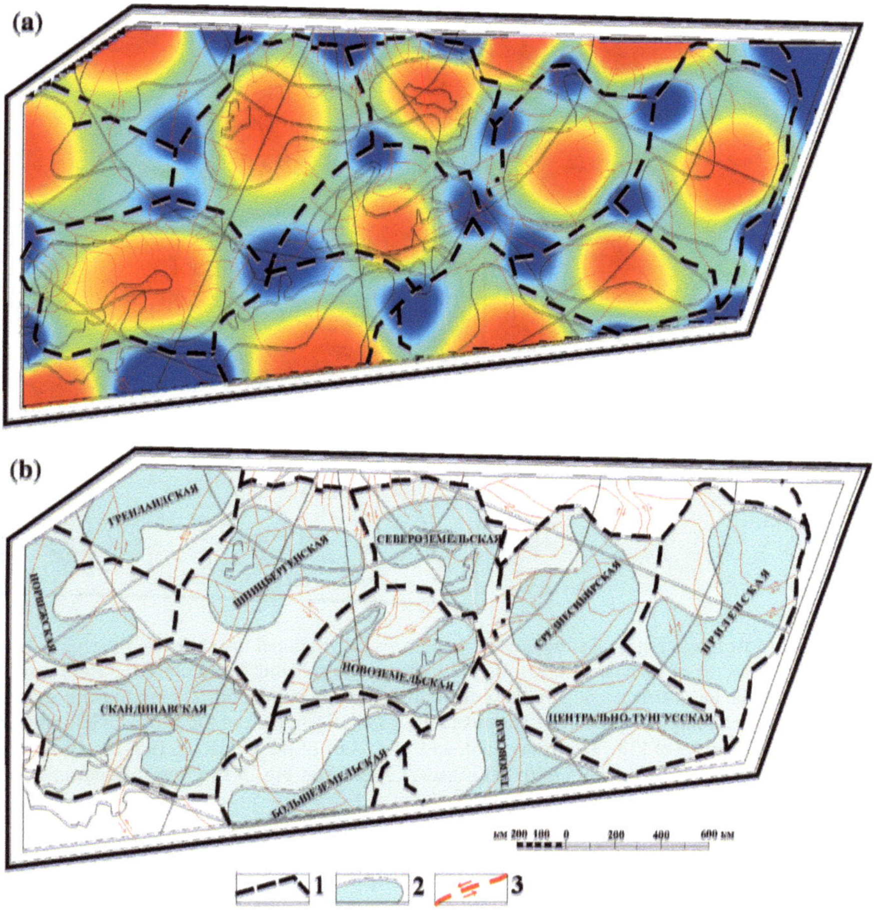

Fig. 3.18 Most recent supra-regional structures of the Earth's gravitational instability of the seventh order of the central sector of Russian Arctics and adjacent territories. **a** base surfaces of the isobase and goniobase lines of the seventh order; **b** cellular structures. 1—structure boundaries; 2—immersing areas of gravitational instability structures; 3—faultings and the directions of displacements along them (normal azimuthal stereographic projection). {Legends on Fig. 3.18b, left to right, top to bottom: Greenlandic, Spitsbergen, Northland, mid-Siberian, Lena Region, Norwegian, Scandinavian, Newland, Central Tungus, Bolshezemelskaya, Tazovskaya}

morphological blocks, including at the global level (Figs. 3.4, 3.5, 3.6, 3.7 and 3.8); therefore, these features can be attributed to their typomorphic characteristics.

The morphology of multi-ordinal morphostructural blocks and their typomorphic structural elements, exposed on the land surface, is modified, distorted, levelled or emphasised by endogenous (tectonic, magmatic, etc.) and exogenous (denudational and accumulative) processes. The resulting relief forms can, with varying degrees of

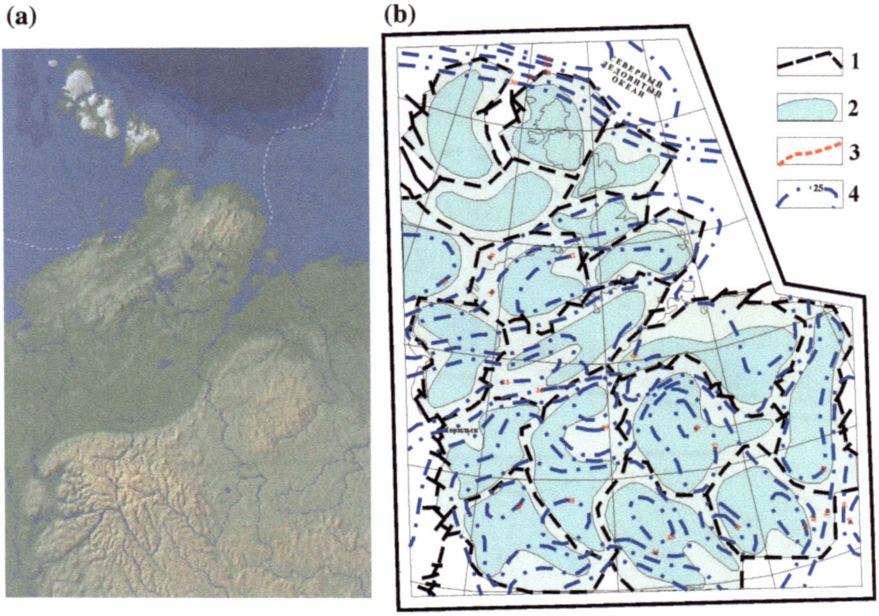

Fig. 3.19 Most recent regional structures of the Earth's gravitational instability of the eighth order of the central sector of Russian Arctic **a** digital hypsometric scheme; **b** morphostructural forms of the relief and the base surfaces of goniobase lines. 1—structure boundaries; 2—immersing areas of gravitational instability structures; 3—various faultings; 4—isohypses of the Mohorovičić surface, in km form the sea level [Sib. Platform. Geol. and Min. Res. of the USSR…, 1987. V.4]. Normal equidistant conical projection. 1:5,000,000 scale. {Legends on Fig. 3.19b, top to bottom: Arctic Ocean, Norilsk}

similarity, outline the morphology of typomorphic elements or only roughly indicate their location.

The comparison of surface elements of multi-ordinal morphostructures with geotectonic structures and morphological characteristics of different discontinuities of the Earth's crust allowed the following patterns to be established. Firstly, all the morphostructures are expressed at depth levels corresponding to their order. Secondly, for morphological blocks of all levels, the conformal nature of the correlations between surface and depth structures is noted. At the deep horizons, the typomorphic elements of morphostructures have an inverse character with respect to the external geomorphological surface, i.e., the central regions of the above-mentioned type of cellular morphostructures at the deep horizons are characterised by the immersion of various crust-mantle boundaries and their marginal parts—by the rising (emerging) of the corresponding depth horizons. According to gravimetry and deep seismic probing, the morphostructures of the regional level are reflected in the lower part of the Earth's crust and in the upper mantle in form of isometric depressions (Fig. 3.23).

Their boundaries in the upper horizons of the Earth's crust are confined to the echelon-conjugated, swell-like or dome rises, the axial parts of which are often complicated by the subsiding grabens. The structural boundaries of the morphological blocks of the local level at deep horizons are represented by the echelon-conjugated, anticlinal, dome or flexural folds. The laying depths of local morphological structures correspond to the sedimentary cover and the basement. Within the southern hemisphere, the morphostructural blocks of the regional level (scales of 1:2,500,000 and 1:1,000,000) have, as a rule, a structural character inversed with respect to the morphological blocks of the northern hemisphere (Figs. 3.21 and 3.24).

Instead of the positive forms of relief, which is typical for the morphostructural blocks of the northern hemisphere, the central parts of the uni-ordinal morphological blocks are represented by the large isometric depression (Figs. 3.21 and 3.24). The large watersheds framing the depression comprise the orographic morphoblock boundaries. The conformity of the depth and surface elements of the structure is retained for the morphostructures of this type. At deep horizons, the central regions of morphostructures are characterised by isometric uplift of various deep horizons, while their marginal parts are characterised by the immersion of the corresponding crust-mantle discontinuities. At the regional level—within South Africa, as well as the Siberian platform and its margins—in addition to the two types of morphostructures considered above, linearly prolate swell-shaped morphological blocks are also rather widespread (Figs. 3.20, 3.21, 3.22, 3.23, 3.24 and 3.25). This type of morphostructure is at first glance characterised by a fairly simple morphological, tectonic and deep structure. However, when the scale is varied, it turns out that they consist, on the one hand, of lower-order isometric morphostructural blocks by themselves and, on the other, of the larger cellular morphological blocks of the regional and supra-regional levels which comprise them as component elements (Figs. 3.19, 3.20, 3.21, 3.22, 3.23, 3.24 and 3.25).

Thus, a direct analogy can be drawn between the revealed hierarchy of the Earth's morphostructural blocks and the structures of gravitational instability (Rayleigh-Taylor instability), based on both geological and geophysical data and on the typomorphic features of their structure. In both cases, there are three main structural types: (1) cellular (hexagonal) structures with emersion in the centre and immersion in the marginal parts of the L-hexagon type; (2) hexagonal cells with immersion in the centre and emersion in the edge parts of the G-hexagon type, and (3) linearly extended, two-dimensional, swell-like Rolls. Both for natural formations and for the structures obtained in the experiments, the conformity of the topographic surface with an internal (deep) structure and the reversed nature of the correlations between them can be observed.

Within this approach, the morphostructures identified above can be considered as a new class of geodynamic objects, which were not detected earlier in traditional geomorphological and geological studies. The contemporary science of thermo-gravitational instability phenomena underlying the formation of these geodynamic objects also predetermines the most appropriate name for them: "the most recent multi-ordinal swell-like and cellular dissipative structures of the Earth's

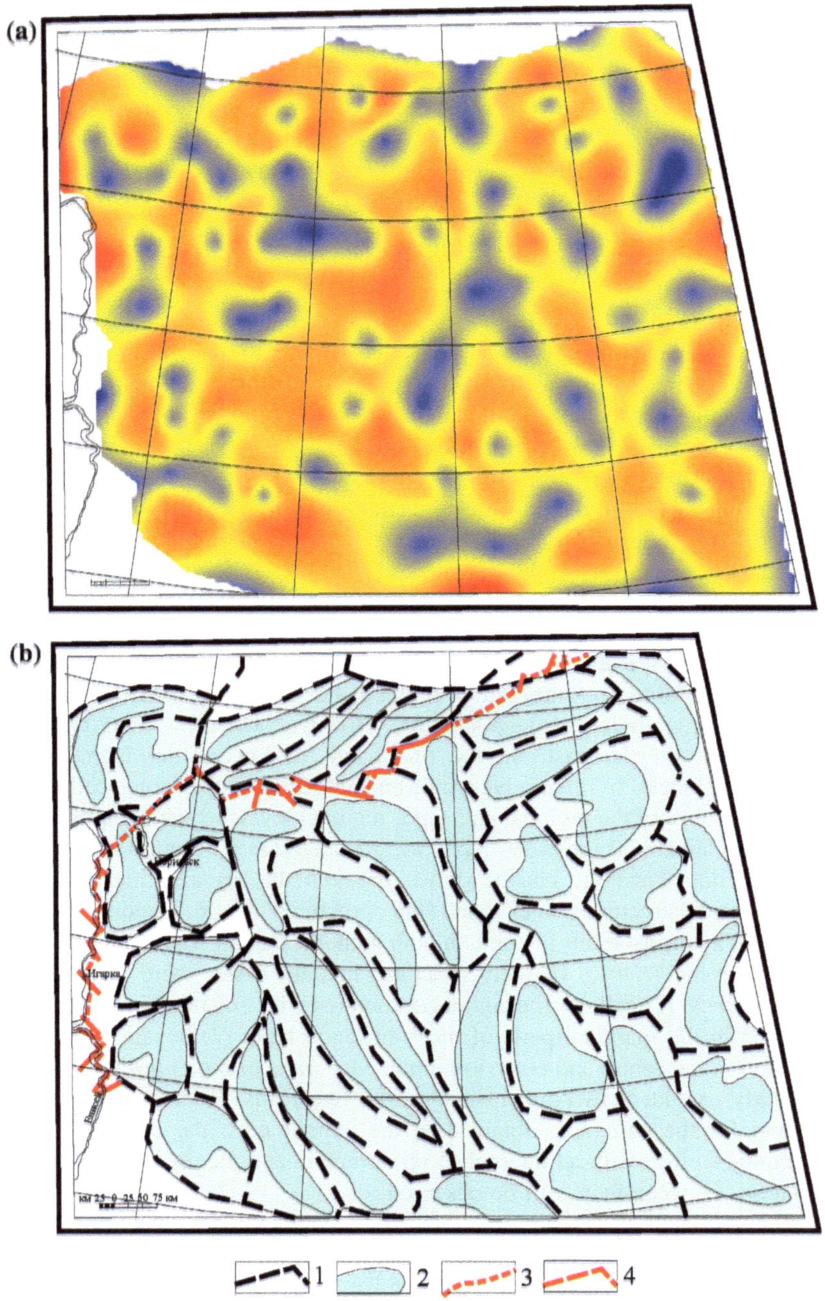
(a)
(b)
1
2
3
4

◄**Fig. 3.20** Most recent regional structures of the Earth's gravitational instability of the ninth order of the Central Siberian Plateau **a** base surfaces of isobase and goniobaselines of the ninth order; **b** cellular structures. 1—structure boundaries; 2—immersing areas of gravitational instability structures; 3—lava plateau boundaries; 4—various faultings. Normal equidistant conical projection. 1:2,500,000 scale. {Legends on Fig. 3.20b, top to bottom: Norilsk, Igarka, Yenisei River}

gravitational instability" (Table 3.1). The determination of the age of these structures as most recent is justified both by the methods that are used to identify them, as well as by the age of the associated alignment surfaces and geological formations. At the same time, the morphometric methods suggest a close correlation between the age of morphostructures and the age of the relief. In this regard, Schultz (1932), who coined the term "neotectonics", repeatedly stressed that there is no united low age boundary of the neotectonic movements that created the main features of the modern relief for the whole Earth in general. This boundary is sliding, although it does not fall below the upper Oligocene—lower Miocene boundary. Further study of the neotectonic movements indicated that the beginning of the neotectonic stage was really asynchronous, and, first of all, it took place in the Pacific and Atlantic segments of the lithosphere (Grachev et al. 1970). According to A.F. Grachev, in the Pacific segment this boundary belongs to the Pliocene; in the Atlantic segment—to the Upper Oligocene—Lower Miocene (Neotectonics … 1998).

On the other hand, the study of geological formations and the presence of relicts of ancient planation surfaces, the earliest age of which is today defined as the Triassic-Jurassic, led to the introduction of the concept of a geomorphological stage in the evolution of the Earth (Gerasimov and Meshcheryakov 1964), the beginning of which coincides with the disintegration of Pangea. At the geomorphological stage of the Earth's evolution, neotectonics designates only one of its cycles. Therefore, it seems appropriate to define a sliding age lower boundary for both the global Pacific and Atlantic morphostructures and for various hierarchies in their composition, in which development individual tectonic cycles can be noted, including the neotectogenesis.

During the development of this cycle of tectogenesis, an explicit confinement of the most recent hierarchies of the structures of the Earth's gravitational instability in noted for the G-hexagon type with immersion in the centre and emersion in the marginal parts of the deep substrate up to the Eurasian continents of the eastern hemisphere, while for the L-hexagon type with emersion in the centre and immersion in the marginal parts of the deep substrate down to the Gondwana continents of the western hemisphere. The most extensive development of two-dimensional swell-like structures is established at the regional level including their regular correlation with the uni-ordinal cellular structures.

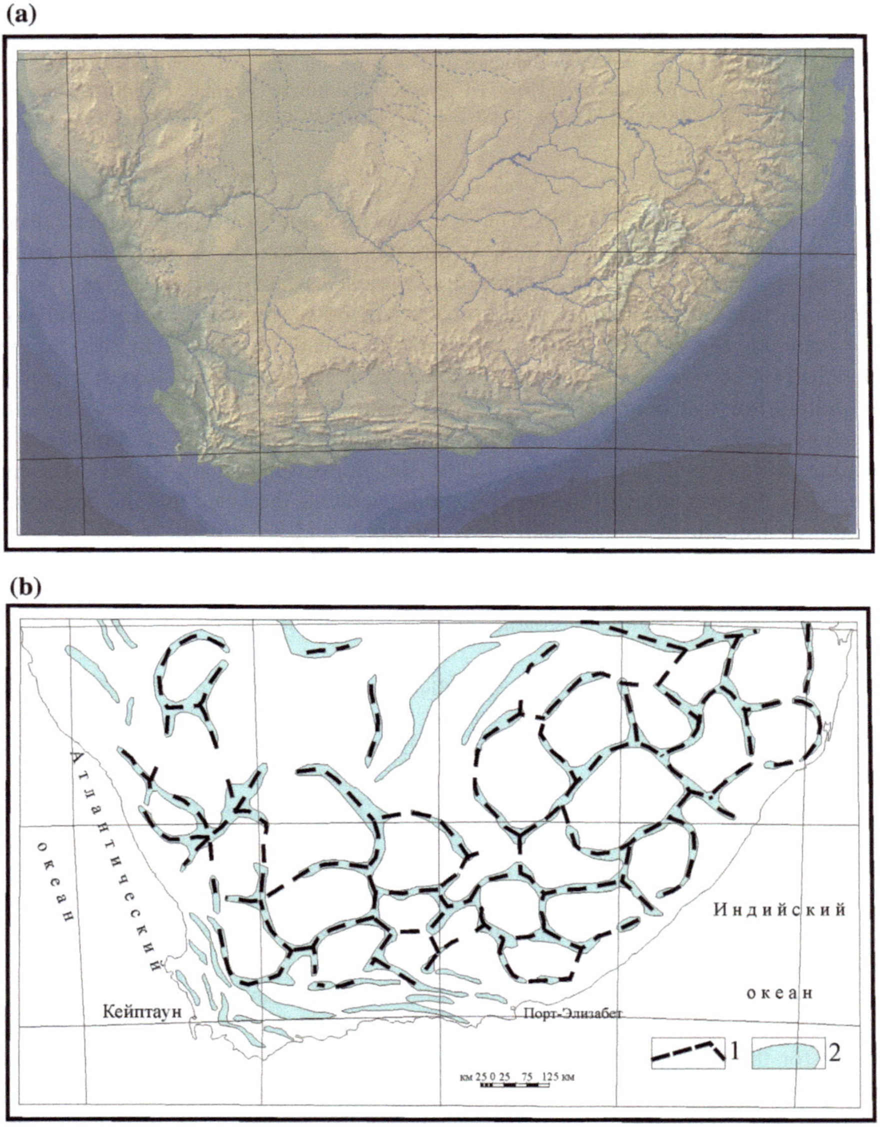

Fig. 3.21 Most recent regional structures of the Earth's gravitational instability of the ninth order of the South Africa **a** hypsometric scheme; **b** cellular-block structure. 1—structure boundaries; 2—immersing (blue) and emersing (white) areas of the Earth's gravitational instability structures. Normal equidistant conical projection. {Legends on Fig. 3.21b, left to right, top to bottom: Atlantic Ocean, Indian Ocean, Cape Town, Port Elizabeth}

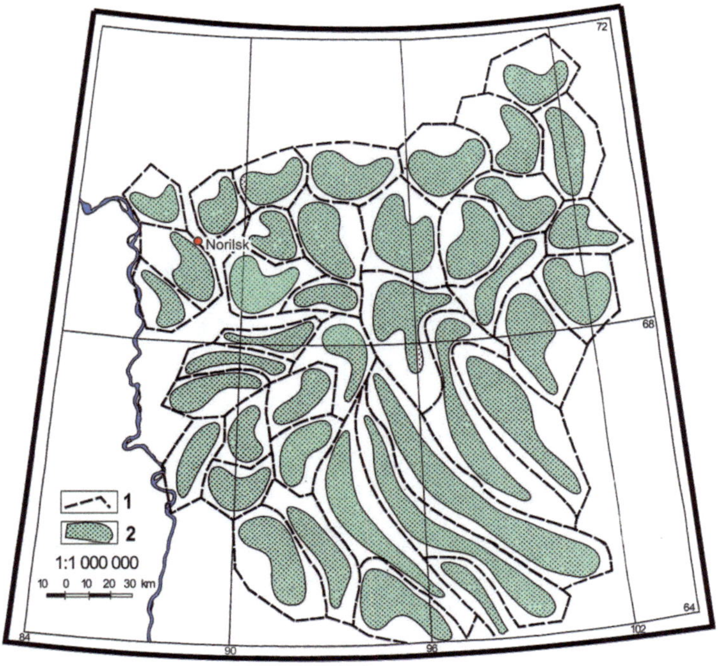

Fig. 3.22 Most recent regional structures of the Earth's gravitational instability of the tenth order within the north-western part of the Siberian platform. 1—structure boundaries; 2—immersing areas of gravitational instability structures. Normal equiangular conical projection. 1:1,000,000 scale

It is important to emphasise that since the most recent multi-ordinal swell-like and cellular structures of the Earth's gravitational instability cover the entire surface of the planet, they can consequently serve as a means of tectonic zoning of the Earth's surface on a geodynamic basis. Morphometric, geophysical and geological data have been widely used in the selection of these structures.

The global structures of the gravitational instability of the Earth, such as the G-hexagon type in the eastern hemisphere and L-hexagons in the western, are associated with the main planetary geo-discontinuities, as well as the distribution of the continental and oceanic crust. As can be seen in Figs. 3.4, 3.5 and 3.6, these deep inhomogeneities of the Earth influence not only the shape of the Earth, but also determine the Earth's global tectonic asymmetry, in other words, its division into the Pacific and Indo-Atlantic tectonic segments. According to seismic tomography data, these global structures reflect the main heterogeneity of the Earth—a liquid core with a radius of almost half of the earth, overlapped with a crystalline mantle—and have remarkable geometric and dynamic properties. Within a structure, each point forms a limit (i.e., there are set points in any of its vicinities).

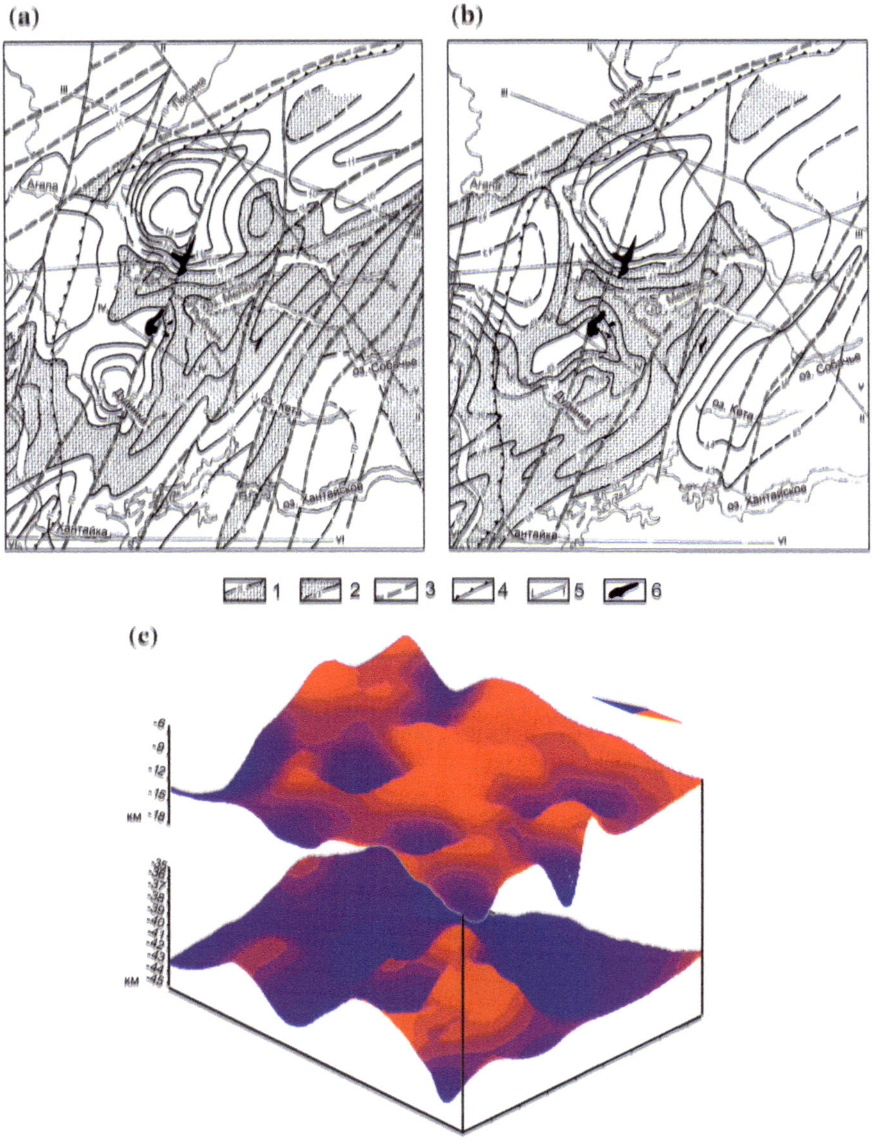

Fig. 3.23 Structural scheme of the Norilsk ore district: **a** crystalline basement roof; **b** Earth's crustal base. Compiled according to the data of Chernyshev et al. (1978); Avetisov and Golubkov (1984); Egorkin et al. (1984); **c** 3D-diagram. 1—isohypses of the sedimentary cover sole >9 and <9 km; 2—isohypses of the consolidated crust sole >42 and <42 m; 3—deep faults; 4—contact of Mezo-Cenosoic cover transgressive bedding; 5—profiles of earthquake converted wave method (ECWM)-deep seismic sounding (DSS) and DSS; 6—spreading area of workable ore-bearing intrusives

In this regard, these structures consist of interconjugated and interpenetrating formations, and it is not always possible to draw the boundary of the structure in the form of a dividing line (Figs. 3.4, 3.5 and 3.6).

These global structures have a scale invariance, that is, they repeat themselves on a smaller scale, forming a complex hierarchical ensemble. The predominant confinement of structures of the L-hexagon type to the continents of the western hemisphere is revealed, as well as the G-hexagon type structures are confined to the Eurasian continents of the eastern hemisphere (Figs. 3.7, 3.8, 3.9, 3.10, 3.11, 3.12, 3.13, 3.14, 3.15, 3.16, 3.17, 3.18, 3.19, 3.20, 3.21, 3.22, 3.23, 3.24, 3.25, 3.26, 3.27, 3.28, 3.29, 3.30, 3.31 and 3.32). It can be seen that such structures possess a fractal property of self-similar sets, the existence of which was first theoretically established by Mandelbrot (1972). Such sets are analogous to the perfect, dense-throughout Cantor set of points that fills the entire segment with the total mass of its points and, at the same time, is equal to zero.

In fact, in this case we are faced with a phenomenon familiar to many researchers when global and supraregional cellular structures, identified on the basis of geophysical field analysis or the deciphering of satellite images at increasing scales, break up into hierarchies of diverse and often completely disconnected tectonic elements while these global or supraregional structures somewhat dissolve and disappear. Therefore, when using the analysis of seismic tomography data, satellite data on geoid and global gravitational potential anomalies, the results of quantitative studies of the relief, the interpretation of aerial and satellite images or structural geological and geophysical studies, these new objects of geodynamics can be identified only if the researcher who undertakes such a task has an idea of fractals and their dimensionalities.

3.3 Calculation of the Fractal Dimensionality of the Most Recent Dissipative Structures of the Earth

Fractal geometry, the new methodological principle that takes into account the wave properties of matter in spontaneous structuring processes of density-unstable media, defines a new paradigm for studying chaotic natural systems whose construction is entirely irreversible. Such systems under certain conditions have the ability of self-organisation (Nicolis and Prigogine 1979), which is based on cooperative processes of stabilisation of collective modes (Haken 1980). Studies have shown that, while reflecting the wave properties of matter, the fractals serve as the approximate models of random media in a number of cases, and a fractal structure and dimensionality act as the parameters of the order of nonequilibrium chaotic phenomena. Theoretically, a fractal can be distinguished in any—even absolutely chaotic—structure, where the element of structuredness may be completely absent at first glance. That kind of approach has become widespread in various fields of natural science (Fractals in Science 1995), including geology

(a)

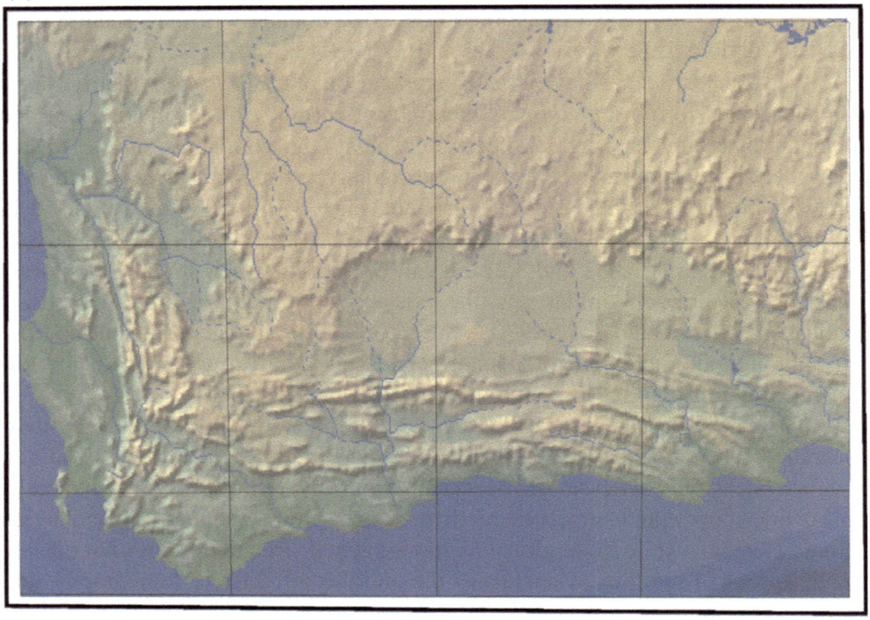

(b)

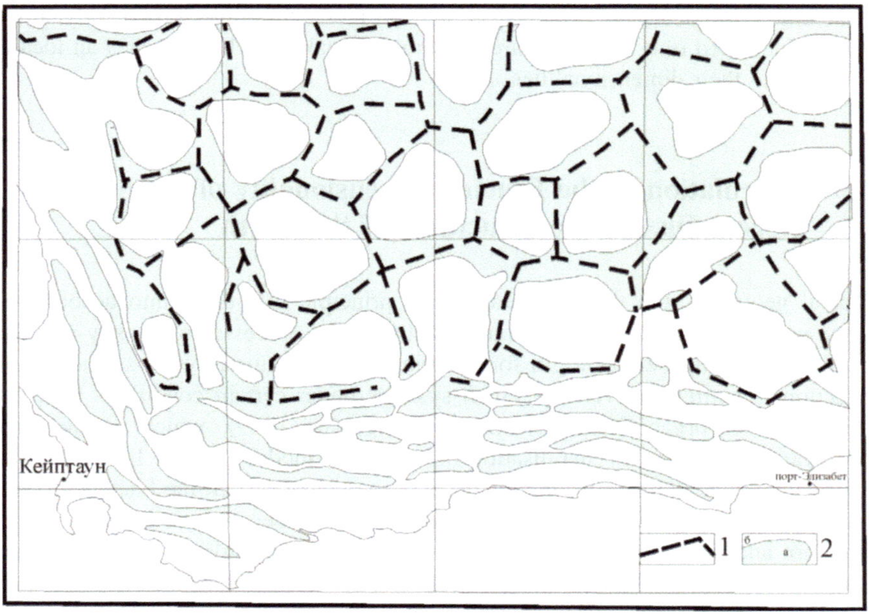

◀**Fig. 3.24** Most recent regional structures of the Earth's gravitational instability of the tenth order of South Africa **a** hypsometric scheme; **b** scheme of polygonal cellular structuring of the tenth order of southern Africa. For explanation refer to Fig. 3.21

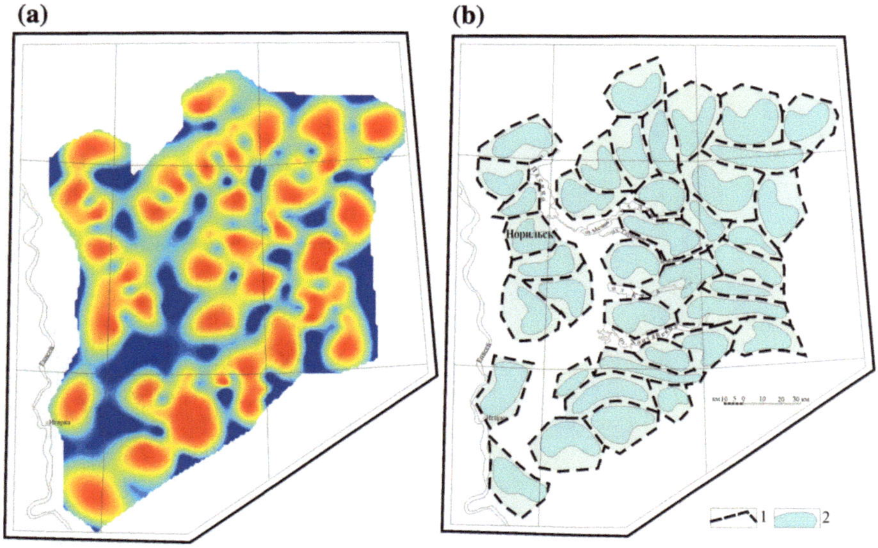

Fig. 3.25 Most recent local structures of the Earth's gravitational instability of the eleventh order within the Norilsk ore district and adjacent territories **a** isobase and goniobase line scheme; **b** polygonal cellular structuring. For explanation refer to Fig. 3.21. {Legend on Fig. 3.25b: Norilsk}

(Fractals … 1993; Ivanyuk et al. 1996; Turcotte 1997). Thus, in our understanding, the empirically established fact of the existence of the fractal as a form of approximation of chaotic structures reveals new wave properties of matter along with new fundamental laws of the evolution of nature.

In mathematics, fractals are by definition self-similar objects, whose dimensionality is strictly different from that of the embedding space and possesses a fractional value. In this connection, let us consider elementary objects that have a homogeneous density distribution: a segment, a square, a cube whose dimensionality d characterises the dependency of the mass of the system M (L) on its linear dimension L (Fig. 3.33). Obviously, for a fragment of the system of the dimension bL (b < ; 1)

$$M(bL) = b^d M(L), \tag{3.1}$$

where d determines the topological dimensionality of the figures: d = 1 for the segment, d = 2 for the square and d = 3 for the cube.

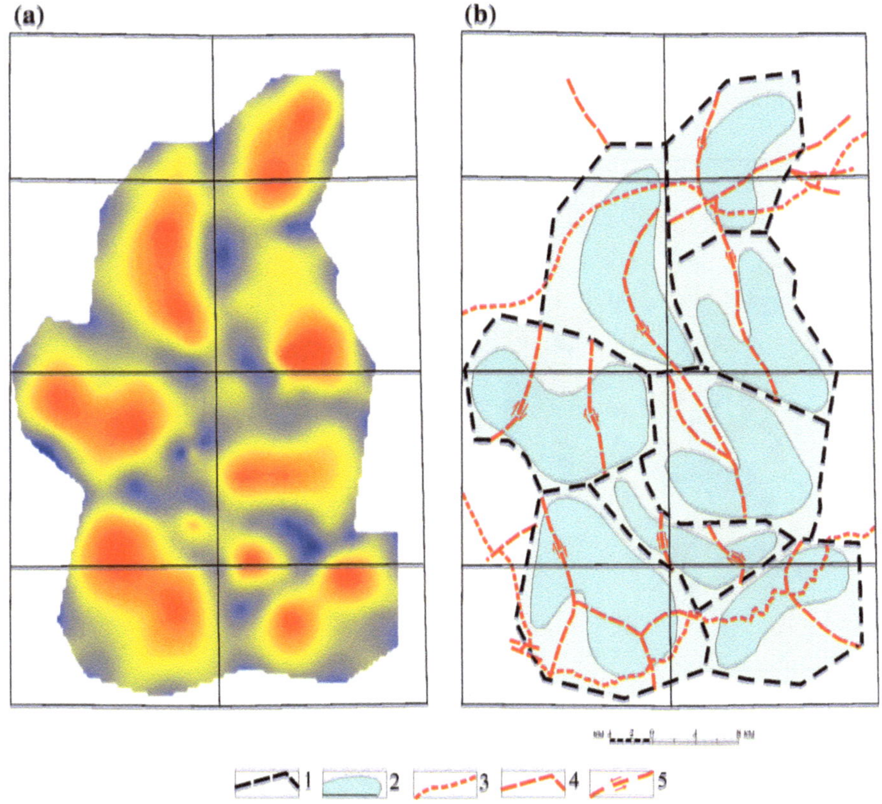

Fig. 3.26 Most recent local structures of the Earth's gravitational instability of the twelfth order within the Kharaelakhsk trap plateau: **a** isobase and goniobase line scheme; **b** polygonal cellular structuring scheme. 1—structure boundaries; 2—immersing and emersing areas of gravitational instability structures; 3—the outcrops of industrial nickel-bearing intrusives under the Quaternary deposits, and their boundaries on deep horizons; 4—the boundaries of lava plateaus; 5—some faultings and the directions of tectonic displacements along them. Normal equiangular conical projection. 1:200,000 scale

Let us divide the unit segment into three equal parts and delete the central part. We repeat the described procedure towards each of the remaining parts of the segment, then towards the remaining parts and so on. As a result, after four iterations, we obtain the pattern indicated in Fig. 3.34. The mass of its fragment with length b = 1/3 is expressed by the ratio

$$M(1/3L) = 1/2M(L) \tag{3.2}$$

from which $1/3 = (1/2)^d_f$ and $d_f = \log_2/\log_3 \cong 0.631$. The fractional exponent obtained is referred to as a fractal dimensionality. Objects that possess such dimensionality, then, are fractals. In the example considered above, an algorithm for

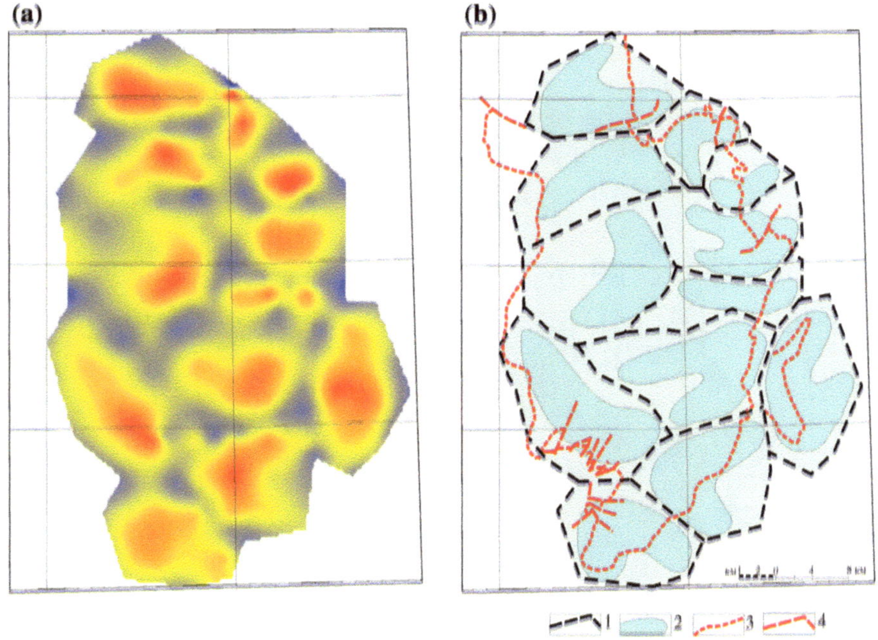

Fig. 3.27 Most recent local structures of the Earth's gravitational instability of the twelfth order within the Norilsk trap plateau **a** isobase and goniobase line scheme; **b** polygonal cellular structuring. 1—structure boundaries; 2—immersing areas of gravitational instability structures; 3—boundaries of lava plateaus; 4—various faultings. Normal equiangular conical projection. 1:200,000 scale

constructing a fractal, known as the Cantor triad set, is described. Another example of a fractal set is the Sierpinski triangle, the method of construction of which can be clearly seen from Fig. 3.35. It can be shown that the fractal dimensionality of both sets is at the same time the dimension of similarity. As the illustration of this fact, both considered examples should be presented as a sequence of repeated compressed representation, in which the image is placed inside the prototype (Feder 1991; Makarenko 2003; Egorov 1997).

For the trivial case of a unit segment, such a representation consists of covering the original segment by its N-reduced copies of the length

$$r(N) = 1/N, \tag{3.3}$$

where N is any integer. In the general case, the scale factor is

$$r(N) = (1/N)^{1/d}, \tag{3.4}$$

where d is the topological dimensionality of a homogeneous figure.

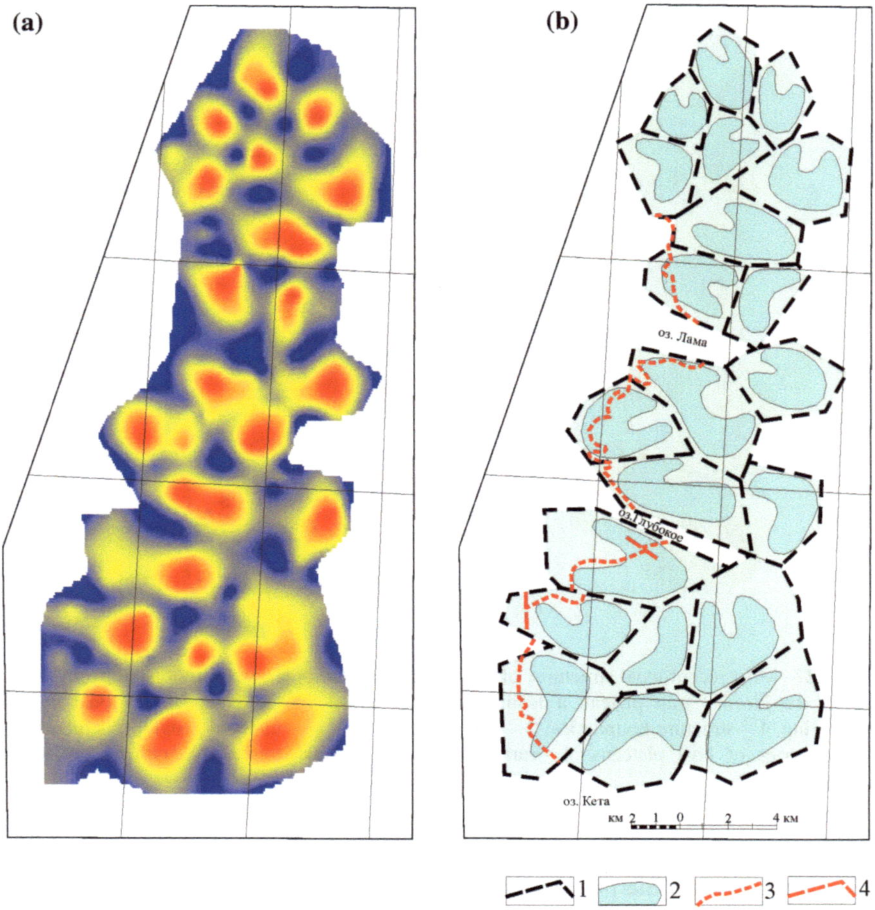

Fig. 3.28 Most recent local structures of the Earth's gravitational instability of the twelfth order within the Syverminsk trap plateau **a** isobase and goniobase line scheme; **b** polygonal cellular structuring. For explanation refer to Fig. 3.20

For the Cantor set at the first iteration, the analogous representation consists in decreasing the unit segment in the ratio r = 1/3 and its coverage by its N = 2 reduced copies. At the second iteration, r = 1/9 = $(1/3)^2$, N = 4 = 2^2, for the third iteration r = 1/27 = $(1/3)^3$, N = 8 = 2^3, at n-th r = (1/3) n = $(1/3)^3$, N = 2^n. Hence

$$r(N) = (1/N)^{1/D_S} \tag{3.5}$$

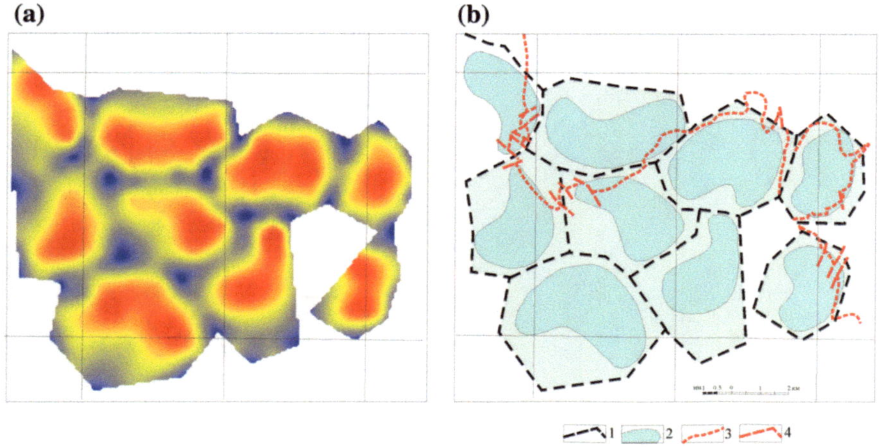

Fig. 3.29 Most recent local structures of the Earth's gravitational instability of the thirteenth order within the northern part of the Noril'sk trap plateau **a** isobase and goniobase line scheme; **b** polygonal cellular structuring. 1—structure boundaries; 2—immersing areas of gravitational instability structures; 3—the outcrops of industrial nickel-bearing intrusives under the Quaternary deposits, and their boundaries on deep horizons; 4—some faultings. Normal equiangular conical projection. 1:50,000 scale

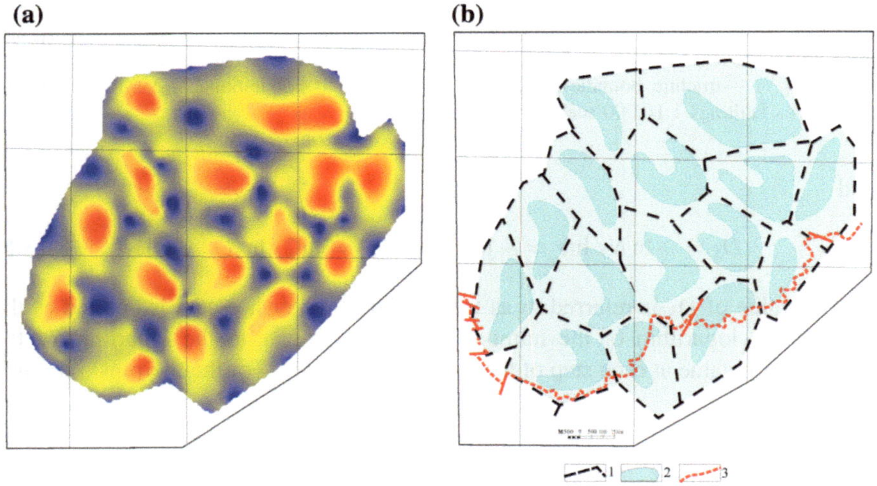

Fig. 3.30 Most recent local structures of the Earth's gravitational instability of the thirteenth order within the Kharaelakh trap plateau **a** isobase and goniobase line scheme; **b** polygonal cellular structuring. 1:50,000 scale. For explanation refer to Fig. 3.26

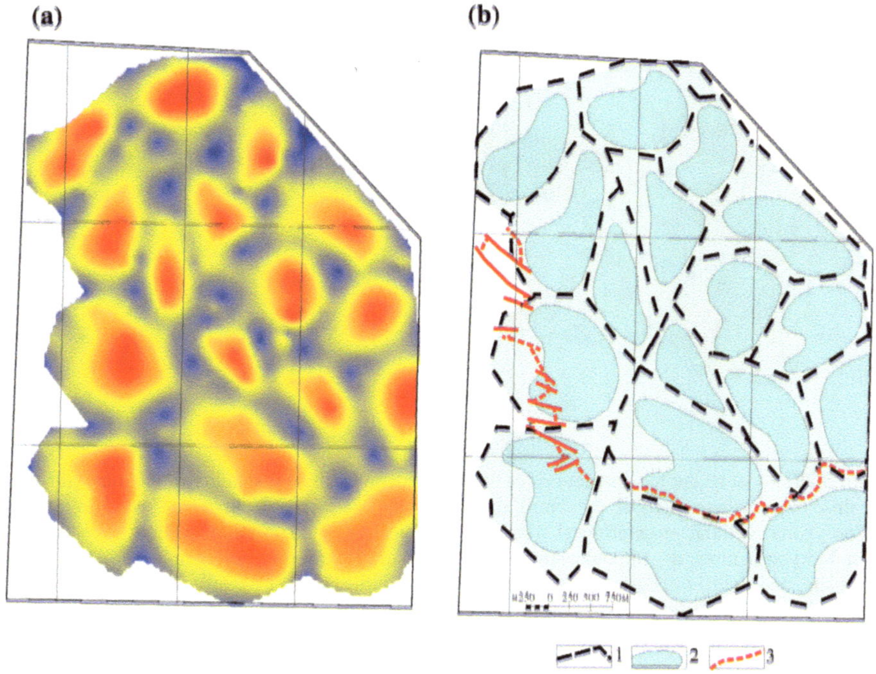

Fig. 3.31 Most recent local structures of the Earth's gravitational instability of the fourteenth order within the Kharaelakh trap plateau **a** isobase and goniobase line scheme; **b** polygonal cellular structuring. 1—structure boundaries; 2—immersing areas of gravitational instability structures; 3—various faultings 1:10,000 scale

and

$$D_S = -\ln N / \ln r = -\ln 2 / \ln(1/3) = \ln 2 / \ln 3 = 0.631. \tag{3.6}$$

The value of D_S is referred to as the similarity dimension and is defined as the ratio of the logarithm of the number of reduced copies to the logarithm of the compression value at each step of the figure construction. For the Sierpinski triangle

$$D_S = \ln 3 / \ln 2 = 1.585. \tag{3.7}$$

The self-similar sets considered above belong to the class of deterministic fractals. They are based on a strict algorithm that establishes the rules according to which the image is self-copied at different scales of observations. By introducing a stochastic component into the self-copying algorithm, the above can also be generalised to the case of probabilistic fractals. It turned out that for a number of natural phenomena possessing wave properties, similar fractal models describe the geometry of structures more adequately as compared with conventional Euclidean models.

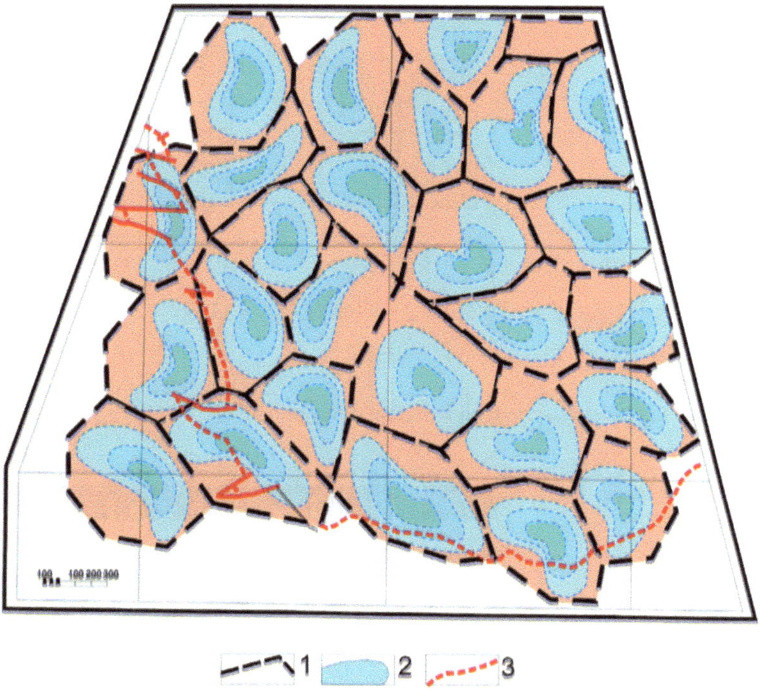

Fig. 3.32 Most recent local structures of the Earth's gravitational instability of the fifteenth order within the Talnakh ore field and adjacent territories. 1—structure boundaries; 2—immersing areas of gravitational instability structures; 3—various faultings. Rectangular coordinate system. 1:10,000 scale

3.3.1 *Calculation of Fractal Dimensionality*

At the phenomenological level, the fractal models describe the features of the geometry of objects with the aid of power equations that define the magnitude measure of point sets. The most general approach to calculating fractal dimensionality is as follows. The test structure is represented as a point set embedded in the observation space. This space is divided into cells with an edge r, and then a number of cells K containing at least one point of the set is counted. The procedure is repeated many times for continuously decreasing r, with the approximation of the obtained dependency by the power equation

$$(K) = ar^{-D}, a = \text{const}. \quad (3.8)$$

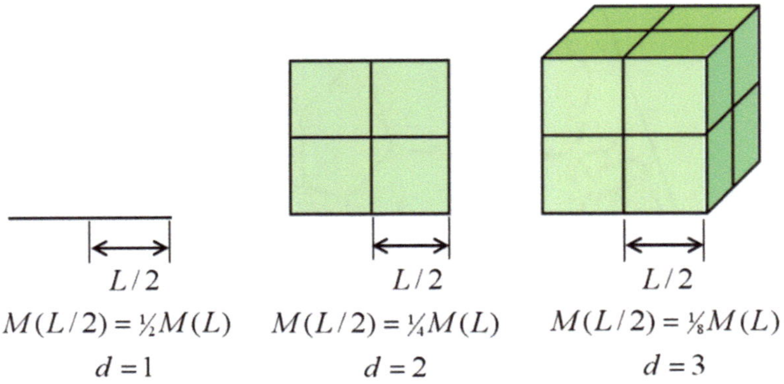

Fig. 3.33 Mass dependency system on its *L* line dimension *M(L)*

Fig. 3.34 Cantor triadic set construction algorithm

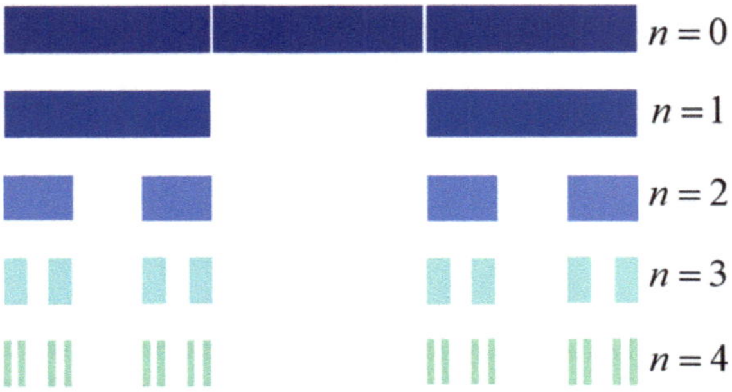

The value of the fractal dimension *D* is derived from the regression equation

$$\log K(r) = \log a - D \cdot \log r \tag{3.9}$$

as the opposite sign slope coefficient of the linear dependency between the logarithms of the variables. The described computational procedure is referred to as the method of cell counting (box-counting method). Fractal dimensionality, defined in this way, is referred to as cellular.

For one-dimensional curves on a plane, the fractal dimensionality varies from 1 (smooth Euclidean lines) to 2 (infinitely sinuous curves filling the plane). With increasing complexity (bifurcation) of the structure, the value of *D* increases. Similarly, for the case of two-dimensional surfaces $2 < D < 3$.

Cellular fractal dimensionality should be distinguished from the dimensionality of a mass or a fractal cluster (Mandelbrot 1988). A fractal cluster should be understood as any inhomogeneous spatial distribution that can be expressed by a power function connecting the mass of some object *M* with its linear dimension *R* (for example, with the radius of the circumsphere) upon the extrapolation to the region of larger *R*:

Fig. 3.35 Sierpinski triangle construction algorithm

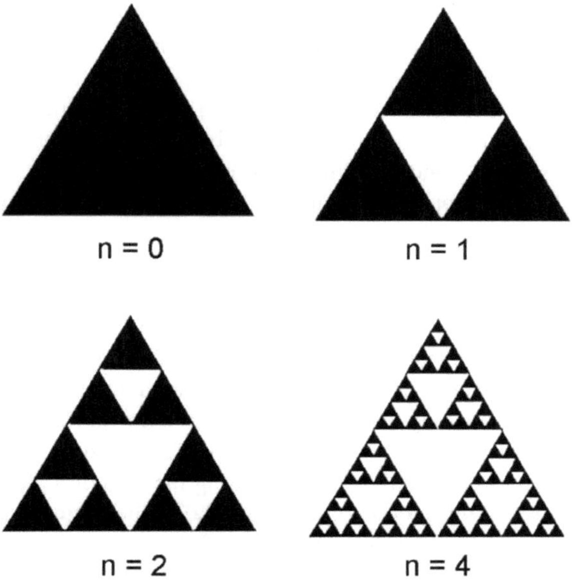

$$M \in R^D \tag{3.10}$$

The exponent of D, if it has a fractional value, is referred to as the mass dimensionality. This value characterises the degree of the space filling by cluster particles or, which is the same, the law of density alteration. Obviously, for homogeneous distributions the value of D will correspond to the topological dimensionality of the embedding space, that is, 2 in the case of a plane and 3 in the case of a volume. The calculation method of D, using the recorded equation, is referred to in the literature as the sandpile method.

The method of a correlation (two-particle) function is closely correlated with the sandpile method (Smirnov 1991; Polikarpov 1995). In general, this function describes the law of change in mass or density within a fractal cluster. For a given cluster particle, it provides the probability of finding an additional particle at a distance r from the first. Imagine an object consisting of a large number of points, somehow located in a D-dimensional Euclidean space. Calculate the average number of points that fit into a sphere of radius R. To do this, consider the function

$$f(R) = 1/\mathrm{N}(\mathrm{N}-1)\sum_{i,j}\theta\left(R - \left|x_i - x_j\right|\right), \tag{3.11}$$

where N is the total number of points, x_i are their coordinates, θ—is the unit function (the Heaviside function that takes a value of 1 if the distance $|x_i \_ x_j|$ is less than R, and the value of 0 if the specified distance is greater than R). When the object under consideration is fractal, $f(R)\in R^D$ and the power D can be considered as the fractal dimensionality of the object.

In the case of the dimensionality of cells, as well as that of mass, it is assumed that the range of definition of the fractal approximation corresponds to an interval of linear dimensions intermediate between the dimension of the smallest element of the object and the dimension of the object itself. At the same time, the value of D does not depend on the calculation method, but only because we perform interpolation or extrapolation, focusing on different aspects of the structure. These aspects are either local inhomogeneity that is visualised in the region of small linear quantities or global heterogeneity comparable to the dimension of the cluster itself. The definition of fractal dimensionality in both cases is closely related to the concept of intermediate asymptotics (Zeldovich and Sokolov 1985).

In connection with a one-dimensional case, the procedure for calculating the cell dimensionality can be understood in terms of the process of measuring the length of the $L(r)$ curve by means of a linear measure r. Similarly, for flat figures it is admissible to talk about measuring the area of a plane figure $S(r)$ by square cells of linear dimension r. In this case, the studied figure may, in general, not have fractal properties. To estimate the fractal dimensionality (within a given range of r), it is significant whether the studied figure is compact or, on the contrary, "loosened", i.e. how sharply or smoothly its area increases with a decrease in the linear measure. The more compact the figure, the higher its fractal dimensionality, and vice versa.

As an example, Fig. 3.36 indicates the results of calculating the fractal dimensionality of prefractals of several generations of a Sierpinski triangle. The images of prefractals are obtained with the help of a computer program with a resolution of 350×400. Measurements are performed in the range $r = 2$–70 using the cell counting method employed by the FractShop software package for structural image analysis. As can be seen, only for prefractal of the seventh generation the calculated value of D coincides with the theoretical value. As the image becomes more simple and compact, the fractal dimension increases.

3.3.2 *Analysis of Seismic Tomograms*

The considered regularities were used to study the structure of seismic tomograms of the crust and upper mantle. During the analysis, the colour images of the seismic tomograms were converted into black and white (bitmap) format using the numerical algorithms of the Photoshop graphics editor. The calculation of the fractal dimensionality of the image was based on the cell counting method. The selection of intervals for calculating D was made taking into account the type of fractal plots and the accuracy of the obtained estimates of the angular regression coefficients. In the capacity of the latter, interval estimates were used, covering the target values with 95% reliability.

As a result of the analysis of the negative anomalies of the SAW12A model, a slight increase in the fractal dimensionality at a depth of 1.83–1.87 is recorded,

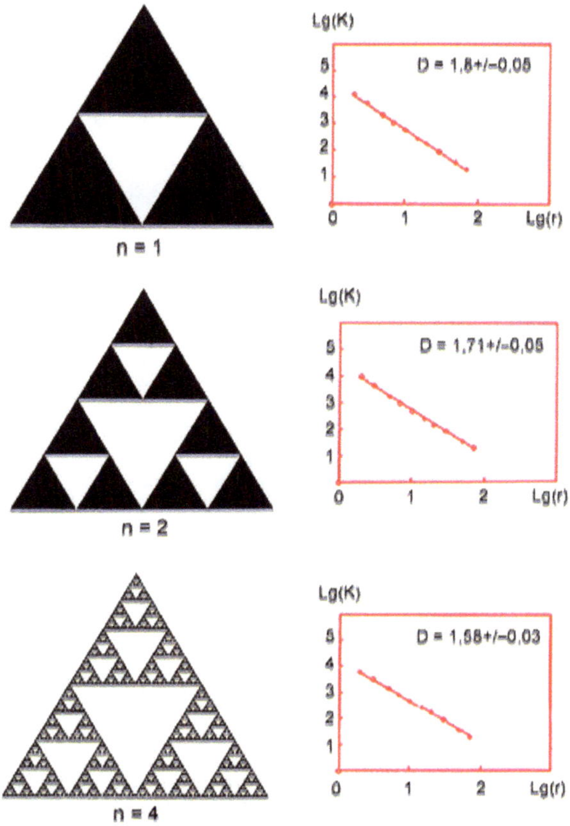

Fig. 3.36 Fractal dimensionality calculation results for the pre-fractals of several generations of a Sierpinski triangle

indicating that the structure at the lower horizons becomes more homogeneous in comparison with the upper horizons (Fig. 3.37). In the case of positive anomalies, a similar trend occurs when comparing the level of 2800 km with the levels of 620 and 150 km (Fig. 3.38).

Resovsky Model. Similar patterns were obtained for a series of tomograms borrowed from (Resovsky and Ritzwoller 1999). Black-and-white images of positive maxima with a resolution of 350 × 700 (Fig. 3.39) were investigated. For the calculations of fractal dimensionality, the cell counting and correlation function methods were used. The data obtained from the analysis are given in Table 3.2.

In Fig. 3.40 estimates of the fractal dimensionality of the structure of high-velocity seismic sections are presented. The circumference representing the outer surface of the sections corresponds to the Moho surface. A dash line indicates a depth of 670 km, at which a significant jump in temperatures and densities (the surface of discontinuity) is observed. The maximum depth of the section is 2700 km. The figure indicates the fractal branching averaged for each of the sections:

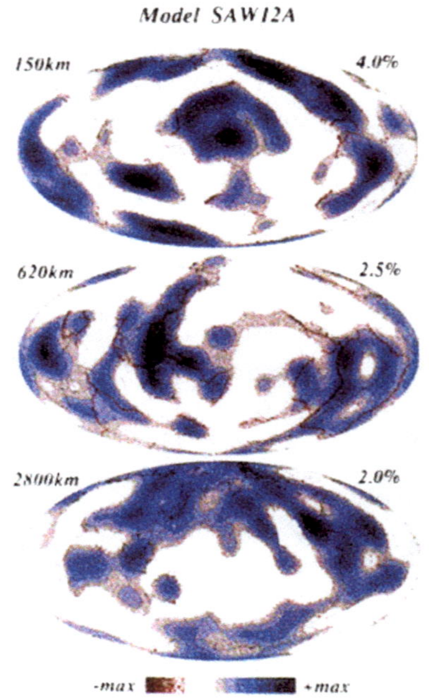

r	K	N	Lg(r)	S	Lg(N)
2	72193	231168	0,301	4,824	5,364
3	38735	102816	0,477	4,533	5,012
5	15805	36850	0,699	4,151	4,566
7	8586	18816	0,845	3,873	4,274
10	4535	9179	1	3,572	3,963
15	2224	4004	1,176	3,232	3,602
20	1369	2244	1,301	2,995	3,351
30	672	990	1,477	2,666	2,996
50	278	351	1,699	2,266	2,545
70	148	171	1,845	2,005	2,233
100	73	78	2	1,736	1,892
150	34	36	2,176	1,452	1,556
200	18	18	2,301	1,193	1,255

r	K	N	Lg(r)	S	Lg(N)
2	74028	233856	0,301	4,832	5,369
3	40655	103936	0,477	4,548	5,017
5	16796	37252	0,699	4,175	4,571
7	9071	19104	0,845	3,898	4,281
10	4760	9313	1	3,596	3,969
15	2305	4048	1,176	3,255	3,607
20	1406	2277	1,301	3,016	3,357
30	684	1012	1,477	2,686	3,005
50	280	351	1,699	2,279	2,545
70	147	171	1,845	2,014	2,233
100	74	78	2	1,721	1,892
150	35	36	2,176	1,421	1,556
200	18	18	2,301	1,148	1,255

r	K	N	Lg(r)	S	Lg(N)
2	75896	230460	0,301	4,847	5,363
3	40978	102580	0,477	4,556	5,011
5	16528	36708	0,699	4,178	4,565
7	8850	18715	0,845	3,899	4,272
10	4612	9108	1	3,597	3,959
15	2239	4048	1,176	3,255	3,607
20	1361	2277	1,301	3,015	3,357
30	659	1012	1,477	2,682	3,005
50	269	351	1,699	2,267	2,545
70	144	171	1,845	1,995	2,233
100	73	78	2	1,699	1,892
150	35	36	2,176	1,389	1,556
200	18	18	2,301	1,139	1,255

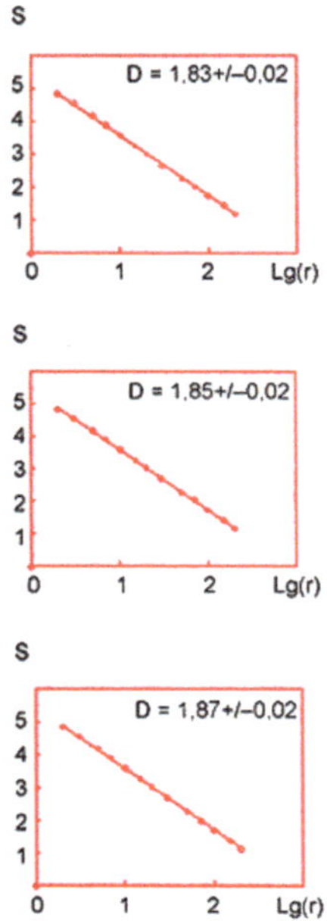

Fig. 3.37 Fractal dimensionality increase with depth during the analysis of the mantle negative anomalies obtained by transversal seismic waves

- the section dissecting Europe, Antarctica and North America in the meridional direction, corresponds to the average fractal dimensionality $D = 1.66$ (*a*);
- the section dissecting Africa and Eurasia has the fractal dimensionality $D = 1.48$ (*b*);
- the section dissecting South America, Africa and India has the fractal dimensionality $D = 2.16$ (*c*);
- the section, dissecting South America, Australia and Eurasia has the fractal dimensionality $D = 1.45$ (*d*).

The numerical series of D values fits into a narrow interval in which the value of fractal dimensionality falls, determined earlier by lateral representations of seismic tomography materials ($D = 1.7$–1.8). It can be assumed that the observed scatter of D values is due, on the one hand, to the approximate nature of the estimation of the fractal structure topology, and, on the other hand, to the substantial lateral and vertical anisotropy of the upper mantle material. The latter can be associated with different dynamics of heating and cooling (internal sources and external sinks) and causes different wave dynamics for each of the sections.

The method for determining the mean fractal dimensionality is the same as for the lateral representations of seismic tomography data, namely:

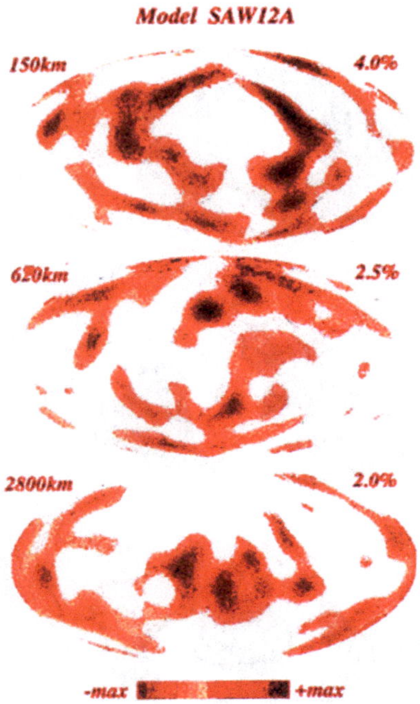

r	K	N	Lg(r)	S	Lg(N)
3	29131	95251	0,477	4,396	4,979
5	12601	34314	0,699	4,04	4,535
7	6912	17490	0,845	3,767	4,242
10	3666	8514	1	3,467	3,93
15	1807	3784	1,176	3,13	3,578
20	1104	2112	1,301	2,893	3,325
30	556	946	1,477	2,569	2,976
50	234	325	1,699	2,172	2,512
70	130	162	1,845	1,927	2,209
100	66	72	2	1,649	1,857
150	31	32	2,176	1,376	1,505
200	18	18	2,301	1,148	1,255

r	K	N	Lg(r)	S	Lg(N)
3	26826	102795	0,477	4,354	5,012
5	12160	36846	0,699	4,02	4,566
7	6721	18909	0,845	3,753	4,277
10	3622	9177	1	3,455	3,963
15	1805	4094	1,176	3,119	3,612
20	1130	2244	1,301	2,883	3,351
30	574	1012	1,477	2,56	3,005
50	238	338	1,699	2,161	2,529
70	131	171	1,845	1,9	2,233
100	72	78	2	1,627	1,892
150	32	32	2,176	1,329	1,505
200	18	18	2,301	1,117	1,255

r	K	N	Lg(r)	S	Lg(N)
3	25721	98368	0,477	4,337	4,993
5	11494	35306	0,699	4	4,548
7	6267	18109	0,845	3,731	4,258
10	3314	8757	1	3,431	3,942
15	1638	3864	1,176	3,094	3,587
20	1022	2139	1,301	2,857	3,33
30	527	966	1,477	2,532	2,985
50	228	324	1,699	2,137	2,511
70	136	171	1,845	1,878	2,233
100	73	78	2	1,632	1,892
150	35	36	2,176	1,372	1,556
200	18	18	2,301	1,116	1,255

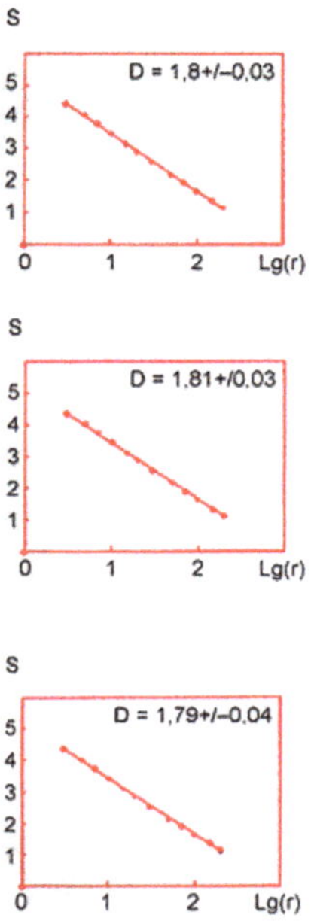

Fig. 3.38 Fractal dimensionality increase with the depth during the analysis of the mantle positive anomalies obtained by transversal seismic waves

- this section is divided by a uniform grid into a finite number of cells (clusters) with a fixed linear dimension (Fig. 3.41). For example, if the cells have a square shape, then the length L of their side is determined;
- the total number of N clusters falling within the given object, for example, in the region of increased velocities, is counted (depicted in blue in Fig. 3.41);
- then increase the linear dimension of the cell, and repeat the counting procedure.
- Finally, we try to approximate the dependency *N* from *L* by the dependency of the form

$$\log(N) = D \cdot \log(L) + \text{const or} N = \text{const} \cdot L^D.$$

Additional analysis shows that the shape of the cells can be arbitrary (rectangles, hexagons, circles, triangles), as well as the geometry of the grid. To simplify the algorithm, it was chosen as a rectangular (Fig. 3.42).

In addition, the program for calculating the fractal dimensionality can evaluate the effect of lateral anisotropy of the upper mantle on the structure of fractal branching of internal gravitational waves. Figure 3.43 indicates the transformation of a high-velocity seismic section, meridionally intersecting Europe, North America

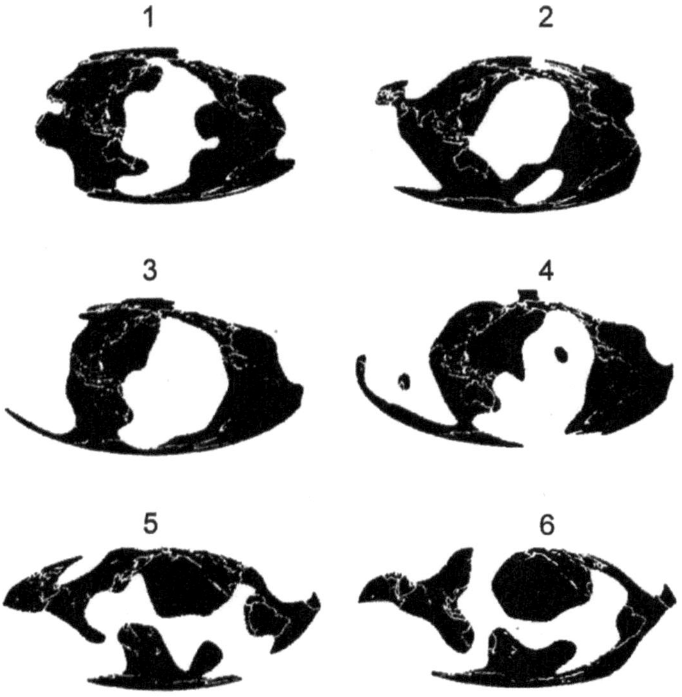

Fig. 3.39 Fractal dimensionality decrease in the tomogram (accounting for the positive velocity maxima of lateral seismic waves) during the transition from the deeper layers to the sub-surface. Source data from Resovsky and Ritzwoller (1999)

Table 3.2 Fractal dimensionality of tomogram in the Resovsky model

Tomogram number	D^a	D^b
1	1.83 ± 0.03	1.72 ± 0.02
2	1.84 ± 0.03	1.76 ± 0.03
3	1.83 ± 0.04	1.68 ± 0.03
4	1.82 ± 0.04	1.64 ± 0.02
5	1.79 ± 0.03	1.63 ± 0.03
6	1.80 ± 0.03	1.60 ± 0.03

[a]Cell count method, r = 1–150
[b]Correlation function method, r = 50–180

and Antarctica, into a system of standing internal gravitational waves formed at a fixed depth. The analysis of the spatial correlation of deep and medium standing waves with each other allows us to split the section plane into regions of spatial stationarity—the regions where the constant amplitude-frequency composition of the wave process takes place. In our case, the areas of spatial stationarity have the

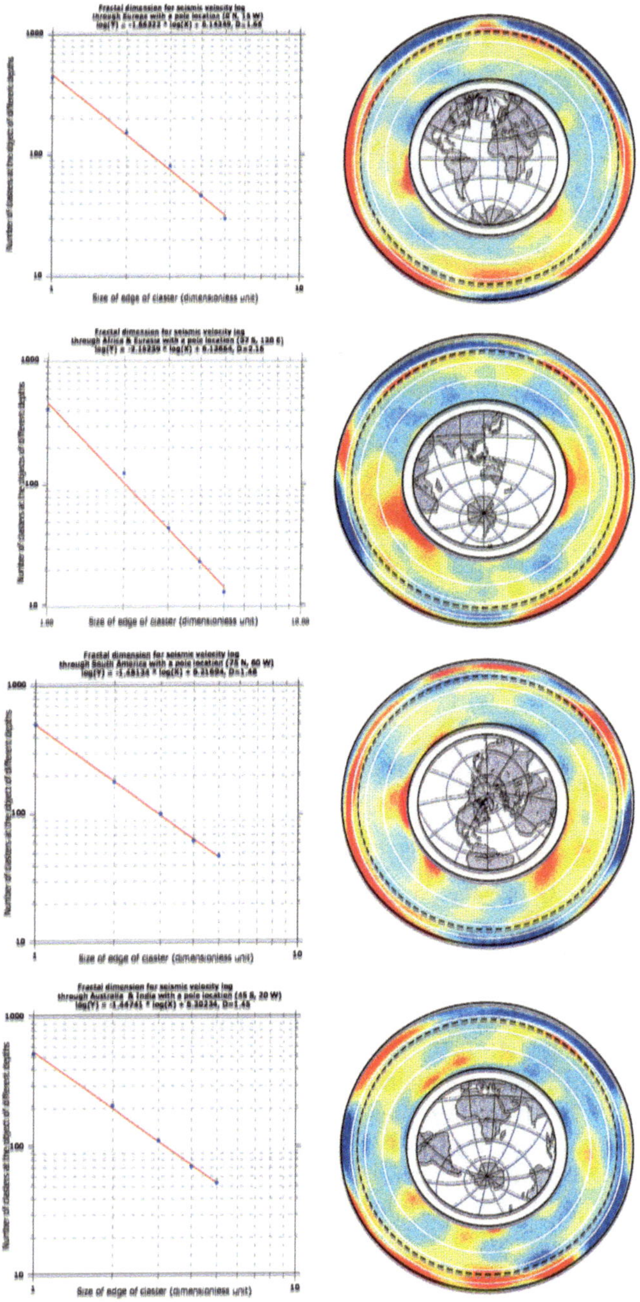

Fig. 3.40 Fractal dimensionality reconstruction in the planetary sections down to 2700 km. Detail explanations are given in the text

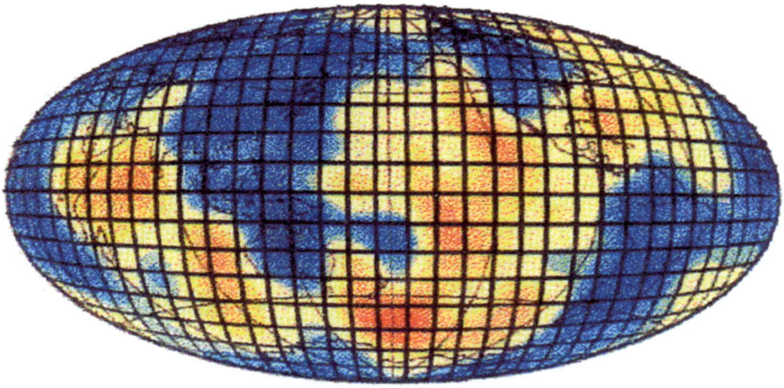

Fig. 3.41 Image fractionation of lateral high-velocity distribution by homogeneous clusters

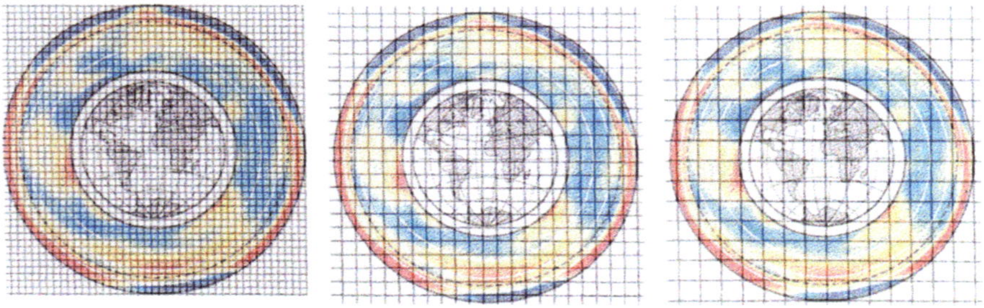

Fig. 3.42 Rectangular cells for the determination of fractal dimensionality

form of three angular sectors, including the Pacific Ocean, North America and Eurasia, the Indian Ocean and Antarctica. An investigation of the dependency of the length of the internal gravity waves λ on depth h shows the existence of a fractal inheritance in the assemblage of these waves, which is violated at depths of 1800 km (sectors 1, 2) and 1350 km (sector 3). From the analysis of plots $\lambda(h)$ it follows that in depth intervals where fractal inheritance exists, the fractal dimensionality varies from 1.18 to 2.77. For these values, the average fractal dimension over the entire section was obtained and estimated with an error of $\Delta D = 0.63$. Therefore, since the error is less than 50% of the mean, our numerical estimates are conditional. The average fractal dimensionality along the section, estimated above ($D = 1.66$), and the fractal dimensionality of the lateral seismic tomography data ($D = 1.7$–$1.8°$) falls in this range of values.

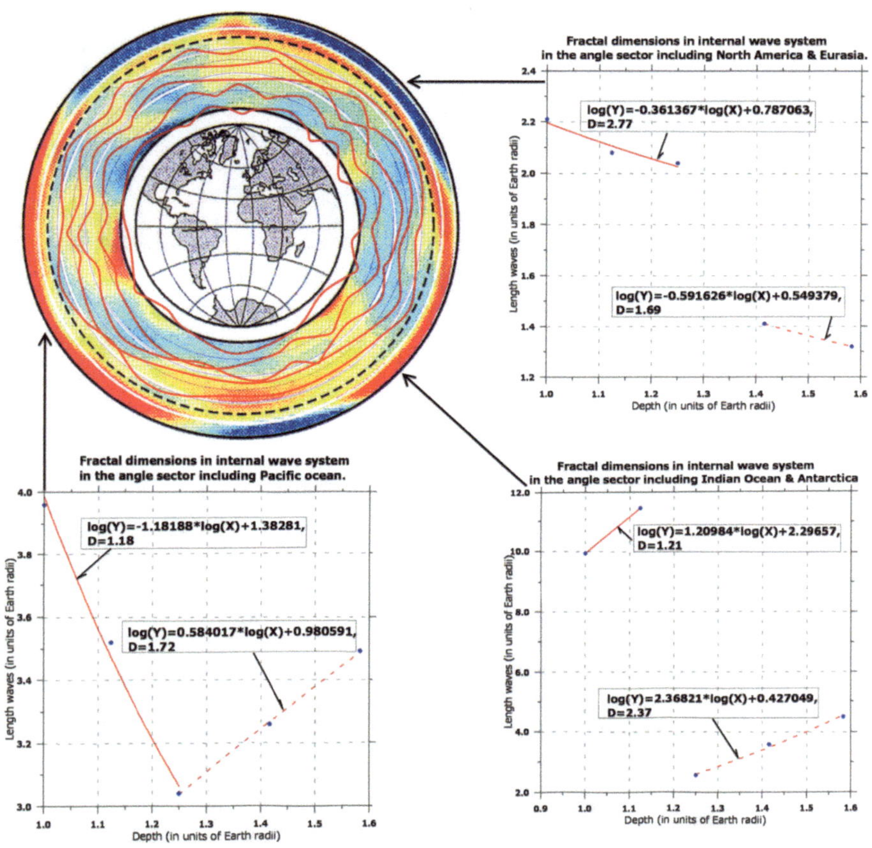

Fig. 3.43 Transformation of high-velocity seismic section meredionally transecting Europe, North America and Antarctic into the system of standing internal gravitational waves

3.3.3 *Analysis of Morphometric Surfaces of Gravitational Instability Structures*

For the evaluation of the spatial distribution of the morphometric surfaces of gravitational instability (dissipative) structures, the images with a resolution of 800 × 1000, 800 × 700, 800 × 1100 (Figs. 3.44 and 3.45) were studied. The results of calculating their fractal dimensionality are given in Table 3.3.

Taking into account the chosen range of *r*, the calculated values of the fractal dimensionality characterise the distribution density of the anomalies within the morphostructures of different scales. The comparative similarity of the obtained *D* estimates confirms the idea of the anomalies studied as the multi-ordinal constituents of a single hierarchically constructed self-similar structure of the Earth's

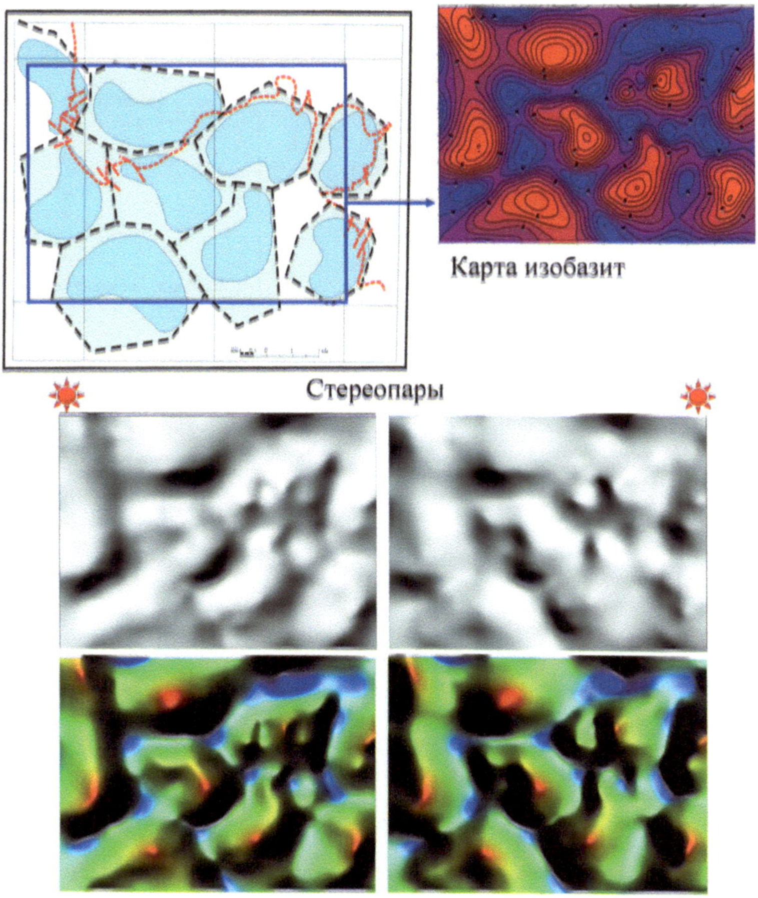

Fig. 3.44 Determination of the boundaries of dissipative polygonal structures of the thirteenth order within the northern part of the Norilsk trap plateau, and the reflection of the typomorphic features of their fractal structure by stereopairs. {Legends on Fig. 3.44, top to bottom: Isobase line map, Stereopairs}

crust. A slight increase in D for fourteenth-order morphostructures reflects a slight increase in density, that is, in turn, confirmed by a visual assessment of structural images.

The data of numerical experiments of the convection process in viscous media were used to interpret the obtained results (Ten et al. 1997). According to these data, the degree of fragmentation (dispersity) of the convective system increases during mixing, and its fractal dimensionality decreases. A similar dependency for the Newtonian liquid is shown in Fig. 3.46. It is characteristic that at the early stages of the process the variations of the D values are insignificant (1.84–1.82) in comparison with late stages, where the decrease in the fractal dimensionality

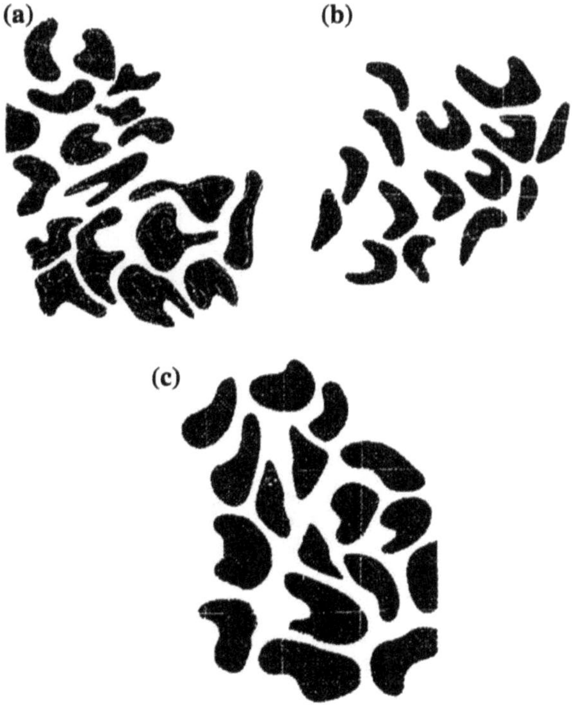

Fig. 3.45 Geomorphological expression of the gravitational instability structures **a** of the eighth order (the central sector of the Russian Arctic area); **b** of the thirteenth order (Kharaelakhsk plateau); **c** of the fourteenth order (Kharaelakhsk plateau). Additional explanations are given in the text

Table 3.3 Fractal dimensionality of dissipative structures

Structure	D^a	D^b
a—of the eighth order, central sector of the Russian Arctic	1.83 ± 0.03	1.72 ± 0.02
b—of the thirteenth order, Kharaelakh plateau	1.84 ± 0.03	1.76 ± 0.03
c—of the fourteenth order, Kharaelakh plateau	1.83 ± 0.04	1.68 ± 0.03

[a]Cell count method, $r = 5–50$
[b]Correlation function method, $r = 30–100$

becomes more noticeable (up to 1.43). In the case of the SAW12A model, the evolution of the structure proceeds in a similar way, assuming that at different depths we are dealing with successive "time sections" of the initial stages of the mantle convection process. The structure seems to be "split" into more fragments, more elongated in shape, with the depth decreased.

As follows from the consideration of gravitational waves, as we move to a region with a higher viscosity, the wavelength decreases, the set becomes more compact and its fractal dimensionality increases. The same regularity is indicated by the processing of the seismic SAW12A model. As we move toward the core, into the high temperature region, the viscosity of the mantle material decreases, the fractal

r	K	N	Lg(r)	S	Lg(N)
2	9992	131524	0,301	3,958	5,119
3	5554	58263	0,477	3,665	4,765
5	2693	20900	0,699	3,306	4,32
7	1714	10579	0,845	3,001	4,024
10	1099	5200	1	2,87	3,716
15	658	2277	1,176	2,624	3,357
20	462	1300	1,301	2,467	3,114
30	270	544	1,477	2,212	2,736
50	143	200	1,699	1,937	2,301
70	85	98	1,845	1,72	1,991
100	50	50	2	1,505	1,699
150	18	18	2,176	1,174	1,255

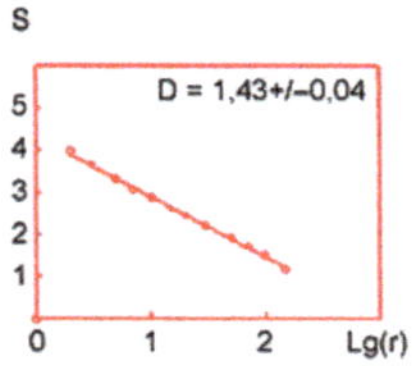

r	K	N	Lg(r)	S	Lg(N)
2	27482	137808	0,301	4,428	5,139
3	12733	61248	0,477	4,086	4,787
5	4939	21945	0,699	3,656	4,341
7	2676	11175	0,845	3,379	4,048
10	1415	5408	1	3,086	3,733
15	702	2415	1,176	2,766	3,383
20	428	1352	1,301	2,54	3,131
30	216	578	1,477	2,23	2,762
50	87	200	1,699	1,939	2,301
70	46	98	1,845	1,573	1,991
100	27	50	2	1,338	1,699
150	12	18	2,176	0,995	1,255

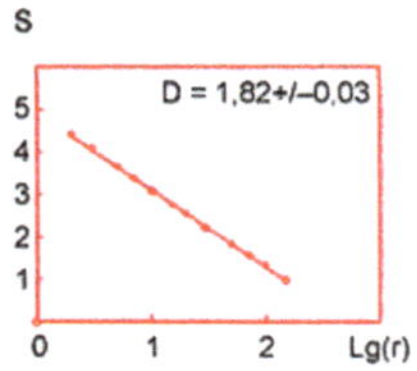

r	K	N	Lg(r)	S	Lg(N)
2	26677	133875	0,301	4,419	5,127
3	12103	59500	0,477	4,073	4,774
5	4545	21420	0,699	3,636	4,331
7	2410	10950	0,845	3,352	4,039
10	1243	5355	1	3,051	3,729
15	599	2380	1,176	2,716	3,377
20	356	1300	1,301	2,476	3,114
30	180	595	1,477	2,168	2,774
50	81	210	1,699	1,78	2,322
70	47	105	1,845	1,541	2,021
100	26	50	2	1,262	1,699
150	16	21	2,176	0,99	1,322

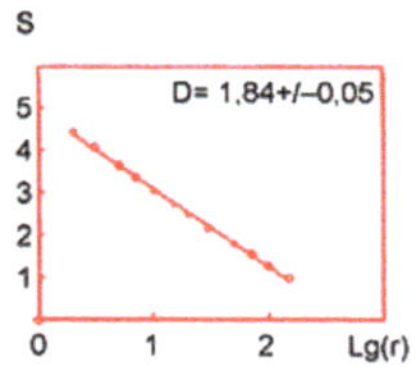

Fig. 3.46 Calculation results of the fractal dimensionality of convection model structures in viscous media (Tan et al. 1997) for the early (bottom), intermediate (centre) and late (upper) agitation stages

dimensionality of the structures displayed in the lateral velocity distribution of shear waves increases.

Thus, the easily implementable methodology of V.P. Filosofov made it possible to distinguish a family of polygonal forms, geometrically similar to Bénard cells, in the structure of topographical relief. A detailed comparison of these surface forms with depth interfaces confirms their absolute conformity. At the same time, as in the structure of the topographical relief, the lines of equal values of seismic wave velocity form polygonal structures similar to Bénard cells. The approximation of the seismic model structure by a random fractal indicates an increase in the fractal dimensionality as the depth increases, i.e. when moving further into the into the high-temperature region of the Earth's interior, where the mantle material has decreasing viscosity. A similar regularity, as is well known, is also observed in gravitational waves.

In summarising the listed facts, we come to our conclusion concerning the significant displacement of matter inside the planet, which arises due to gravitational (mechanical) disequilibrium. It is commonly known that this displacement of matter in continuous media creates cellular standing internal gravitational waves at the interfaces of nonequilibrium media. The presence of such waves is indicated by fractal approximations. Both seismic data and the structure of the topographical relief reveal polygonal forms that constitute a geometric representation of gravitational waves, or dissipative structures of the Earth.

Thus, seismic tomography data provide an essential addition to our understanding of the distribution of density inhomogeneities in the Earth's interior. The authors of the recent tomographic generalisations almost unanimously note the comparability of the results obtained by different researchers. In particular, this applies to long-period anomalies. During the layer-by-layer comparison of the spectra of high-velocity anomalies, an undoubted tendency of "redshift" is appeared with the transition to greater depths. The dimensions of the anomalies at the core-mantle boundary and in the lower areas of the lower mantle are larger than in the upper mantle (Kennet and Bowman 1990; Su et al. 1994). Nevertheless, the alteration of the anomalous spectrum is not monotonous. The structure of high-velocity seismic sections reveals a fractal dimensionality $D \approx 1.45$–2.15. Into this narrow range of values D falls both the value of the fractal dimensionality, determined for surface structures, as well as the numerical estimate of D, performed on the lateral representations of seismic tomography materials ($D = 1.7$–1.8).

The presence of such fractal dimensionality, as well as the presence of self-similar elements in morphogenesis and structuration, is indirectly indicative of the wave nature of geological processes. Thus, the calculation of the fractal dimensionality of such natural formations turns out to be equivalent to the parameterisation of the wave process, which is physically realised on the basis of a set of modes having their numerical correlations determined by the alteration of the wavelength (or a wave number), i.e., by the scale factor. The oscillation frequency of these cellular standing waves (or structures) cannot exceed the buoyancy frequency of the density-unstable geological medium, and, in this sense, is comparable with the duration of geological periods in the Earth's evolution. Here, from the standpoint of wave theory, the geological structures must also be taken to represent the historical categories, since, if they did not, a complete erasure of the boundaries between them and the cellular standing internal gravitational waves would have occurred.

References

Gerasimov N.P. Experience in the geomorphological interpretation of the general scheme of the USSR geological structure. Problemy fizicheskoi geografii. AN SSSR. 1946; 12:33–46. (In Russ.).

Meshcheryakov Yu. A. The main elements of the Earth's morphological structure and the problem of its origination. Izv. AN SSSR. Ser. geogr. 1957; 4:3–15. (In Russ.).

Khudyakov G.I. Principal foundations of morphotectonic studies (on the example of studying the southern mainland part of the Soviet Far East). Published summary of the Doctor's dissertation. Novosibirsk; 1974. (In Russ.).

Khudyakov G.I., Ezhov B.V., Ishchenko A.A. On the content of the concept of "geological-geomorphological conformal complex" (to the creation of geomorphological structure classification). Morphostructure studies in the Far East. Vladivostok: DVNTs AN SSSR; 1983:40–47. (In Russ.).

Aki K., Lee W.H.K. Determination of three dimensional velocity anomalies under a seismic array using first Parrival times from local earthquakes. 1. A homogeneous initial model. J. Geophys. Res. 1976; 81(23):4381–4399.

Dziewonski A.M., Hager B.H., O'Connell R.J. Largescale heterogeneities in the lower mantle. J. Geophys. Res. 1977; 82(2):239–255.

Ritzwoller M.H., Levshin A.L. Eurasian surface wave tomography; group velocities. J. Geophys. Res. Solid earth. 1998; 103(3):4839–4878.

Fukao Y., Maruyama S., Obayashi M., Inove H. Geologic implication of the whole mantle P-wave tomography. J. Geol. Soc. Japan. 1994; 100(1):4–23.

Su W.J., Woodward R.L., Dziewonski A.M. Degree 12 model of shear velocity heterogeneity in the mantle. J. Geoph.Res. 1994; 99(4):6945–6980.

Vasco D.W., Johnson L.R. Whole earth structure estimated from seismic arrival times. J. Geophys. Res. Solid earth. 1998; 103(2):2633–2672.

Resovsky J.S., Ritzwoller M.H. A degree 8 mantle shear velocity model from normal mode observations below 3 mHz. J. Geoph. Res. 1999:104(131): 993–1104.

Jordan P. The rheology of polimineralic rocks - an approach. Geol. Runschau. 1988; 77(1):285–294.

Durek J.J., Ritzwoller M.H., Woodhouse J.H. Constraining upper mantle anelasticity using surface wave amplitude anomalies. Geophys. J. Int. 1993; 114(2):249–272.

Cadek O., Yuen D.A., Steinbach V., Matyska C. Lower mantle thermal structure deduced from seismic tomography, mineral physics and numerical modelling. Earth and Planet. Sci. Lett. 1994; 121(3–4):385–402.

Cadek O., Kyvalova H., Yuen D.A. Geodynamical implications from the correlation of surface geology and seismic tomographic structure. Earth and Planetary Sci. Letters. 1995; 136(3-)4: 615–627.

Romanowicz B. A global tomographic model of shear attenuation in the upper mantle. J. Geophys. Res. Solid earth. 1995; 100(7):12,375–12,394.

Perrot K., Tarits P., Goslin J. Regionalized comparison of global Earth models of seismic velocity, geoid and topography; implications on the geodynamics of ridges and subductions. Physics of the Earth and Planet. Inter. 1997; 99(1–2):91–102.

Yuen D.A., Cadek O., Chopelas A., Matyska C. Geophysical inferences of thermal-chemical structures in the lower mantle. Geophys. Res. Letters. 1993; 20(10):899–902.

Kawakatsu H., Niu F. Seismic evidence for a 920 km discontinuity in the mantle. Nature. 1994; 371(6495):301–305.

Kennet B.L.N., Bowman J.R. The velocity structure and heterogeneity of the upper mantle. Physics of the Earth and Planetary Interiors. 1990; 59(3):134–144.

Maruyama S., Kumazawa M., Kawakami S. Towards a new paradigm on the Earth's dynamics. J. Geol. Soc. Japan. 1994; 100(1):1–3.

Anderson O.L. The temperature profile of the upper mantle. J. Geophys. Res. 1980; 85(12): 7003–7010.

Turcotte D., Schubert J. Geodynamics: geological applications of the physics of continuous media. M.: Mir; 1985. (In Russ.).

Dziewonski A.M., Woodhouse J.H. Threedimensional Earth structure and mantle convection. Abstracts. 1989; 1(28th IGC):427–428.

Nikolaev N.I. Neotectonics and geodynamics of the lithosphere. M.: Nedra; 1988. (In Russ.).

Morelli A., Dziewonski A. The harmonic expansion approach to the retrieval of deep Earth structure. Seismic tomography. With applications in global seismology and exploration geophysics. Ed. Nolet G. Dordrecht. 1987; 251–274.

Zhivago A.V. Morphostructure of the bottom of the southeast Pacific. Metalliferous sediments of the South-Eastern part of the Pacific Ocean. AN SSSR. In-t okeanologii. M.; 1979:8–47. (In Russ.).

Aplonov S.V. Paleodynamics of the West Siberian Plate. Sov. geologiya. 1989; 7:27–36. (In Russ.).

Kropotkin P.N., Efremov V.N. New evidence of plate tectonics theory. Geotectonics. 1994; 1:16–24. (In Russ.).

Holmes A. Principles of physical geology. London; New York: T. Nelson and sons; 1944.

Holmes A. Principles of physical geology. 2nd new and fully rev. ed. N/Y: Roland press; 1965.

King L. Morphology of the Earth. M.: Progress; 1967. (In Russ.).

Krasnyi L.I. Global divisibility of the lithosphere in terms of the geoblock concept. Sov. geologiya. 1984; 7:17–31. (In Russ.).

Carey S.W. Searching for the laws of the development of the Earth and the Universe. M.: Mir; 1991. (In Russ.).

Filosofov V.P. Remarks on the theory and practice of the morphometric method. Morphometric method in geological studies. Saratov: Sarat. SU; 1963:249–261. (In Russ.).

Panov B.P. Quantitative characteristic of the river drainage. Tr. GGI. 1948; 4:122–149. (In Russ.).

Filosofov V.P. Experience in the genetic classification of continental plains. Scientific notes. V. 65. Saratov: Sarat. SU; 1959a:189–200. (In Russ.).

Filosofov V.P. A short guide to the morphometric method of searching for tectonic structures. Saratov: Sarat. SU; 1980. (In Russ.).

Filosofov V.P. The order of valleys and their use in geological research. Scientific annual collection for 1955. Saratov: Sarat. SU; 1959b:38–40. (In Russ.).

Egorkin A.V., Zyuganov S.K., Chernyshev N.M. The upper mantle of Syberia. 27-i Mezhdunar. geol. kongr. Moskva, 4-14 avg., 1984, sekts. S. 08, Geofizika. Dokl. M. 1984; 8:27–42. (In Russ.).

Avetisov G.P., Golubkov V.S. The deep structure of the central part of Norilsk ore region according to earthquake converted-wave method and DSS data. Sov. geologiya. 1984; 10:86–94. (In Russ.).

Grachev A.F., Demenetskaya R.M., Karasik A.M. The Mid-Atlantic Ridge and its continental continuation. Geomorphology RAS. 1970; 1:42–45. (In Russ.).

Mandelbrot B.B. Statistical methodology for nonperiodic cycles: from the covariance to R/S analysis. Annals of Economic and Social Measurement. 1972; 1(3):257–288.

Haken H. Synergetics. M.: Mir; 1980. (In Russ.).

Fractals in Science. Berlin.: Springer Verlag; 1995.

Fractals and Chaos in the Earth Sciences. Ed. by Sammis C.G., Saito M., King G.C.P. Basel.: Birkhauser; 1993.

Turcotte D.L. Fractal and chaos in geology and geophysics. 2nd edition. N/Y: Cambridge University Press; 1997.

Feder J. Fractals. M.: Mir; 1991. (In Russ.).

Makarenko N.G. Fractals, multifractal measures and attractors. Nonlinear waves. 2002: collection of articles. N. Novgorod: In-t prikl. fiziki RAN. 2003:381–394. (In Russ.).

Egorov D.G. Information measures for the analysis of geological self-organising systems. SPb.: Nauka; 1997. (In Russ.).

Mandelbrot B. Self-affine fractal sets. Fractals in physics. M.: Mir. 1988:9–47. (In Russ.).

Smirnov B.M. Physics of fractal clusters. M.: Nauka; 1991. (In Russ.).

Polikarpov M.I. Fractals, topological defects and confinement in lattice calibration theories. Uspekhi Fizicheskikh Nauk. 1995; 165:627. (In Russ.).

Zel'dovich Ya.S., Sokolov D.D. Fractals, similarity, intermediate asymptotics. Physics-Uspekhi (Advances in Physical Sciences). 1985; 146(3):493–506. (In Russ.).

Ten A., Yuen D.A. et al. Fractal features in miing of nonNewtonian and Newtonian mantle convection. Earth and Planet. Sci. Lett. 1997; 148(1–4):401–414.

Schulz S.S. Report on the work of the Dzharkent geological party in the summer of 1930 in the Dzhungarian Alatau. L.: Geolgiz; 1932. (In Russ.).

Gerasimov I.P., Meshcheryakov Yu.A. Morphostructure and morphosculpture of the Earth's surface. Modern problems of geography. M.: Science; 1964: pp. 225–230.
P.M. Goryainov, G.Yu. Ivanyuk, N.V. Sharov, (1997) Fractal analysis of seismic and geological data. Tectonophysics 269 (3-4):247–257
Chernyshev N.M. et al. Deep structure of the north West Siberian Plate from seismic data. Soviet Geology. 1978; 9: pp. 46–58. (In Russ.).
T. YUKUTAKE, (1968) Two Methods of Estimating the Drift Rate of the Earth's Magnetic Field. Journal of geomagnetism and geoelectricity 20 (4):427–428

Chapter 4
Fractal Hierarchies and Paleogeodynamic Reconstructions

From the standpoint of the material manifestation of wave properties in the processes of spontaneous structuring of unstable masses, the essence of the new general geological paradigm follows from the discovery of the colossal global structures of the Earth's gravitational instability of G-hexagon type, exhibiting immersion in the centre and emersion in the marginal parts of the deep mantle substrate in the eastern hemisphere, and L-hexagon type, with emersion in the centre and immersion in the marginal parts of the deep mantle substrate in the western hemisphere, the basis of which development is explained in terms of wave processes. On the one hand, according to seismic tomography and satellite data on global anomalies of the geoid and gravitational potential, these global structures comprise interconnected and interpenetrating entities, which reflect the inextricable dynamics of energy transfer in wave processes. Their constituent parts are common to any of these structures and each point within the structure itself comprises the limit (that is, there are points of the set in any of its neighbourhoods). In connection with this fact, it is not always possible here to draw the boundary of the structure in the form of a dividing line. On the other hand, the results of quantitative studies of the relief, anomalies of the geoid and gravitational potential, as well as the interpretation of satellite and aerial pictures demonstrate that these global structures, or cellular standing internal gravity waves repeat themselves at different scales in accordance with the mechanism of wave decomposition into modes. Such behaviour determines the predominance of different orders of L-hexagons in the continents of the western hemisphere, and hierarchies of G-hexagon structures in the Eurasian continent forming the eastern hemisphere. In connection with these circumstances, these structures themselves acquire the remarkable geometric and dynamic properties of fractal (or fractional) dimensionality. Following Prigogine, we also referred to the structures having such an origin as dissipative, thereby striving to preserve scientific continuity as well as emphasise the constructive morphogenetic role of dissipative wave processes in their formation.

O. V. Petrov, *The Earth's Dissipative Structures*, Springer Geophysics,
https://doi.org/10.1007/978-3-319-93614-7_4

It is this author's opinion that these fractal hierarchies of dissipative structures (or internal standing gravitational waves) generated by heat and mass transfer processes within the Earth's interior deserve special study. This idea comes from the fact that they control the development of open, nonequilibrium natural systems, which are characterised by the appearance of a special kind of structural regularity and the wave-based principles of spatial and temporary organisation.

The potential of dissipative structures for creating new models for the evolution of natural or thermodynamically open systems is very significant. One of the first to exploit this potential was the English mathematician A. Turing, who did not at that time propose their wave nature. His work, published as early as 1952, initially went practically unnoticed. Turing established, in particular, that in open natural systems, under certain conditions, two dissipative structures can emerge simultaneously. One of these structures can be associated with internal sources of matter and energy, and the other—with their external sinks. In such systems, bifurcation or oscillatory regimes are established. Later, Turing's bifurcation was found in models of chemistry and ecology, in plasma physics and fluid dynamics. The emergence of such dissipative structures and the oscillatory regimes of their evolution turned out to be a common property of the most diverse natural systems. And how incredible at first the spontaneous emergence of temporal regularity in such natural systems seemed to many specialists is evidenced by the already classic history of the discovery of oscillatory reactions in chemical systems. While the first reports of reactions in which stationary oscillations are observed appeared in the works of (Morgan 1916) and (Bray 1921), contemporaneous chemists attached little significance to them. Another periodic chemical reaction was discovered in 1951 by B.P. Belousov. Despite the fact that this reaction itself looks very beautiful—in the course of the process the blue colour of the solution periodically changes to red– for a long while, he was not able to publish his work since everyone believed that such a reaction would be simply impossible. The first short report was published in 1959 in a book of abstracts on radiation medicine. Now the Belousov-Zhabotinsky periodic reaction has received global recognition, with hundreds of articles and monographs being devoted to it. All this became a powerful stimulus for the study of such natural systems and the discovery of the determining role of the wave properties of matter in their evolution forms a new paradigm in modern natural science.

Since, in terms of thermodynamics, our planet also refers to open natural systems, it becomes obvious that the phenomena of its density instability can be associated not only with internal sources of energy and matter (the energy of decay of radioactive isotopes, the gravitational energy released during formation and the growth of the Earth's core, the internal energy of tidal interaction with the Moon, the energy of chemical-density, crystallisation, thermodiffusion and other processes), but also by external heat sinks (nonequilibrium cooling through the surface of the Earth). In order to assess the role of internal sources and external sinks of heat and matter in the formation of fractal hierarchies of dissipative structures of the

Earth (cellular standing internal gravity waves) throughout its geological history, we obviously need to turn to paleogeodynamic reconstructions. These reconstructions were compiled in the works of V. E. Khain, Ch. B. Borukaev, N. A. Bozhko and others, published in the last three decades of the 20th century. They are intended to illustrate the manifestation of the wave properties of matter in geodynamic processes. Time intervals are given in accordance with the "Stratigraphic Code of Russia" (2006).

4.1 Dissipative Structures of the Earth and Paleogeodynamic Reconstructions

According to geological maps of our planet, the most ancient Archaean rocks are found on the Earth's surface in the form of remnants of isolated cratons. They are separated by Proterozoic and younger belts, as well as by ocean depressions (Fig. 4.1a).

However, taking into account the palinspastic reconstructions that remove the influence of global structures of the Earth's gravitational instability, and numerous direct geological and indirect data (Early history of the Earth 1980; Borukaev 1985; Windley 1985; Veevers 1989), an analysis of the spatial distribution of the sialic protocrust in the Early Archean confirms the formation of the protocrust by the time period of 4000–3800 Ma with the emergence of the Protopangean crescent (Fig. 4.1b). According to these data, the asymmetric (one-sided) sickle-shaped distribution of the sialic crust of Protopangea and its spatial confinement to the central part of the global structure of the Earth's gravitational instability of the L-hexagon type, with emersion in the centre and immersion in the marginal parts of the deep mantle substrate, were inherent in the Precambrian Earth. Morphological features of the central part of this global structure, established on the basis of seismic tomography core relief data, convincingly explain the sickle-shaped distribution of the Archaean sialic protocrust in the western hemisphere. At the same time, in the eastern part of the Pantalassa, as follows from Fig. 4.1, similar, but significantly smaller in volume, parts of the protosialic crust could also exist within the marginal part of the G-hexagon global structure of the Earth's gravitational instability.

The association of the bulk mass of the protosialic crust with the arcuate-curved central part of the global structure of the Earth's gravitational instability of the L-hexagon type, with the emersion in the centre and immersion in the marginal parts of the deep mantle substrate, leads us, following Glukhovsky et al. (1994), to the conclusion that the Earth's hot belt, formed by the system of ascending mantle plumes, originated in the early stages of its evolution.

Mantle plumes, in accordance with modern concepts, extend spatially over the offsets of the layer D, located at the "core-mantle" boundary (Zonenshain and

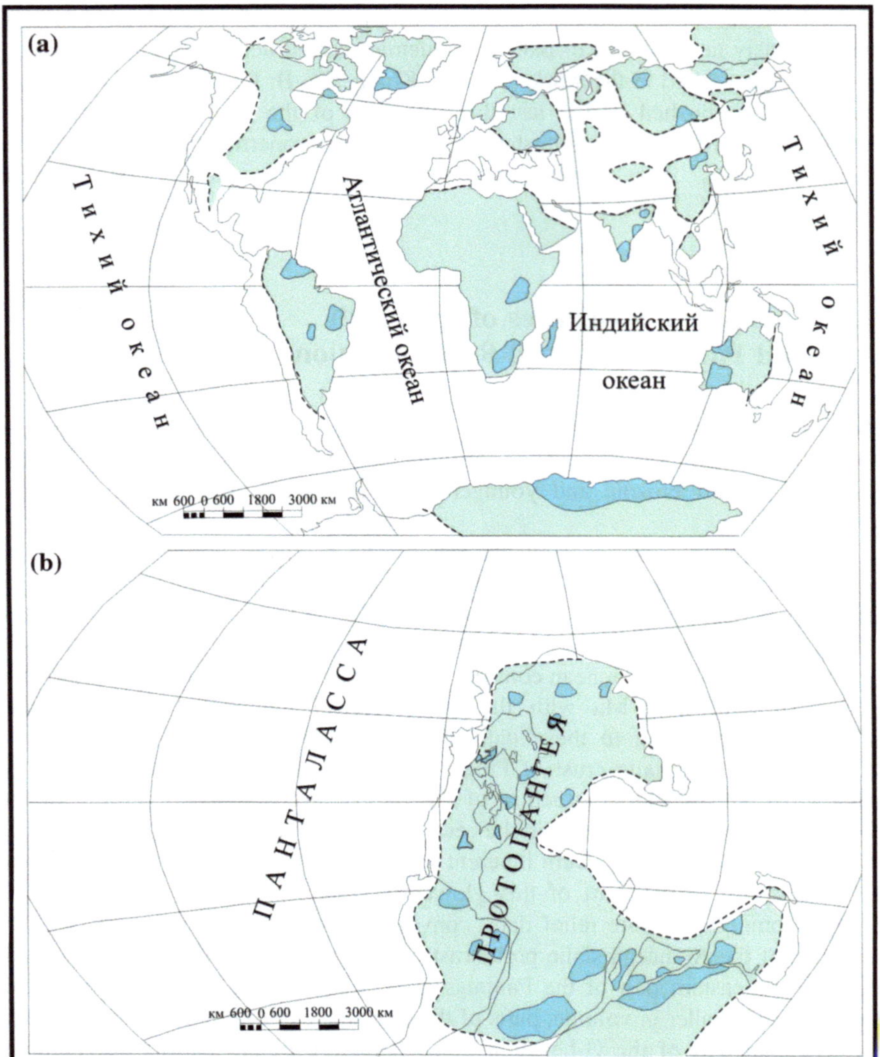

Fig. 4.1 Global paleogeodynamic scheme-reconstruction, Archaean (3500–3200 Ma). Polyconic projection of the Central Institute of geodesy, cartography and aerial laser scanning, version of the Great Soviet Encyclopaedia. Contemporary arrangement of the sections of **a** the Archaean crust on the continents of the Earth; **b** in the reconstruction of Protopangea. 1—identified sections of the ancient (up to 3500 Ma) Archaean crust; 2—estimated boundaries of the Archaean crust distribution

Kuzmin 1993; Larson and Olson 1991; Kumazawa and Maruyama 1994), and are therefore closely related to the central, emerging part of the L-hexagon global structure. These same global structures of gravitational instability control the distribution of currently active mantle plumes, which indicators on the Earth's surface are presented by hot spots. A deformation structure, caused by plastic flow during the ascent of mantle diapir, is typical for mantle xenoliths in volcanic rocks poured into such hot spots (Brooks 1973). Moreover, according to the data of the global seismic tomography (Maruyama et al. 1994; Dziewonski and Woodhouse 1989), within the boundaries of the global structures of the Earth's gravitational instability, the superplume corresponding to the areas of the emergence of the lower mantle is split into several smaller plumes in the upper mantle, which, in turn, divide into even smaller ones at the base of the plate, and so on. Such a successive split of plumes towards the surface of the Earth leads to the emergence of fractal clusters that resemble the structure of viscous fingers in petroleum geology.

In the earliest stages of the Earth's evolution, in accordance with the main provisions of the nuclear model of the evolution of the Precambrian lithosphere (Glukhovsky 1990), the formation of a thick and mature continental crust (up to 50 km) occurred in the areas of influence of mantle plumes. Otherwise, in the inter-nuclear spaces beyond the plumes, an enderbite-basite and basite crust was formed with a relatively lower thickness (up to 30 km). The sections of the Archaean crust, identified in the ancient shields of the continents, are predominantly "grey-gneiss". For these rocks, the oldest radiometric ages of 3500–3800 Ma were obtained. In that way the prevailing opinion in the literature today was formed that the oldest crust of the Earth is precisely a complex of "grey gneisses". According to published data, the rocks of this complex are intrusive, effusive, pyroclastic and sometimes sedimentary formations also. The characteristic features of the rocks among them, in addition to only external features (grey colour and gneiss-like character), are the composition corresponding to the average andesite, dacite, tonalite, trondjemite, plagiogranite; the apparent predominance of sodium over potassium, the low content of Rb, U, Th, Ti, F, Li, Nb, Ba, B and the enrichment with Ni, Cr, V compared to normal granitoids; very low ratio Sr^{87}/Sr^{86} (0.699–0.701). Although the main criterion for classifying these rocks as a complex of "grey gneisses" is typically the circumstance that they form the basement of the Archean greenstone belts, in reality, the ratios of the "grey gneisses" with the earliest generation of greenstone belts are highly ambiguous.

A comparison of the geochemistry of the Archaean metabasites of the Aldan shield with the basalts of the Deccan and Siberia trappean formation shows that the metabasites of nuclears were formed under the influence of deep mantle matter flows, as well as the products of Phanerozoic intraplate magmatism (Glukhovsky et al. 1994). Such continental intraplate plumes produce domes with a diameter of up to 1000 km, the central sections of which rise in 1–2 km above the surrounding terrain (Kinsman 1975). An example of an intraplate location in the oceanic region is a plume under the Hawaiian Islands. Plumes of this type generate an intraplate hot point, or Hawaiian hot spot (Vogt 1976). An example of a plume located on the divergent boundary of plates is that located near Iceland. Plumes of this type generate mid-ocean hot spots, or hot spots of the Icelandic type (Vogt 1976).

Vogt (1972, 1979) noted that a certain synchrony is observed in volcanic eruptions in a number of hot spots above the plumes. These include, in particular, plumes near the Hawaiian Islands, Iceland, the Azores, the Afar triangle and the Cape Verde Islands, Madeira and the Galapagos. In addition to the present, peaks of their activity fall on dates of 15, 42 and 65 Ma ago. They correspond to peaks of volcanic activity on island arcs. Apparently, there is a direct connection between the activity of various plumes and the activity of global structures of the Earth's gravitational instability.

Thus, the data of paleogeodynamic reconstructions on the nature of the distribution of the sialic protocrust with respect to the global structures of gravitational instability of the L-hexagon type in the western hemisphere and the G-hexagon type in the east indicate the emergence of these global structures. Its formation had apparently already occurred by the Early Archean era, immediately following the isolation of the liquid core, which, according to the data of (Kumazawa and Maruyama 1994), took place almost simultaneously with the accretionary formation of the Earth over a period of no more than 10 Ma. These structures, in turn, control the spatial distribution of mantle plumes and determine ultimately the fundamental asymmetry of the Earth (its division into the Indo-Atlantic and Pacific tectonic segments), occurred almost simultaneously with the accretionary formation of the Earth over a period of no more than 10 million years (Kumazawa and Maruyama 1994). How, having emerged at the earliest stage of the Earth's development, could these global structures be maintained for more than 4500 million years? Obviously, the answer to this question should be sought in the thermal history of the Earth.

As is known, the heat flux, which is currently lost from the surface of the Earth, is due, on the one hand, to the decay of radioactive isotopes U^{238} and U^{235}, Th^{232} and K^{40}, and, on the other hand, to the cooling of the Earth. The amount of heat lost per time unit by the internal regions of the Earth is determined by the value of the total heat flux from the entire surface of the Earth. According to Turcotte and Schubert (1985), the cooling of the Earth provides a heat flux on the surface equal to 21 mW/m^2, or 25–30% of the global average surface heat flux of 75 mW/m^2. The remaining 70–75% of the heat flux is of radiogenic origin. At present, the main exothermic reactions due to which heat is generated in the mantle are the U^{238} decay reactions. But in the remote past the U^{235} and K^{40} decay reactions dominated, since the half-lives of these isotopes are shorter. In this case, the decrease in the temperature of the Earth's interior with time is a consequence, on the one hand, of the general cooling of the Earth through its surface, and on the other hand, of the almost complete decay of some of the radioactive elements having short half-lives.

Direct and indirect data indicate that the heat emission rate in the Earth was twice or three times higher 3800 Ma ago than it is now. With a decrease in heat generation, the overall temperature of the core, mantle and lithosphere decreases, making the main contribution to the secular cooling of the Earth. Thus, as a result of the secular cooling of the Earth with the passage of geological time, a nonequilibrium increase in the density of mantle and lithospheric rocks occurs. Simultaneously with—and as a consequence of—the increasing pressure from the crystalline mantle and cooling of the outer liquid core, a crystallisation-gravitational differentiation occurs, leading to the precipitation of iron-nickel crystals, of which an internal solid core is formed. The crystallisation-gravitational differentiation of the outer liquid core also determines the activity of mantle plumes localised, generally, in the areas of the emersion of the deep mantle substrate at the boundary of the lower mantle and core. Obviously, as a result of these processes, the boundary of the crystalline mantle and outer liquid core will remain gravitationally unstable throughout almost the entire geological history, thereby ensuring a stable existence of the global G-hexagon structures in the eastern hemisphere and L-hexagons in the western hemisphere, originated at the moment of formation of liquid core.

4.2 Late Archaean–Early Proterozoic (3200–2500 Ma)

The global structural plan of the Archaean level within Protopangean boundaries is defined by structural patterns of three types of tectonic regions. For the northern hemisphere, a cellular structure plan of the Laurasian (North Asian) type is typical and is formed with large polygonal-rounded sections of proto-cratons corresponding to the most stable, relatively rigid blocks of ancient crustal formation included in younger melanocratic (greenstone) zones of folded metamorphic regions. These zones form complex branching systems that have structural correlations with proto-cratons of the frame type, comprising, as it were, the melanocratic matrix into which the latter are incorporated (Fig. 4.2).

Polygonal cratons themselves, in turn, are often dissected by narrow troughs and graben-synclinoria of the second order. Usually, these structures have the structure of greenstone formations. Their length is in the order of hundreds—and the width, tens—of kilometres. Among the granite-greenstone regions of the northern hemisphere, bimodal (rift induced) and calcium-alkaline (plate tectonic) belts are distinguished. In general, belts of full development with calcium-alkaline differentiated volcanism of the middle part of the section are more typical for the Laurasian continents (Khain and Bozhko 1988).

Based on rift-induced and plate tectonic models, the greenstone belts of the Laurasian continents are formed over the older tonalitic ("grey gneiss") sialic crust of Protopangea in the process of rifting and subsidence over mantle diapirs (Grachev and Fedorovsky 1980). Calcium-alkaline volcanism is explained either by partial melting of early tholeiites during their immersion, or by melting of an older sialic material (Borukaev 1985). These models do not exclude each other, since the embedding of greenstone belts occurs onto the previously formed sialic

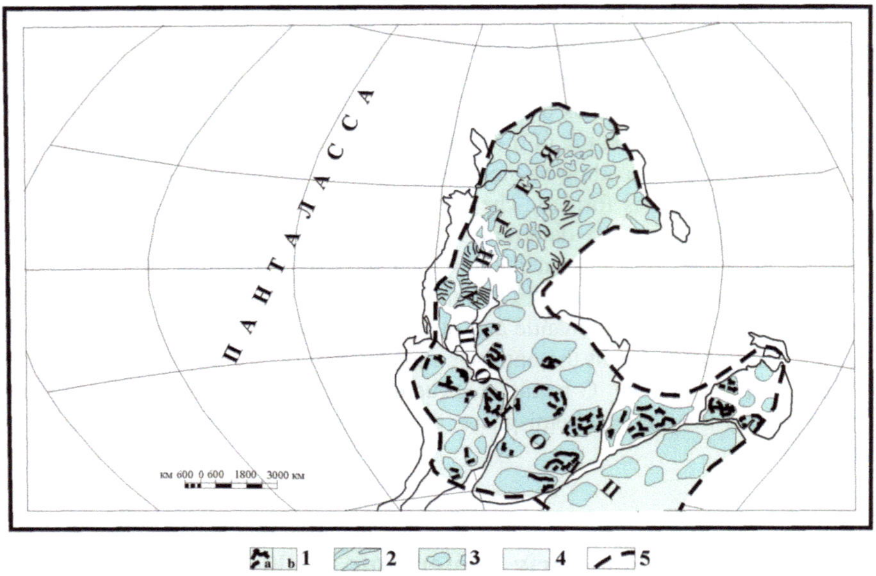

Fig. 4.2 Global paleogeodynamic scheme-reconstruction, Late Archaean—Early Proterozoic (2500–1650 Ma). 1—sections of the Archean crust of the Gondwana (Rhodesian) type (a); extrapolated areas of the non-dissected Archean cratons, some parts of which were processed in the post-Archaean period (b); 2, 3—sections of the Archaean crust of the Laurasian (North Canadian and North American) type: 2—with the predominance of linearly extended greenstone belts (North Canadian type), 3—with the predominance of isometric lithoplinths (North Asian type), 4—assumed area of the spread of the Archaean crust in the reconstruction of Protopangea; 5—approximate boundary of expansion of the Archaean continental crust

crust in the rifting process in both cases. The destruction of the sialic crust in the northern hemisphere was more extensive and, apparently, often led to new oceanic crust formations. Thus, Borukaev (1985) distinguishes the distribution areas of the Archaean oceanic crust (a region of the Timpton type) within the platforms of the Laurasian series. On the Gondwana shields, the areas of the Timpton type are not identified. As many researchers note, the most important feature of the Laurasian (North Asian) tectonic regions is the long-term preservation of the structural plan even after the completion of the formation of Archean complexes (Figs. 4.3 and 4.4).

Generally, this feature is inherent to the shields of Laurasia, but does not pertain to the shields of Gondwana, which developed somewhat differently. Obviously, as a result of the lack of structural inheritance, the Archaean structural plan of the Gondwana (Rhodesian) type is determined in the modern level by the unsystematic arrangement of the synforms created by greenstone formations, among the granite-gneiss complexes (Fig. 4.2).

In general, areas of this type are characterised by a wide manifestation of granite-dome structures. A classic example is the Zimbabwe massif, a region which

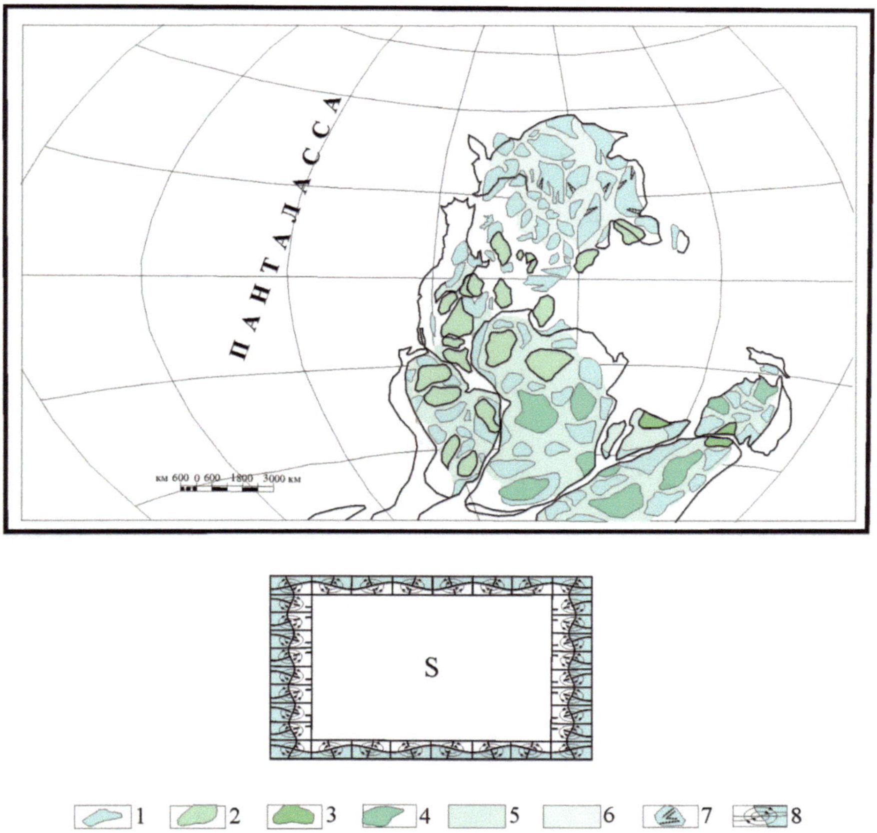

Fig. 4.3 Global paleogeodynamic scheme-reconstruction, Early Proterozoic (2500–1650 Ma). 1—residual cratons of Epiarchaean cratonisation (more ancient than 2500 Ma); 2—intracontinental basins with the formation of an oceanic crust, or crust of a transitional type; 3—pools with the development of the Late Proterozoic greenstone belts; 4—basins (depressions) made with unfolded platform complexes of the Early—Late Proterozoic; 5—ensialic intracratonic protogeosynclines (miogeosynclines) and zones of the non-geosynclinal tectonic processing of the basement; 6—esimatic intracratonic protogeosynclines (eugeosynclines); 7—protoaulacogens; 8—displacements and streamlines in cellular Early Proterozoic structures of the Earth's gravitational instability

served as an example to A. McGregor for developing of his notions about "herds of domes". Between the domes, narrow-keeled and intensely compressed synclinoria are clamped. In the modern structural plan, the synforms are extremely diverse. They have the form of arcs, twisting and straight lines, loops, etc. (Fig. 4.2). Greenstone complexes of granite-greenstone areas of the Gondwana (Rhodesian) type are characterised only by komatiite-tholeiitic formations with an explicit predominance of contrasting varieties, both mono- and polycyclic (Bozhko 1992; Sivoronov 1988).

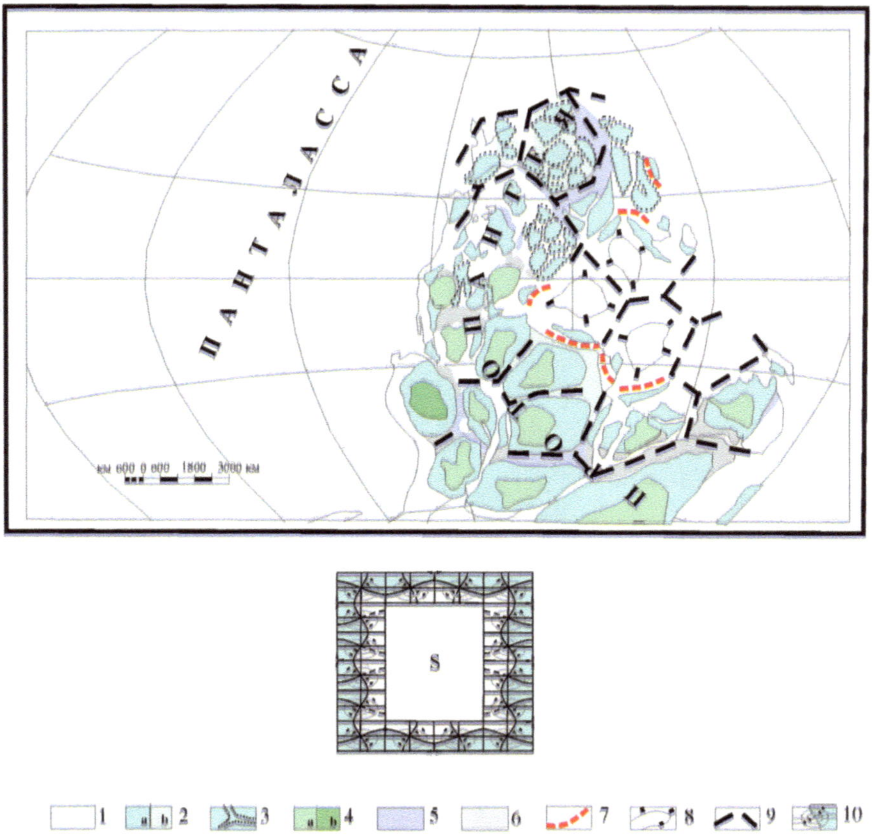

Fig. 4.4 Global paleogeodynamic scheme-reconstruction, Early Riphean (1350–1030 Ma). In the polyconic projection of the Central Institute of geodesy, cartography and aerial laser scanning, version of the Great Soviet Encyclopaedia. 1—oceanic crust; 2—continental Early Precambrian crust (a) and areas of passive margins (b); 3—aulacogens; 4—depressions, performed by the unfolded complexes of the Early Riphean (a) and acid volcano-plutonic associations (b); 5—ensimatic intracratonic geosynclines; 6—zones of non-geosynclinal tectonothermal basement processing; 7—zones of subduction; 8—directions of spreading of the Earth's crust; 9—boundaries of Earth's gravitational instability structures of the fifth generation; 10—displacements and streamlines in the Early Riphean gravitational instability structures of the Earth

The formation of a mature continental crust—and, respectively, the oldest proto-platform—covers over the future Gondwana continents began earlier than in Laurasia, and resulted in the formation of giant layered dyke-like intrusions that are absent at the northern continents. Archean basite-ultrabasite hypabyssal associations along with granite-greenstone associations are widespread in the Gondwana type tectonic regions. These associations are aligned with well-sorted and relatively hollow-lying orthoquartzites, limestones and shales with a source consisted of plausible inland and shelf seas. In South Africa, sediments of similar sedimentary basins evolved, ranging in the age from the Pongola basin (about 3000 Ma) to the youngest sections of the Witwatersrand Triad near the Archaean–Proterozoic boundary (about 2800 Ma) within the Transvaal syneclise. The largest Bushveld layered complex (350 km across) is dated to the Transvaal syneclise with an age of 2095–1930 Ma. In Australia, the proto-platform deposits of the Hammersley and Nabberu basins were accumulated in the interval 2770–2670 Ma, i.e., from the late Archean. Relatively synchronously, the formation of giant dyke-like complexes of the north-eastern continuation of Western Australia (2420 Ma) occurred. Around these basins, one can find areas marked by granulite metamorphism, which affected, at least in some cases, the rocks of the orthoquartzite-limestone-basite-ultrabasite association. In this regard, it should be noted that for the Archean crust of the Gondwana type, in addition to the granite-greenstone areas, another type of structural elements is characteristic—the tectonothermal processing belts (TPB), or granulite belts. The earliest of these is apparently the Limpopo belt, the beginning of the evolution of which refers to 3350 Ma, and the first granulite metamorphism —to 3100–1900 Ma. In the western granulite belt of the Yilgarn block (Australia), this metamorphism can be even older—about 3300 Ma; the same applies to the Antarctic craton. In the Atlantic belt of Brazil and in the belt that encompasses Sri Lanka, the south of India, and the Eastern Ghats, granulite metamorphism refers to the end of the Archean only, but perhaps it was already a repeated one (Friend 1984).

Unlike the cellular structural patterns of the Archaean crust of the Gondwana and Laurasia (North Asian) types, the structural plan of the Archaean crust of the Canadian type, intermediate between them, is linear, though not always clearly pronounced (Fig. 4.2). Linearity is defined here by a quasi-regular concentration in some zones of the synforms located either echelon-like to each other or forming chains. Element spatial orientation of a regular pattern is diverse. At the Canadian shield, it constitutes a fan opened to the northeast; at the Ukrainian shield granite-gneiss and greenstone "belts" are elongated in the meridional direction.

Analysis of the formation series over the Canadian Shield shows that within its boundaries there were two different tectonic regions: *Slave*, within which are widely distributed both separate synforms and entire zones, implemented mainly of the "sedimentary" (North Canadian) type of the section, and the *Superior*, which is dominated by the synforms composed of the "volcanogenic" (South Canadian) type of the section (Early history of the Earth 1980).

V. Schwerdtner and others describe two orders of diapirism in the granite-greenstone region of the Archean province of Lake Superior in Ontario (Schwerdtner et al. 1978). In the process of remobilisation of early stratum batholiths in the layered and gneissic tonalite-granodiorites giant diapirs of the first order formed, responsible for the global structure of the upper metavolcanic—metasedimentary formations (greenstone belts). These diapirs, 100 km in diameter, contain a plenty of diapirs of the second order, 15–20 km in diameter, in the form of domes and embedded conformably within the metavolcanic shells. Schwerdtner (1982) also pointed out the similarity between the polydiapiric nature of the

Archaean gneiss domes and the structural characteristics of the individual domes of the Zechstein evaporites in Central Europe. It is known that layers of evaporite deposits form linear ridges, which later evolved in culminations (domes) and saddles (Dixon 1991). Brun proposed a similar mechanism for explaining the swarm of uplifts characteristic of the Karelian domes of Finland (Brun 1980).

Talbot interpreted the very complex mushroom-shaped gneiss-migmatite complexes of East Greenland in the similar manner, considering them as a product of thermal convection (Talbot 1971). At the same time, he proceeded from the assumption that there is no jump in density with a transition from the upper horizons of the overlying mass to its twisted inverted forms, which indeed resemble convective cells.

Thus, a comparative analysis of the Archean paleotectonics of the northern and southern hemispheres reveals all three types of structural form associated with gravitational instability phenomena:

- cellular structures with emersion in the centre and immersion in the marginal parts of the L-hexagon type, i.e. structures of the Gondwana (Rhodesian) type;
- cellular structures with immersion in the centre and emersion in the edge parts of the G-hexagon type, i.e. structures of the Laurasian (North Asian) type;
- linearly extended swell-like structures of the Rolls type, i.e. structures of the Canadian type.

There is an obvious confinedness of L-hexagon structures to the southern hemisphere, and G-hexagon structures to the Eurasian continents of the northern hemisphere (Fig. 4.2). In this case, the linearly extended swell-like structures of the Rolls type, which are especially widely manifested in the composition of greenstone belts occupy an intermediate position in the early stage, when the heat flux was still 3–4 times higher than the modern one (Fig. 4.2).

At the same time, the synchronicity of the main global events is noted in the Archaean aeon. To date, no more than three generations of greenstone belts have been identified in any of the tectonic regions in the late Archaean. In general, Middle Lopian (more than 3000–2900 Ma) and two Late Lopian (2800–2600 Ma) are established among them. Over the course of the Archaean, the maximum dimension of successive greenstone belts, as they were preserved today, progressively increased. At the same time, bundles of simultaneous greenstone belts within the supercontinent of the Protopangea, with the passage of the Archaean time, occupied less and less regions, until at the beginning of the Proterozoic (more than 2000 Ma) they ceased to be a characteristic feature of the Earth's crust.

The remaining fragments of the Late Archaean crust testify that the system of small cells (with a diameter of hundreds of kilometres) at the end of the Archaean has acquired a certain ordering. Its reflection is the formation of a polygonal (predominantly hexagonal) lattice of granulite belts with domes of granite-greenstone domains embedded in it (Khain and Bozhko 1988). The final formation of this structure refers already to the Early Proterozoic.

4.3 Early Proterozoic (2500–1650 Ma)

In the Proterozoic aeon, the difference between the structural plans of the Laurasian and Gondwana groups of continents is preserved (Fig. 4.3). Within the northern series of continents, Proterozoic ensimatic intracratonic protogeosynclines (eugeosynclines) form a geometrically regular network. In the cells of the network there are protoplatforms comprising round-angular, mostly isometric blocks of the continental crust—fragments of the Epiarchaean supercontinent from hundreds to more than a thousand kilometres across. In turn, the protoplatforms are characterised by a wide spread of protoaulacogens, often forming triple junctions with protogeosynclines (Fig. 4.3). It is clear that, in the Early Proterozoic, the plate tectonics mechanism has already become apparent within Laurasia in its present form. First of all, this is evidenced by the discoveries of the present ophiolitic associations of Jormua and Outokumpu in Finland (Kontinen 1987), 1960 and 1970 Ma respectively, and Purtunnck in the Cape-Smith belt of Canada of about the same age (Picard et al. 1990).

Passive margins of the Early Proterozoic are reliably reconstructed in the Wopmei belt (Canada) and in the south-west of Sweden (Kalevian). Models with a subduction mechanism at the level of 1880 Ma were developed by Barbey et al. (1984) for svecofennides and "united plates of America" (Khain and Bozhko 1988). In Laurasia, a few Lower Proterozoic protoplatform covers, as a rule, are associated with sedimentary-volcanogenic formation of protogeosynclines (the Huron group on the edge of the Penokyi protogeosyncline) or belong to the paleorifts (the structures of the Baltic Shield, the Kodaro-Udokan Depression). As a result, geological data testify that during the Early Proterozoic, the supercontinent of Protopangea in the northern hemisphere experienced a successive destruction with disintegration into separate polygonal blocks—protoplatforms, in the gapping between which protogeosynclines emerged over the crust of the suboceanic or even oceanic type (Khain and Bozhko 1988). However, the unity of the trajectory of the apparent pole migration, obtained from paleomagnetic data, shows that the gapping scale could not exceed 1000–2000 km (Piper 1982).

In the southern hemisphere, the Early Proterozoic destruction was incomparably less major, and the process of formation of the mature continental crust was carried out at the Gondwana continents more widely and productively. Here, at the large sialic blocks (protoplatforms) in the Early Proterozoic, sub-continental type depressions were formed, where an accumulation of powerful volcanogenic-sedimentary formations took place in the calm environment. N. A. Bozhko and E. V. Pavlovsky attributed these formations to the protoplatform cover, while Borukaev (1979) explained them in terms of geosynclinal unfolded complexes, which, in general, perform independent syneclises, which take into account the specificity of their composition. Some of these were formed in the inherited structures (syneclise Rand, Hammersley), while others were newly formed. The latter include the Franceville group at the base of the Congo syneclise. In northern Australia, in the Kimberley Basin, similar sedimentary-volcanic sediments began to form 1900 Ma ago. At the Guiana Shield, in the area of Epiarchaean consolidation of Pakaraim, sedimentary-volcanogenic deposits with the age of 1900–1750 Ma are developed. Later, in the interval 1750–1650 Ma, a clastic cover of the Roraima group was formed at the Guiana shield.

Along with the Early Proterozoic depressions, which are constructed from unfolded platform complexes, Early Proterozoic greenstone complexes were formed in some Gondwana basins. The thickness of formation of these basins reaches many kilometres. This type includes the province of Maroni–Itakayunas of the Guiana Shield and Birim of West Africa. Like their Archaean analogues, deposits of Early Proterozoic greenstone complexes are complexly dislocated, metamorphosed in greenshales and amphibolite, sometimes even granulite facies, and are breached by numerous plutons of granitoids. In the Early Proterozoic, we are dealing almost exclusively with the subtype of greenstone regions, which was described by Khain and Bozhko (1988) in terms of belts of incomplete development with bimodal volcanism concerning the Archean. For these belts, the layering is supposed to have occurred over the strained, thinned and processed continental crust without breaking its continuity.

The Early Proterozoic Gondwana belts are the ensialic intracratonic proto-geosynclines (miogeosynclines) and the zones of non-geosynclinal tectonic basement processing. There are no ophiolites, i.e. island-arc associations corresponding lateral geochemical trends in the protogeosynclinal belts. The structure of such belts is characterised by the symmetry extrinsic to subduction systems. None of this permits an interpretation of the evolution of these structures in terms of plate tectonic models, of the type clearly exemplified by the Birim belt of West Africa. The picture of the greater ensialicity of the Early Proterozoic Gondwana crust is complemented by the absence in its boundaries of the Yatulian-type protoaulacogens, in which an important role is played by the tholeiitic basalts.

The Early Proterozoic tectonothermal processing zones within Gondwana are composed of Archean material primarily metamorphosed from greenshale to granulite facies. Such are the Australian belts, the Atlantic belt of Brazil, the East Gat belt of the Hindustan Peninsula, where the main era of tectonothermal processing started at 1600–1500 Ma ago. Khain and Bozhko (1988) distinguish, as a sufficiently obvious, high heat flux, tangential compression and active uplifting among the geodynamic conditions for the development of tectonothermal processing zones.

Thus, a comparative analysis of the Early Proterozoic paleotectonics reveals in the northern hemisphere the predominance of thermogravitational (convective) instability cellular structures for the Eurasian continents with immersion in the centre and emersion in the edge parts of the G-hexagon type, and, for Gondwana—cellular structures with emersion in the centre and immersion in the marginal parts of the L-hexagon type. These structures of Earth's thermogravitational instability were formed under conditions of a heat flux, higher than today: according to Hynes (1982), the asthenosphere had a temperature 100 °C higher in the Early Proterozoic than at present.

It can be clearly seen that it is the evolution of various types of structures of the Earth's gravitational instability that determines the asymmetry features of the Early Proterozoic supercontinents of Gondwana and Laurasia, which have been repeatedly emphasised in the literature (Borukaev 1985; Bozhko 1982; Khain and Bozhko 1988; Sivoronov 1988). At the same time, the synchronicity of the main global events during the Early Proterozoic is noted: the closure of the Early Proterozoic protogeosynclines mainly to the 1800–1750 Ma boundary. Also the formation of volcanic-plutonic belts of the era of cratonisation with widely manifested ground volcanism and a peculiar gabbro-anorthosite-rapakivi and sometimes granite plutonism is manifested. This magmatism covers a period of 1800–1600 Ma ago, during the transition from Early to Late Proterozoic. At the same time, the process of "slamming" the Early Proterozoic protogeosynclines, i.e. skimming of their sedimentary-volcanogenic performance in terms of its metamorphism and granitisation, was evidently due to the deposition of much larger Late Proterozoic structures of the Earth's gravitational instability (Figs. 4.4 and 4.5).

4.4 Early Riphean (1650–1350 Ma)

Comparative analysis of paleotectonics reveals the emergence during the Early Riphean of a new generation of cellular structures of the Earth's gravitational instability (Fig. 4.4). The average diameter of these structures is 2500–2000 and 1250–1500 km, their area is 3–4 and 1.0–1.5 thousand km^2. For the Eurasian group of continents, the predominance of cellular structures of gravitational instability of the G-hexagon type is defined; for the Gondwana type, the L-hexagon type is defined (Fig. 4.4).

The geodynamic environment of the formation of G-hexagon structures, with immersion in the centre and emersion in the marginal parts of the underlying mantle substrate, caused the active processes of the Early Riphean destruction of the Protopangea in the northern hemisphere. These processes resulted in the formation of a large number of aulacogens, representing paleorifts. The rifting nature of the structures under consideration is emphasised by their pattern in the plan, the nature of the "graben" performance, bimodal volcanism, dyke zones preceding the depositing of the troughs. Within Protopangea, the aulacogens of North America are characterised by their unusually large depositional thickness (up to 12 km) and terrigenous-carbonate composition. The Lower Riphean parts of the aulacogenic sections of the East European Craton are less thick and characterised by a substantially terrigenous composition with a wide development of redbeds (Khain and Bozhko 1988). The Early Riphean sedimentation of the aulacogens of Siberia distinguishes itself by its transgressive character. The formation of variegated sediments is replaced by overlying terrigenous coastal-marine sediments, while those in turn are replaced by siliceous-carbonate (Milanovsky 1987; The Siberian Platform 1987).

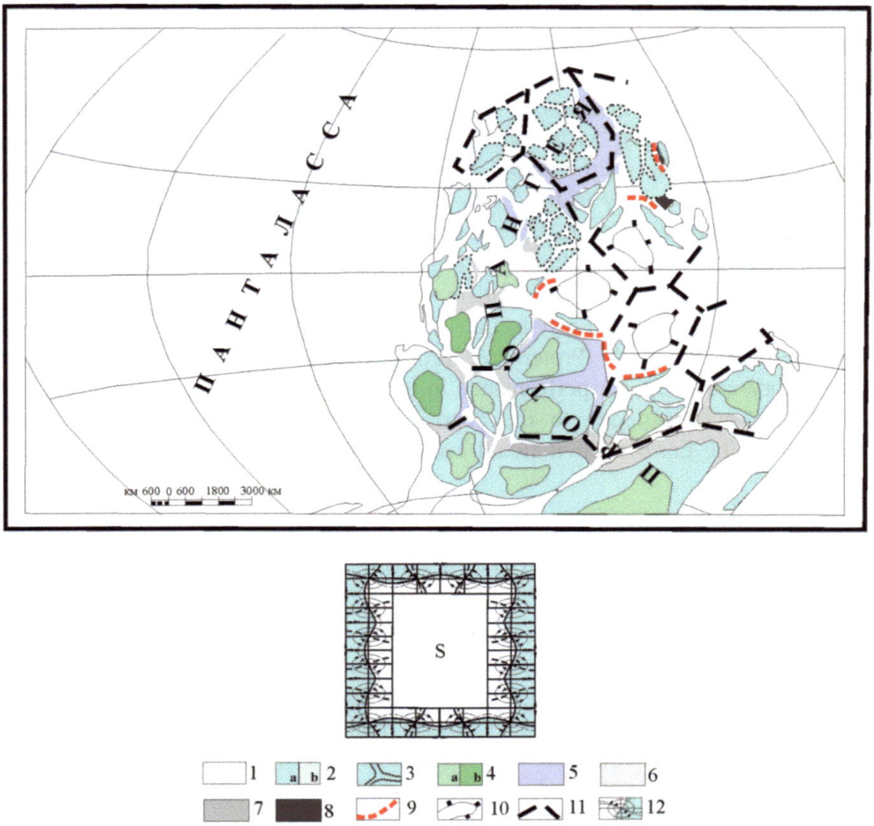

Fig. 4.5 Global paleogeodynamic scheme-reconstruction, Middle Riphean (1030–1030 Ma). In the polyconic projection of the Central Institute of geodesy, cartography and aerial laser scanning, version of the Great Soviet Encyclopaedia. 1—oceanic crust; 2—Early Precambrian continental crust (a), passive margins (b); 3—aulacogens; 4—depressions formed with unfolded complexes of the Middle Riphean (a) and acidic volcanic plutonic associations (b); 5—the same, with the development of acidic volcanic plutonic associations; 6—ensimatic intracratonic geosynclines; 7—zones of non-geosynclinal tectonothermal processing of the basement; 8—folded systems; 9—accretionary complexes and subduction zones; 10—directions of spreading of the Earth's crust; 11—boundaries of the structures of the Earth's thermogravitational instability of the fifth generation; 12—displacements and streamlines in the Middle Riphean structures of the Earth's gravitational instability

According to Khain and Bozhko (1988), for the Eurasian continents within the Yenisei Ridge, the troughs of Kazakhstan, the Altai-Sayan and Baikal-Vitim Regions, the formation of a new generation of intracratonic troughs is marked, along with the formation of the ensimatic intrageosynclines separating the major Early Riphean cratons (Fig. 4.4). The presence of ultrabasites—incomplete ophiolitic associations in ensimatic intracratonic geosynclines—indicate a significant intensity of Early Riphean rifting in these zones and its possible transition to spreading.

The first manifestations of marginal-plate tectonic processes within the boundaries of Southeast Asia, where ophiolite belts and island-arc complexes established, are associated with the broad development of destructive processes in the internal parts of the Eurasian continent (Zhai et al. 1985; Lazko et al. 1982).

In Gondwana, the geodynamic environment of the formation of cellular structures of gravitational (convective) instability of the Earth, such as L-hexagons, determined the almost complete absence of the aulacogenic stage of development of the cratons. During the Early Riphean, extensive syneclises were formed within the individual cratons (Taudenni, Congo, Victoria River, MacArthur, Cuddapah, Vindi and others) having unfolded terrigenous complexes and persilicic volcanic-plutonic associations (Fig. 4.4). The tectonic regime of formation of these deposits is characterised by a continuing high contrast in terms of vertical movements. This is indicated by the considerable thicknesses of the deposits. Thus, 6 km of sediments was accumulated in the MacArthur Basin during the Early Riphean. Sandstones, quartzites, argillites, aleurolites, containing packs of limestones and dolomites, prevail in the composition of sediments that carry out relatively small basins of sedimentation. The development of the trappean bed formation and acidic volcanics is characteristic. These Early Riphean associations were accumulated directly at the ancient basement, bypassing the "aulacogenic stage".

Within the continents of Gondwana, the isolation of individual basement blocks in the Early Riphean occurs mainly as a result of the formation of a new generation of a type of structure contrasting with those pertaining to Laurasia, such as intracratonic ensialic geosynclinal systems and tectonothermal processing belts (TPB). Ensialic intracratonic geosynclines deposit in the form of narrow elongated downfolds discordant with respect to the structures of the basement (Fig. 4.4). Almost everywhere there is a sharply unconformal bedding of the formations that perform the zones under consideration in the Archean or Lower Proterozoic complexes (Khain and Bozhko 1988) or, rarely, inheritance of the Early Precambrian structures.

TPBs correspond to the areas of the Preriphean Protopangea, which are covered by multiple metamorphism (up to the granulite facie), magmatism, isotope rejuvenation, superimposed deformations without significant participation of the Riphean superstructure. In the Early Riphean, there were several such zones in the North and South America, Europe, Hindustan, Australia, Antarctica. They all have a linear form and an extent of several thousand kilometres. Considering the geodynamic situation in manifestation of regimes of the non-geosynclinal TPBs within the framework of the model being developed, it is necessary to start from their confinement to the zones of immersion of the deep mantle substrate during the formation of the Earth's gravitational instability structures of the L-hexagon type. The role of stretching at the first stages of their existence, which causes the formation of small grabens and rifts in the TPB zones should be noted here. In the final stages, stretching is replaced by compression, as evidenced by deformations of the supracrustal strata, the appearance of thick zones of diaphthorites, thrust formation, metamorphism and ultrametamorphism and active uplifting.

Therefore, the results of the paleogeodynamic analysis give grounds to state that for the Early Riphean structures of the Earth's gravitational instability all the main regularities of the structure revealed for their contemporary analogues are preserved: the conformal nature of the correlations of surface and depth structures, the typomorphic elements of their structure and the confinement of G-hexagon structures to the Eurasian continents of the northern hemisphere, and L-hexagons—to the Gondwana continents of the southern hemisphere.

4.5 Middle Riphean (1350–1030 Ma)

The global Middle Riphean paleotectonic situation is characterised by the further development of the Early Riphean generation of the Earth's gravitational instability structures (Fig. 4.5). Having emerged in the Early Riphean, the generation of cellular structures of gravitational instability of the G-hexagon type is still present in the northern hemisphere (Fig. 4.4). The aulacogens continue to play an important role here. The sedimentation pattern intended within Laurasia in the Early Riphean is also preserved. This pattern manifest itself as terrigenous-carbonate in the aulacogens of the Cordillera and north-west of North America, a significantly terrigenous in European and volcanogenic-sedimentary in the Siberian aulacogens. However, universally the Middle Riphean strata correspond to a new cycle of sedimentation. Along with the development of aulacogens deposited in the Early Riphean, new generations arise—the Grand Canyon, Kivino, Pachelmsky, Krestsovsky and others. In some aulacogens of Eastern European and Siberian cratons reconstructions of structural plan and activation of magmatic processes occurred. During the Middle Riphean in Laurasia the ensimatic intracraton geosynclines, which dismember separate cratons, continued development. And the movement of marginal plate processes from the southeast of Asia into the Arabian-Nubian region indicates an increase of the scales of Protopangea's destruction in the northern hemisphere.

In the southern hemisphere within the Gondwana continents, the paleotectonics of the Middle Riphean is determined by the development dynamics of the structures of the Earth's gravitational instability of the L-hexagon type, with emersion in the centre and immersion in the marginal parts of the deep mantle substrate (Fig. 4.5). In common with the Early Riphean, in the active tectonic environment of the rapidly downfolding and relatively small basins in the central parts of the cratons, an accumulation of unfolded sedimentary complexes continues during the Middle Riphean, with the development of acidic volcanic-plutonic associations in some cases (Fig. 4.5). This is indicated by the thickness of the deposits (thousands of meters), sharp facies alterations within the strata and the presence of not only basic, but also acidic effusives.

The largest basins are preserved in Hindustan and Australia. Small (less than hundreds of kilometres across) isolated basins of Abercorn, Kingingloro are characteristic of the Tanzanian block of Africa. As a part of the sediments that perform these basins, we can trace the Early Riphean decline in the scale of the formation of epicraton volcanic-plutonic associations.

The destruction of the continental crust within Gondwana is manifested much weaker than in Laurasia. Here, during the Middle Riphean, the Early Riphean generations of intracratonic geosynclines of both subtypes continued to develop. Common to all intracratonic geosynclines of the Middle Riphean within the continents of the southern group is the predominance of stretching at the initial stage of their evolution. At the same time, the formational series fixes the stage of the continental rift and the expansion impulse of the trough in the ensimatic geosynclines. On the example of Middle Riphean geosynclines, the intracrustal discontinuity has an uneven character, which is well observed along the continuation of the Mavritan-Brazilian belt (Khain and Bozhko 1988). Here the developmental areas of almost complete ophiolitic association are replaced by zones where only alpinotype ultrabasites were introduced. The final stage of the development of intracratonic geosynclines, which takes place against the background of the system compression, is characterised by the development of folding, deep metamorphism, granitisation, the formation of large flat overthrusts, after what many of these geosynclines turned into folded systems at the end of the Middle Riphean (Fig. 4.5).

Belts of the non-geosynclinal TPBs are fixed within the continents of the Gondwana group by the deep-mantle substance immersion zone and warping crust zones in the marginal parts of the Earth's gravitational instability structures of the L-hexagon type. These geostructures continue to exist actively in Eastern Gondwana and arise in the Transantarctic Mountains, on the island of Madagascar, in South Africa and North America (the Grenville belt). The TPB zones are characterised by the presence of regional lines with a pronounced jump in metamorphism, increased dislocations, mylonitisation and migmatisation, as well as numerous intrusions of anorthosites, the association of granites with them, not subdivided into syn- and post-tectonic ones. Considering the geodynamic situation of the manifestation of TPB zones, we should once again note the role of stretching at the first stages of their existence, what causes the formation of small grabens and rifts (in the TPB zones). At the final stages, it is replaced by compression, as evidenced by deformations of the supracrustal strata.

Thus, the dynamics of the development of Earth's gravitational instability structures of the G-hexagon type in the northern hemisphere and the type of L-hexagons in the southern determined the asynchrony of the Middle Riphean paleotectonic events in Laurasia and Gondwana. Within the latter, as a result of diastrophism, which began at 1300–1050 Ma ago, the closure of intracratonic geosynclines occurred, and most of them turned into newly formed folded systems. At the same time, the destruction of the continental crust in geosynclinal troughs continued in Laurasia (Fig. 4.5).

4.6 Late Riphean (1030–600 Ma)

Analysis of paleotectonics reveals in the Late Riphean an accumulation of cellular structures of the Earth's gravitational (convective) instability of the sixth generation (Fig. 4.6). The average diameter of these structures is 4000–6000 km, and their area is up to 10–20 million km^2.

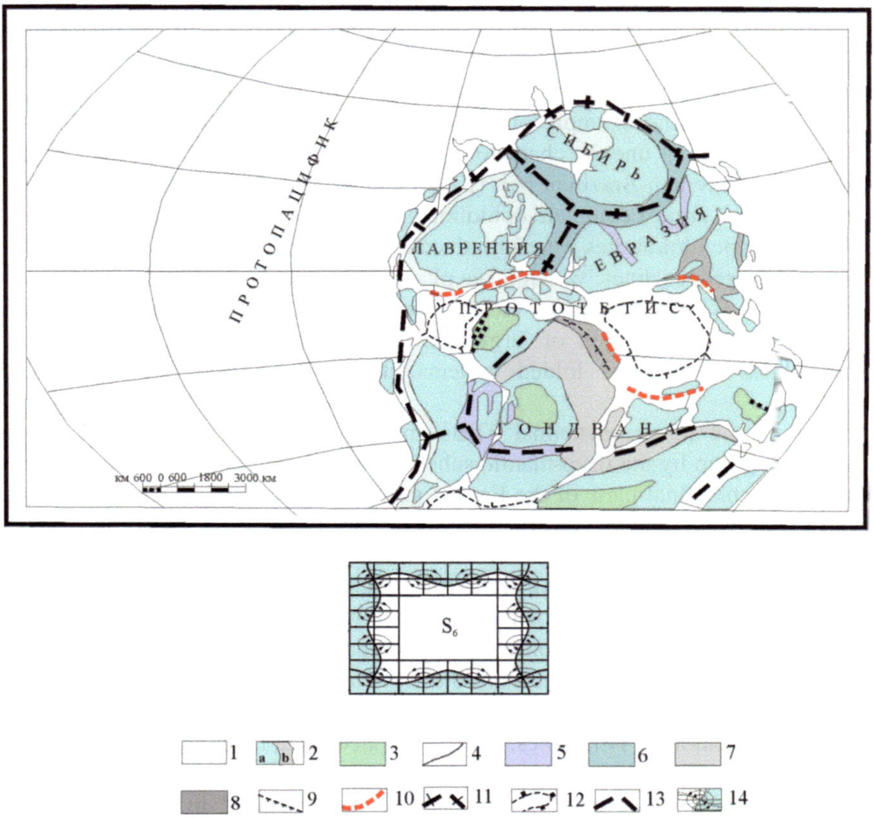

Fig. 4.6 Global paleogeodynamic scheme-reconstruction, Late Riphean (1030–600 Ma). In the polyconic projection of the Central Institute of geodesy, cartography and aerial laser scanning, version of the Great Soviet Encyclopaedia. 1—oceanic crust; 2 (a)—continental Early Precambrian crust and (b) areas of passive margins; 3—depressions, formed by the unfolded platform complexes; 4—aulacogens; 5—ensialic intracratonic geosynclines; 6—ensimatic intracratonic geosynclines; 7—zones of non-geosynclinal tectonothermal basement processing; 8—accretionary complexes and newly formed folded systems; 9—ophiolite belts in accretionary complexes; 10—subduction zones; 11—boundaries of the central zones and directions of spreading of the Earth's crust; 12—directions of diffused spreading of the Earth's crust; 13—boundaries of thermogravitational Earth's instability structures of the sixth generation; 14—displacements and streamlines in the structures of the Earth's gravitational instability of the sixth generation

At Laurasia, this epoch is associated with the formation of a new generation of the ensimatic intracratonic geosynclinal troughs that divided this supercontinent into cellular supercratonic blocks: Laurentia, Siberia and Eurasia (Fig. 4.6). The fragments of these Late Riphean rift structures are now disunited, being established in the form of thick prisms of the clastic Upper Riphean deposits (Khain and Bozhko 1988). The geodynamic environment of the formation of supercratonic blocks of the northern hemisphere allows their deposition to be connected with the cellular structures of the Earth's gravitational instability, such as the G-hexagons. At the same time, the Early-Middle Riphean ensimatic troughs separating the cratons experienced diastrophism and inversion within the Late Riphean cellular supercratonic blocks (Fig. 4.6), with some aulacogens also being closed within Laurentia. In their place, newly formed intracratonic folded structures emerged. Conversely, the aulacogens of Siberia and Europe did not experience folded deformations, with inversion being expressed in these structures in the uplift.

In Gondwana, the geodynamic environment of the formation of the Late Riphean cellular structures of the Earth's gravitational (convective) instability, such as the L-hexagons, determined the laying of large platform basins (syneclises) confined to the central parts of the supercratonic blocks (Fig. 4.6). Structures of this type were formed in the Late Riphean in Africa, South America and Hindustan. One of these, the Taudenni syneclise in the west of Africa, now has an almost rectangular shape (1000 × 1800 km), but in the Riphean it merged with the Volta depression, forming an even larger structure (Khain and Bozhko 1988).

Significant dimensions are typical of the syneclises of Congo, San Francisco, Vindhyan and Central Hindustan. Moreover, in contrast of the Lower and Middle Riphean, sections of these depressions are characterised by moderate thicknesses.

By the Late Riphean, the laying of a new generation of the TPB belts and intracratonic geosynclines in the marginal parts of the L-hexagon supercratonic blocks has already occurred. The tectonic regime of all these structures at early stages of development is characterised by a distinct rift-induced style. The newly formed zones are much more complex in structural plan than in the Early and Middle Riphean and often have longitudinal zonation, which is expressed in the transition along the continuation of the ensimatic zones into the ensialic ones.

The analysis of the global structural plan indicates, following the inversion of the Early Middle-Riphean troughs in Gondwana during the Late Riphean, marginal downfolds were laid down in the separate marginal parts of the supercratonic blocks of this type. These downfolds are usually extended along the folded Lower Middle-Riphean zones containing incomplete ophiolite associations. In the Central Brazilian system, east of the Aramino zone, the downfold of Paraguay–Araguaia is laid, stretching 2500 km from the issue of the Amazon in the north to the midstream of the river Paraguay in the south. Almost symmetrical to it, the downfold of Brasilia, having a length of about 1000 km, is laid in the western frame of the craton of San Francisco.

In the Late Riphean, an almost continuous belt of miogeosynclines emerged, encircling the near-Pacific part of the Earth. The belt itself comprised a global chain of rift-induced ocean basins along the edge of the Protopangea, delimited from

Protopacific and pushed back by masses of the continental crust (Fig. 4.6). In the subsequent complex Phanerozoic history, these masses were fragmented into smaller ones, and, as a result of the closure of these basins, joined to the continents in the form of separate blocks or exotic lands (Saleeby 1983).

The geodynamic environment of the formation of cellular structures of gravitational (convective) instability of the Earth, such as L-hexagons, determined the strengthening of the processes of secondary ocean formation and the development of marginal tectonic regimes in the Arabian-Nubian region, the Anti-Atlas, the Grampian belt and South-East Asia (Fig. 4.6). In the Late Riphean, these processes led to the completion of the split of Protopangea due to the formation of an oceanic depression in the western region of Proto-Tethys between Africa, on the one hand, and North America, and Europe on the other.

The above paleotectonic analysis shows a consistent complication of the global structural plan from the Early Riphean to the Vendian. With the increase in the dimension of the Earth's gravitational instability structures and intensification of the destruction of Protopangea, a gradual replacement of the dominance of the intraplate regime by that of the marginal plate around the middle of the Late Riphean, as well as a transition of the location of the mobile zones from the mosaic structure to the belt structure, occur.

4.7 Vendian–Middle Cambrian (600–500 Ma)

From the Vendian period up to the Middle Cambrian, the generation of cellular structures of the Earth's gravitational instability of the sixth generation, which appeared already in the Late Riphean, is preserved. In the northern hemisphere, the disintegration of Laurasia and the opening of the oceans of Iapetus, Central Asian and Inuit are associated with the geodynamics of the development of G-hexagon structures (Fig. 4.7). At this time, destructive processes also cover the internal parts of the cratons. The structural plan that emerged in its most general form was determined by the existence of the main bands of ocean depressions, on either side of which the fragmented segments of the continental crust were located. In these segments, a series of medial mass-blocks, being preserved from destruction, were separated by complex branches of predominantly ensimatic intracratonic geosynclines, many of which comprised apophyses of the main oceanic branch (Khain and Bozhko 1988).

The Vendian situation brightly illustrates the opposite geodynamic nature of the inner zones of the cellular structures of Laurasia and Gondwana. In Laurasia, a different intensity of processes of gravitational instability determined the absence of platform sedimentation within the Laurentia and the formation of thick Vendian and Cambrian sediments over huge areas of the Siberian and Eurasian cratons.

In Gondwana, in the syneclises of South America, Africa and Hindustan, there was a reverse trend during the Vendian period: some reduction took place in the

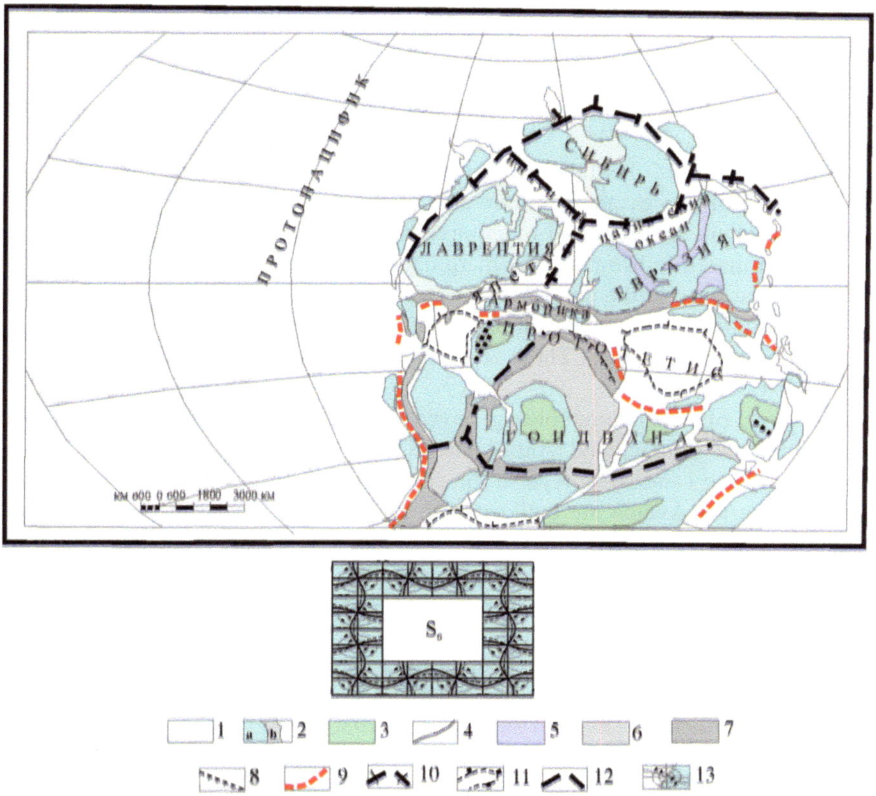

Fig. 4.7 Global paleogeodynamic scheme-reconstruction, Vendian—Middle Cambrian (600–500 Ma) 1—oceanic crust; 2—continental Early Precambrian crust (a), passive margins (b); 3—depressions formed by the unfolded platform complexes; 4—aulacogens; 5—ensialic intracratonic geosynclines; 6—zones of non-geosynclinal tectonothermal basement processing; 7—accretionary complexes and newly formed folded systems; 8—ophiolite belts in accretionary complexes; 9—subduction zones; 10—the boundaries of the central zones and directions of spreading of the Earth's crust; 11—directions of diffused spreading of the Earth's crust; 12—boundaries of the Earth's thermogravitational instability structures of the sixth generation; 13—displacements and streamlines in the structures of the Earth's gravitational instability of the sixth generation

areas of sedimentation and their concentration in the central parts of the depressions. The geodynamic situation, preserved in the southern hemisphere since Late Riphean times, was noted in the Vendian-Cambrian by the diastrophism of the intracratonic zones and by the large-scale manifestation of the ancient TPB substrate in the marginal zones of the structures of the Earth's gravitational instability, such as L-hexagons (Fig. 4.7). In the vast expanses of Africa and South America, the largest Libyan-Nigerian, Mozambique and Brazilian TPB belts had appeared. Processing was resumed in the Rajasthan, East Gat, and Satpuri belts of India, as well as in Antarctica. All of the "Pan-African" processing zones had experienced total isotope rejuvenation, reflected in Rb-Sr and K-Ar datings, mainly in the range of 650–500 Ma. Within these zones, granitoids are distinguished with a synchronisation to the "Pan-African" processing and a formation due to repeated anatexis at various crustal levels.

The major part of the TPB zones is involved in repeated deformations. In the initial stages of processing, a dominance of the situation of tension is observed, being only afterwards replaced by compression. In this case, the immersion zones of the deep mantle substrate in the marginal parts of the structures of gravitational instability are fixed in the upper horizons of the Earth's crust by the manifestations of metamorphism, i.e. the development of the diaphthoresis along the zones of superimposed deformations. These zones undergo folding and granitisation, turning into folded belts.

The origin of the Kadoma folding, covering a significant part of Western Europe, Newfoundland, Britain, and Anti-Atlas, is associated with the geodynamics of structures of the L-hexagon type in the area of Proto-Tethys at 584–540 Ma. It is assumed that the Avalon arc collided with the northern part of the African craton with a precedent formation of a volcano-plutonic belt in Anti-Atlas. Subsequently, this arc separated from Africa again and, together with the newly formed crust of Western Europe, formed the Armorica microcontinent. The formation of the accretionary Arabian complex and the closure of the rear basin with deformation and metamorphism of the strata of Anatolia, Iran, and the Pamirs were completed by the Vendian (Fig. 4.7). Here, the Vendian-Cambrian deposits already have the character of a platform (Belov 1981; Bozhko 1986).

In the marginal parts of Western Gondwana, within certain segments of the Pacific "ring", the paleotectonic setting of the active margins of the Andean type emerges during the Vendian-Cambrian periods. This applies to the Andes itself as well as the Trans-Antarctic Mountains. As follows from the global paleogeodynamic scheme of reconstruction, the marginal zones of structures of the L-hexagon type in the areas of articulation of the oceanic and continental crust fix the passive margin in the Proto-Andean belt (Fig. 4.7).

Concluding the review of the paleotectonics of the Vendian and the Cambrian, it should be noted that at that time in Laurasia, the destruction of the continental crust progressed, leading to the formation of large ocean basins—Iapetus (Proto-Atlantic), Paleo-Asian and Inuit—and to the disentanglement of the continents of Laurentia, Eurasia and Siberia at the beginning of the Paleozoic. Within Gondwana, it was a matter of closing the Late Riphean heaves, transforming them into folded systems, and the new fusion of blocks of the ancient continental crust into a single supercontinent under the influence of the "Pan-African" orogenesis. Geodynamically, in the Vendian-Cambrian time, the development of Laurasia turned out to be subordinate to the features of the geodynamics of the G-hexagon structures of the gravitational instability of the Earth, and Gondwana—to the geodynamical features of the L-hexagon structures.

4.8 Late Cambrian–Early Devonian (500–416 Ma)

Paleogeodynamic reconstructions show that a new seventh generation of the Earth's gravitational instability structures is established in the time interval from the Upper Cambrian to the Lower Devonian. With the increase in the dimension of the structures of this generation compared with the earlier ones, the predominant role here begins to be played by plate- and marginal plate processes. Thus, a significant part of the margins of the pre-existing Laurasian platforms were transformed from passive into collision-based. North America and Eastern Europe (Baltic) merged into a single continent of Laurussia or Euroamerica (Fig. 4.8). The Greenlandic Caledonides and the Northern Appalachians were dragged over the north-eastern margin of North America and the Scandinavian Caledonides, i.e. over the north-western margin of Eastern Europe.

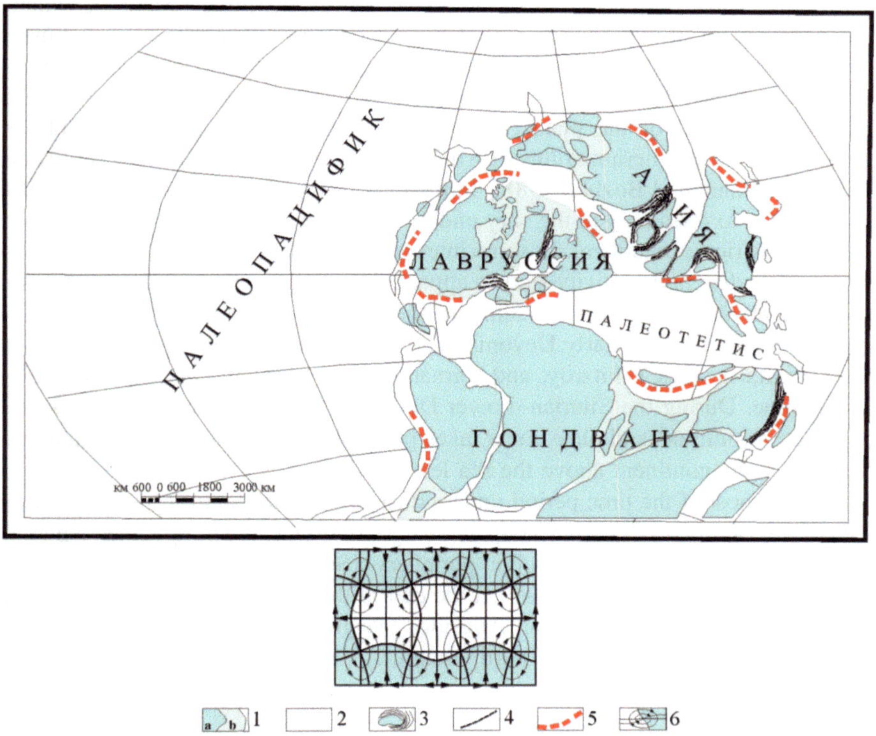

Fig. 4.8 Global paleogeodynamic scheme-reconstruction, Late Cambrian–Early Devonian (500–416 Ma). 1—inland crust; 2—oceanic crust; 3—folded accretionary complexes including continental fragments (microcontinents, median masses); 4—the main ophiolite belts, zones of tectonic melange development and intensive deformations; 5—subduction zones of the Earth's crust including overthrusts and underthrusts; 6—displacements and streamlines in the Earth's gravitational instability structures of the seventh generation

At the beginning of the Ordovician, the destruction of the continental crust that existed at the end of the Baikal stage took place in the Urals. At this time, an ocean type basin that separated the East European continent from the Siberian continent and the Kazakh microcontinent also emerged. Most of Central Kazakhstan, as well as the Northern and Middle Tien Shan, were affected by orogenesis at the time of end of the Silurian to the beginning of the Devonian (Fig. 4.8).

In the Paleo-Asian Ocean, the central part comprised the youngest Ob-Zaisan system, its eastern extension consisted of the Gobi-Hingan, while, to the north-east, the Amur-Okhotsk zone predominated. The eastern branch of the latitudinal continuation through Dunbei emerged into the South Primorye and North-East Korea. Spreading in this part of the Paleo-Asian Ocean began in the Silurian or even Ordovician and continued until the Devonian (Khain 1989). After the Devonian activation, the stage of immersion of the East European, partly of the Siberian and North American platforms; simultaneously the stage of formation of their most important structures, i.e. syneclises, anticlises and plate cover, began.

In Gondwana, during the Late Cambrian—Ordovician, there was a peneplanation of mountain structures and a real platform (oroplatform) regime was established throughout. An earlier manifested decrease in tectonic activity on the periphery of Gondwana then affected its interior regions, as can be seen from the spread of marine formations of the plate cover, which first of all (during the Cambrian) appeared on the northern margins of Africa and Hindustan, as well as in Arabia and Australia (Khain and Bozhko 1988). In the interior regions of Gondwana the plate cover began to form only at the very end of the Ordovician and the Silurian (South America, Africa, Antarctica). In some internal regions of Gondwana, some signs of increased endogenous activity start to appear. These signs are the manifestations of acid and alkaline magmatism mainly in the south and east of the Sahara in the interval of the Silurian–Early Devonian and the Devonian formation aulacogens of Saur-Ugarta in Africa, Fitzroy, and Carnarvon and Perth in Australia at the even earlier time. During the Silurian–Lower Devonian timespan, the maximum spread of marine conditions and plate cover takes place, indicating the minimum height of the Gondwana continent above the sea level.

During most of the time period under consideration, the significant Paleo-Thetis area of the Gondwana's periphery facing the Mediterranean belt remained a passive margin as opposed to the Pacific periphery (South America, Australia) from Venezuela to Hindustan. However, at the end of the Devonian, the situation began to change. The Mauritanians and Atlasides (Maghribides) emerged, having experienced an overthrusting on the northwestern African margin of Gondwana.

4.9 Early Devonian–Middle Jurassic (416–161 Ma)

The Early Devonian–Middle Jurassic is characterised by the development of the structures of the Earth's gravitational instability of the eighth generation (Figs. 4.9, 4.10, 4.11, 4.12, 4.13, 4.14 and 4.15). This global plan for structural instability with

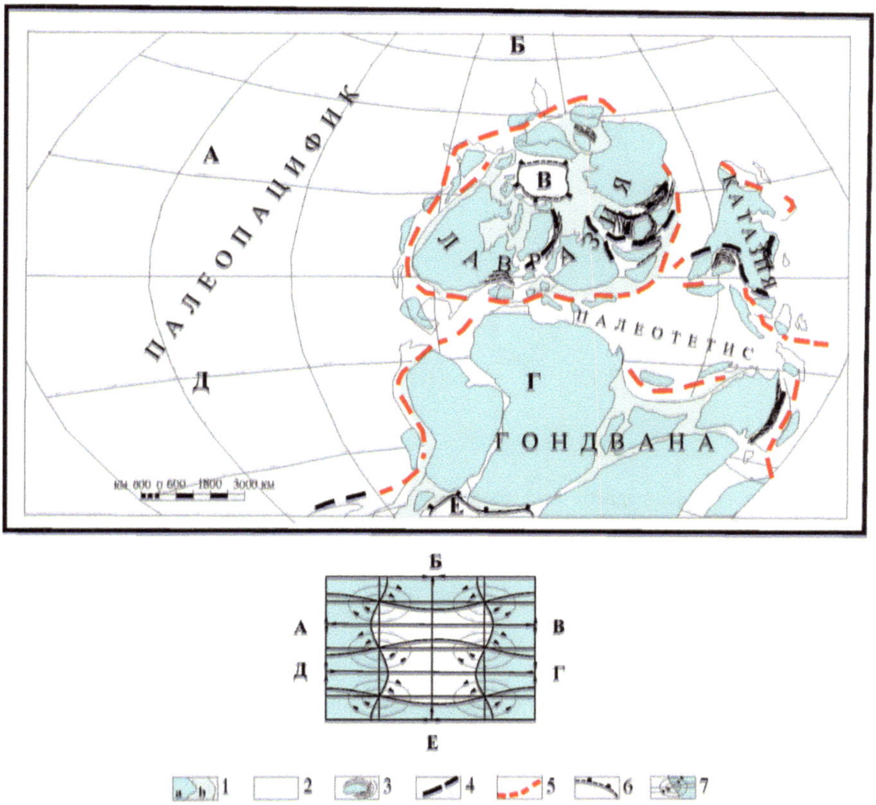

Fig. 4.9 Global paleogeodynamic scheme-reconstruction, Early Devonian—Early Carbonic (416—318 Ma). 1—inland crust (a) and shelf areas (b); 2—oceanic crust; 3—folded accretionary complexes including continental fragments (microcontinents, median masses); 4—main ophiolite belts, zones of tectonic melange development and intensive deformations; 5—subduction zones of the oceanic crust including overthrusts and underthrusts; 6—directions of the spreading of the crust; 7—displacements and streamlines in the Earth's gravitational instability structures of the third order

the emersion of the deep mantle substrate within the South Pole (see the scheme of displacements and streamlines in the structures of the third-order Earth's gravity instability in Figs. 4.9, 4.10, 4.11, 4.12, 4.13, 4.14 and 4.15) determined the displacement to the northern hemisphere (drift of the whole Paleopangea–the Mesopangea) from the Early Devonian to the Middle Jurassic period. Judging by the paleobiogeographical and paleomagnetic data (Fig. 4.10), in the Early Devonian–Early Carboniferous the equator was located at an angle of 55°–65° to the present and passed through the Eastern European platform, Northern Europe and North America (Figs. 4.9, 4.10 and 4.11). In the Middle Jurassic, as a result of

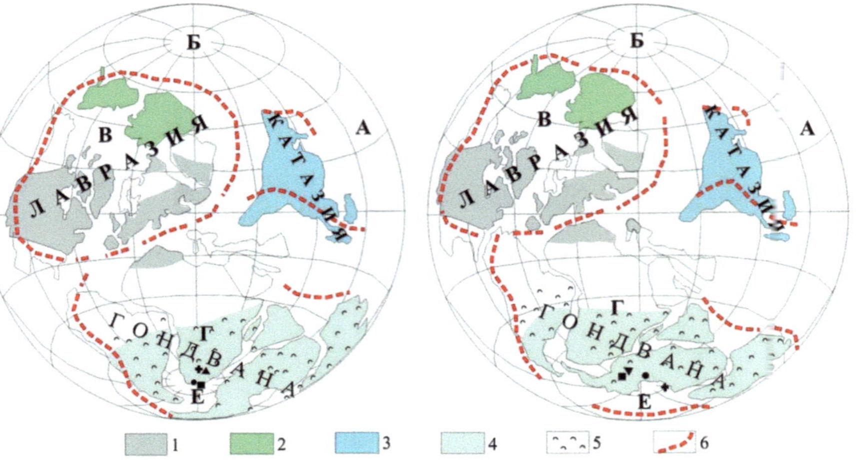

Fig. 4.10 Global paleomagnetic and paleobiogeographic reconstruction for Early, Medium (a) and Late (b) Carbonic. Geodynamic reconstructions (Khain 1989a) with the changes and additions of the author different colours show the position of paleomagnetic poles for different continents. 1–4—expansion of Carboniferous flora: 1—Euramerian (C–P), 2—Angara (C–P), 3—Catasian (C–P), 4—Gondwana (C–P), 5—glacial sediments of the Late Carbonic; 6—subduction zones

the migration of the Mesopangea to the northern hemisphere, the equator was found within Central Africa and South America (Figs. 4.14 and 4.15).

Gondwana entered the late stage of existence at the end of the Early Carboniferous. This was the stage of the unification of Gondwana with Laurasia into a single supercontinent—Paleopangea (Figs. 4.11 and 4.12). The northern half of Africa, located outside the Mauritanids and Atlasides, experienced deformations within a large radius. The bulk of the deformations, occurring in Perm, was imprinted in the pre-Mesozoic structure (Khain 1989a). In the southern half of Africa—as well as in South America, Hindustan, and partly in Australia and Antarctica—the formation of rift systems with the manifestation of trappean volcanism began (Figs. 4.11, 4.12 and 4.13). Rifts developed against the background of a general surface uplift of the continents, covered by the ice cover, which began in the Namurian period (Australia).

Structural reorganisation in the Early Devonian–Early Carboniferous in the northern hemisphere was expressed in the isolation of the Laurasian supercontinent as an independent geodynamic structure of the Earth's gravitational instability of the L-hexagon type. The final closure of the ocean basin separating the East-European continent from the Siberian continent and the Kazakh microcontinent occurred during the Late Paleozoic—specifically, from the end of the Early Carboniferous and Early Triassic inclusive—that is, in the classical Hercynian era of tectogenesis (Figs. 4.11 and 4.12). In the Southern Tien Shan, lying on the southern extension of the eastern zones of the Urals, the relatively calm

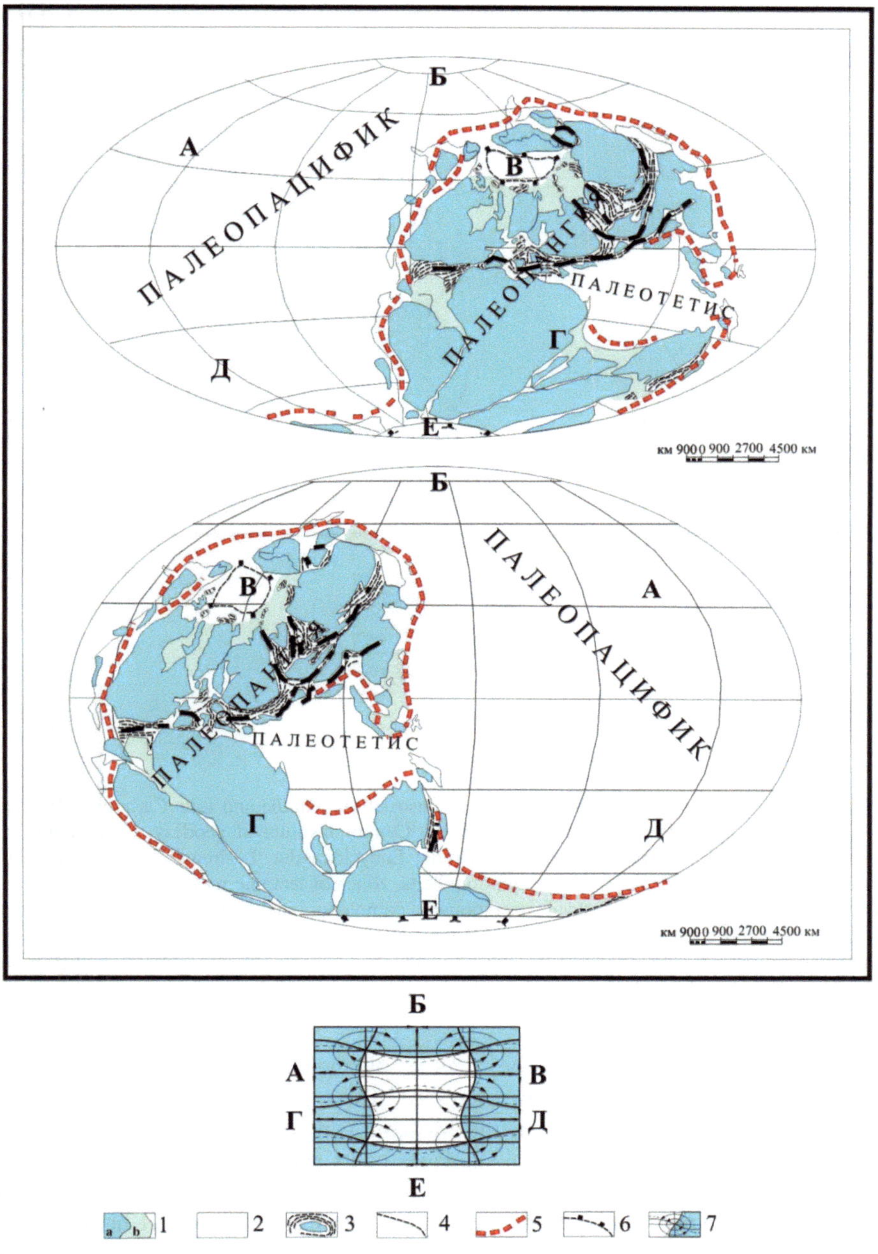

Fig. 4.11 Global paleogeodynamic scheme-reconstruction, Middle and Late carbon—Permian (318—251 Ma) (a) in an arbitrary projection of Aitov—Gammer; (b) in pseudocylindrical elliptic Mollweide homalographic projection. The explanations are given in Fig. 4.9

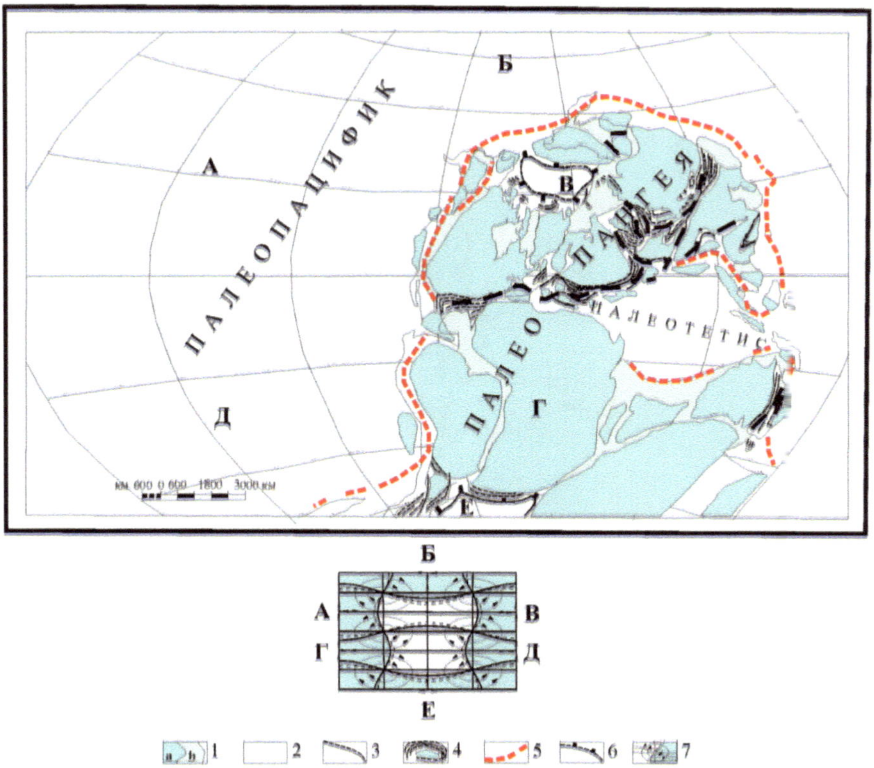

Fig. 4.12 Global paleogeodynamic scheme-reconstruction, Middle and Late Carboniferous—Permian (318—251 Ma ago). Polyconic projection of the Central Institute of geodesy, cartography and aerial laser scanning, version of the Great Soviet Encyclopaedia. 1—inland crust (a) and shelf areas (b); 2—oceanic crust; 3—main ophiolite belts, zones of tectonic melange development and intensive deformations; 4—folded accretionary complexes including continental fragments (microcontinents, median masses; 5—subduction zones of the oceanic crust including overthrusts and underthrusts; 6—directions of the crust spreading; 7—displacements and streamlines in the Earth's gravitational instability structures of the third order

development continued until the middle of the Middle Carboniferous. The island-arc stage here was weakly expressed. The typical island-arc volcanites are known only in the Devonian. In the Moscow century, the Altai microcontinent uplift started and the gravitational shariages of its carbonate cover began to form with their transformation into olistoplacks and olistoliths of a thick flysch-olistostromic strata of the medium–upper Carboniferous (Khain 1989a). Actually, the orogenic stage of the development of the Southern Tien Shan with the formation of molasses and granites had already occurred in Perm (Figs. 4.11 and 4.12).

As follows from the global paleogeodynamic reconstruction scheme (Figs. 4.9 and 4.10), the zone of Laurasia and Catasia joining corresponds to the boundaries of

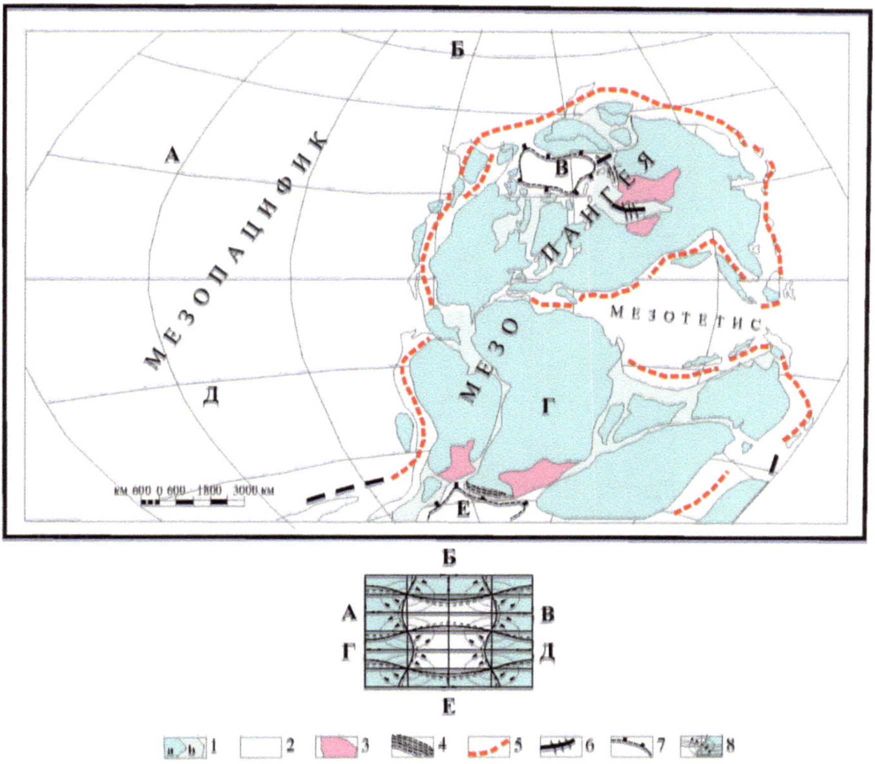

Fig. 4.13 Global paleogeodynamic scheme-reconstruction, Early–Late Triassic (251–200 Ma) 1—inland crust (a) and shelf areas (b); 2—oceanic crust; 3—areas of trappean volcanism; 4—folded accretionary complexes; 5—subduction zones of the oceanic crust including overthrusts and underthrusts; 6—boundaries of the central zones and the spreading direction of the Earth's crust; 7—directions of diffused spreading of the Earth's crust; 8—displacements and streamlines in the Earth's gravitational instability structures of the third order

the junction of structures of the L-hexagon type. Here, on a latitudinal segment of the Paleo-Asian ocean, subduction proceeded in both directions—northward, under the Caledonian frame of the Siberian craton, and southward, under the marginal uplift of the Sino-Korean craton. From the middle Carboniferous the closure of the axial band of the Paleo-Asian basin began, and in the Late Perm its development was completed—the Siberian continent closed with the Sino-Korean one (Figs. 4.11 and 4.12). The exception was the extreme north-eastern Amur-Okhotsk branch of the belt in question, in which Hercynian diastrophism is almost not manifested, and deep-sea sedimentation in the axial zone continued continuously in the Late Paleozoic and Early Mesozoic, up to Middle Jurassic inclusive.

In general, the considered stage of Laurasia's formation is characterised by the development of island-arc stages and diastrophism along its periphery and the development of stretching processes in the region of extensive Arctic geodepression

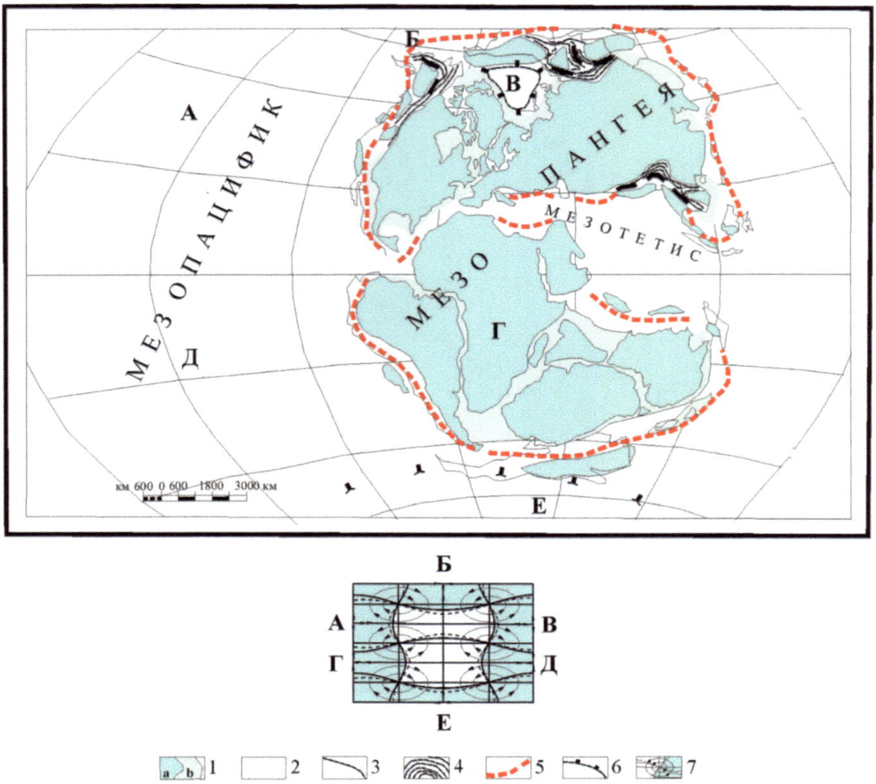

Fig. 4.14 Global paleogeodynamic scheme>;reconstruction, Early—Middle Jurassic (200–161 Ma). Polyconic projection of the Central Institute of geodesy, cartography and aerial laser scanning, version of the Great Soviet Encyclopedia. Explanations are given in Fig. 4.12

and the West Siberian plate (Figs. 4.9, 4.10, 4.11, 4.12, 4.13, 4.13 and 4.15). The Arctic depression itself and its characteristic triple zonality—namely, linear orogens, depression ellipses (sedimentary basins) and shield uplifts—originated in Perm (Pogrebitsky 1976). At the Perm-Jurassic pre-oceanic stage of its development, as shown by the reconstruction of the deviation areas, not the oceanic depressions were located in the core of geodepression, but the arch swelling out of the crust, from which the sediments entered the downfolds located on the site of the current shelves and the Amerasian basin (Ronkina and Vishnevskaya 1982). A similar Perm uplift is also established on the territory of Western Siberia (Surkov and Zhero 1981; Khain 1979). At the turn of the Early and Late Perm, the carbonate and evaporite formations are displaced by terrigenous. In the Late Triassic time, the organisation of giant trappean formations of Siberia is associated with the development of mantle plumes within the Arctic depression and the rift systems of Western Siberia continuing it (Fig. 4.13).

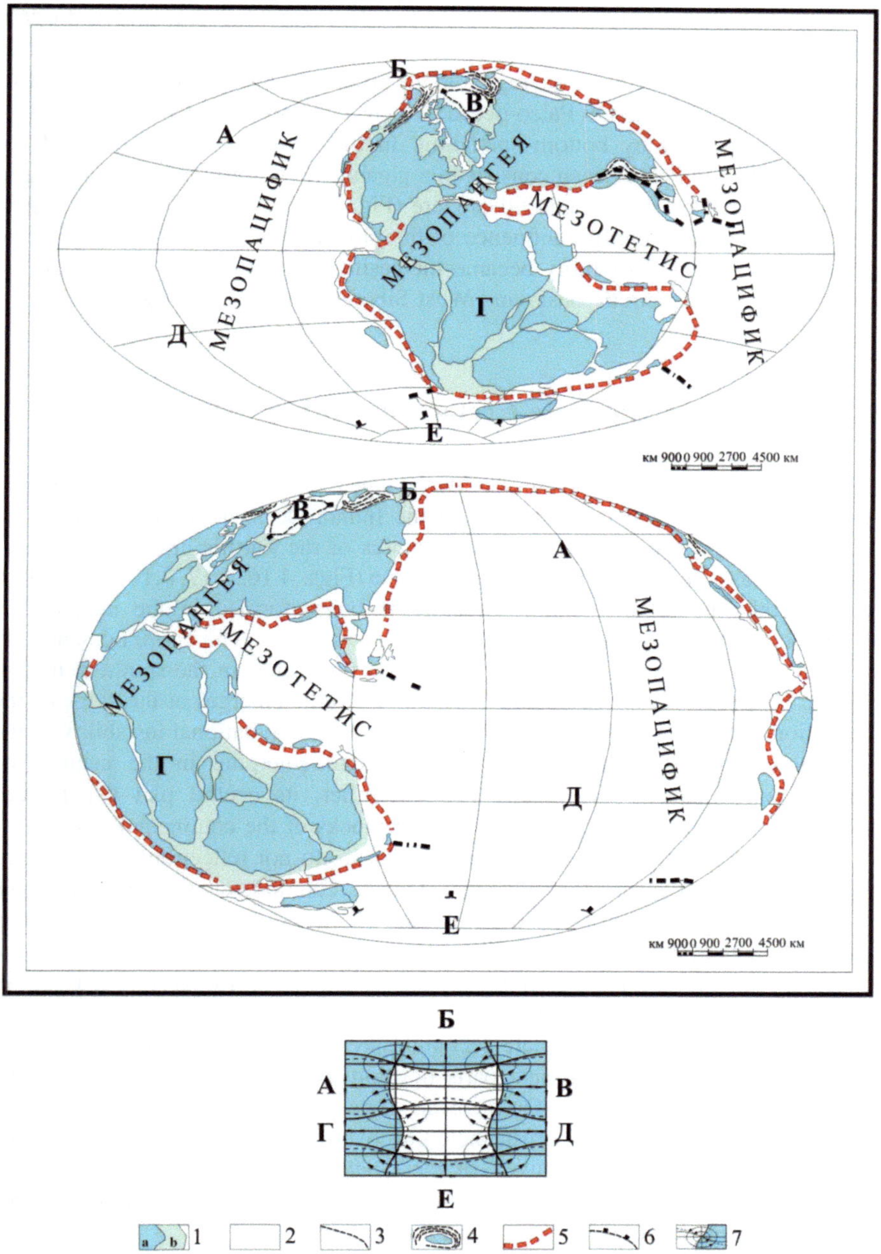

Fig. 4.15 Global paleogeodynamic scheme-reconstruction, Early–Middle Jurassic (200–161 Ma) (a) in an arbitrary projection of Aitov–Gammer; (b) in pseudocylindrical elliptic Mollweide homalographic projection. Explanations are given in Fig. 4.12

Thus, Aplonov (1989) was able to show that the Perm-Triassic rifting in Western Siberia went much further than it was previously thought. In Aplonov's scheme, one of the continental rifts of the region in the process of evolution has turned into a young ocean, called the Ob Paleo-ocean, which for almost two tens of millions of years has expanded by bottom spreading. In the current structural plan of the basement, the Ob Paleo-ocean consists in a gigantic, clinoform gapping tapering to the south, cured by the oceanic lithosphere (Fig. 4.13). It should be noted that in general in the Mesozoic, the influence of the global first-ordered structures of the Earth's gravitational instability became increasingly visible in the features of the geodynamics of the Arctic basin and West Siberian plate (Fig. 3.4).

4.10 Late Jurassic–Late Cretaceous (161–65 Ma)

In the southern hemisphere within the Mesopangea, the beginning of the large-scale split of Gondwana and the formation of the Indian Ocean is associated with the establishment of a new generation of structures of the Earth's gravitational instability, such as L-hexagons, in Late Jurassic time (Figs. 4.16, 4.17, 4.18 and 4.19). It is clear that the second similar structure is being formed in the area of the Mesopacific in the region of the modern East Pacific Rise. The entire branched rift system, accompanied by Early Cretaceous volcanic zones in the southern half of Africa, Hindustan, East Africa and Western Australia, emerges at the stage of the formation of the Gondwana's structure of the Earth's gravitational instability of the L-hexagon type over the deep mantle substrate emerging in the centre (the Paleotectonic maps of Gondwana 1987). Further, its central part experienced spreading immersion, collapse and spread of blocks of the continental crust.

The resulting depressions of the Indian Ocean do not reflect any genetic connection with the banks bordering them. On the contrary, all the signs from the overlap of these depressions on the structural elements of their coasts is that they quite different in origin and age. The edges of the Indian Ocean floor cross the platform structures of the African, Indian and Australian platforms and Madagascar, as well as the structures of their Precambrian folded basal part.

With the immersion and plate tectonic spreading of the central part of the Gondwana structure, its marginal parts develop in the regime of active tectonic margins (Figs. 4.18 and 4.19). On the basis of paleotectonic reconstructions, zones of subduction of the oceanic crust, including overthrusts and underthrusts, can be established here. In the framework of the Gondwana structure, Late-Cimmerian (Australian, etc.) folded accretionary complexes, zones of development of tectonic melange and intensive deformations are noted.

Along with the opening of the Indian Ocean in the Jurassic–Cretaceous, the tectonic processes that established the laying of the Atlantic oceanic depression were activated. The results of deep-sea drilling within the wide zones of the oceanic margins of the Atlantic indicate the fragmentation in the Late Jurassic–Early Cretaceous of ancient platform blocks and associated Paleozoic structures by a system of faults with the

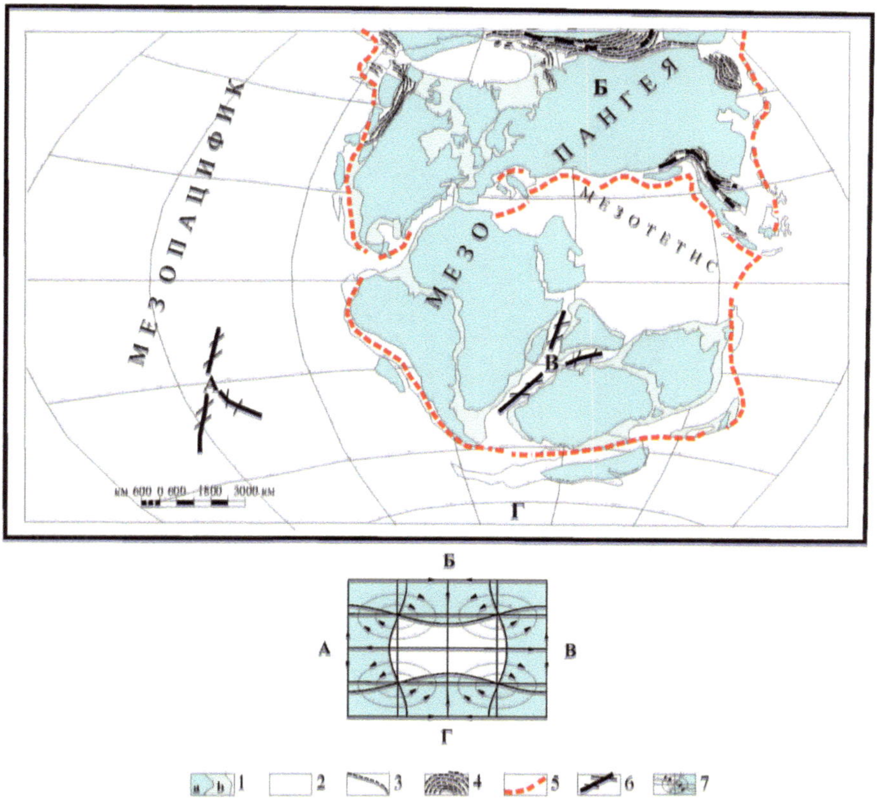

Fig. 4.16 Global Paleogeodynamic scheme-reconstruction, Late Jurassic (161–145 Ma). Polyconic projection of the Central Institute of geodesy, cartography and aerial laser scanning, version of the Great Soviet Encyclopaedia. 1–5—in Fig. 4.12; 6—boundaries of the central zones and directions of spreading of the Earth's crust; 7—displacements and streamlines in the Earth's gravitational instability structures of the second order, the ninth generation

formation of step blocks, as well as systems of horsts and grabens. The fragmentation was accompanied by intensive trappean magmatism, stable emersion and collapse of the continental margin towards the future ocean. It is evident that all these processes are already associated with the influence of a larger global L-hexagon structure of the Earth's gravitational instability of the first orders (Fig. 3.4).

From the Jurassic to the Cretaceous period, the active influence of a larger global structure of the gravitational instability of the Earth, such as the G-hexagons of the first order, also manifested itself within the Arctic geodepression, the West Siberian lowland and in the adjacent territories. Within the Arctic geodepression during the Jurassic–Cretaceous period, the arch swellings of the continental crust in its centre experience the immersion, collapse and thinning, and in their place large downfolds of the crust (sedimentary basins or young plates) appear and the Amerasian oceanic basin is formed.

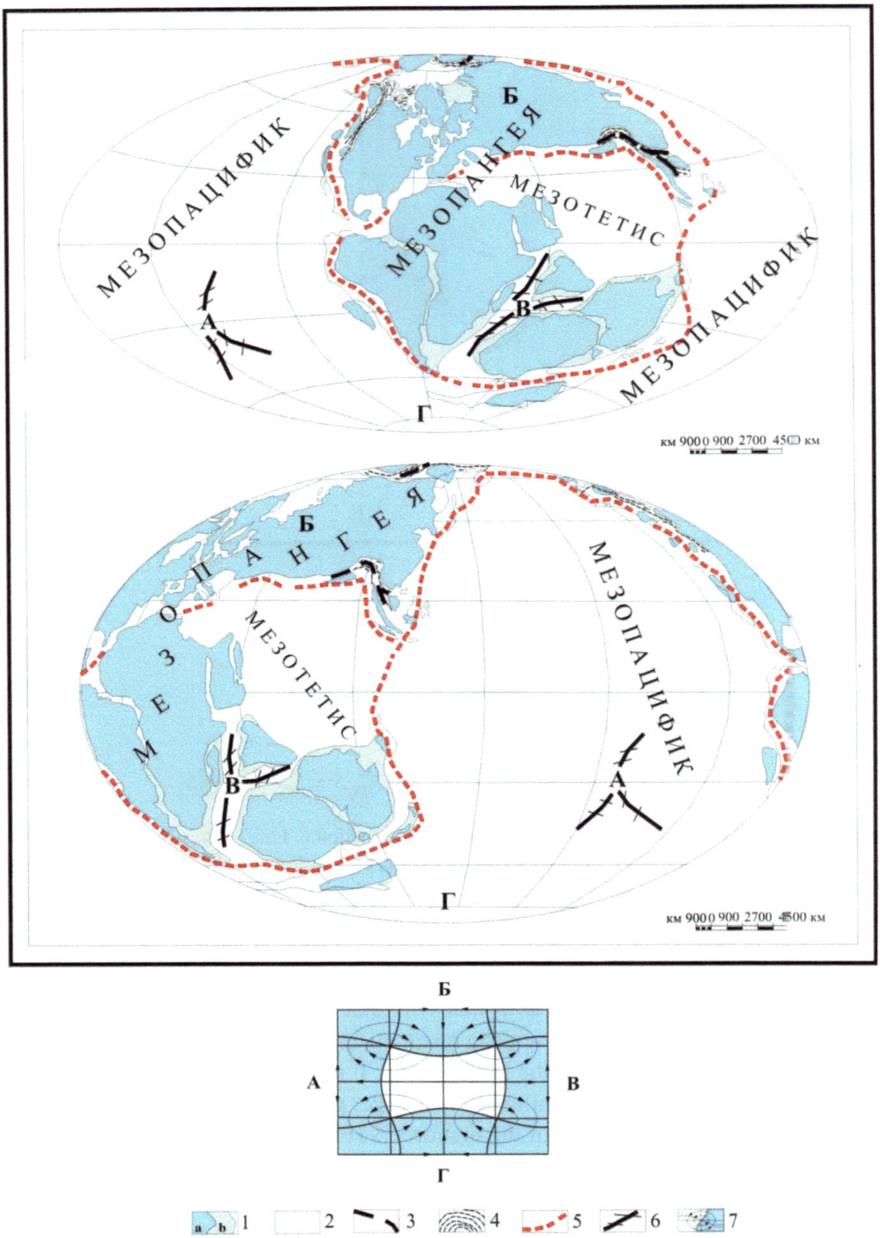

Fig. 4.17 Global paleogeodynamic scheme-reconstruction, Late Jurassic (161–145 Ma) (a) in arbitrary projection of Aitov—Gammer; (b) in pseudocylindrical elliptic Mollweide homalographic projection. Explanations are given in Fig. 4.12 (1–5) and 2.16 (6–7)

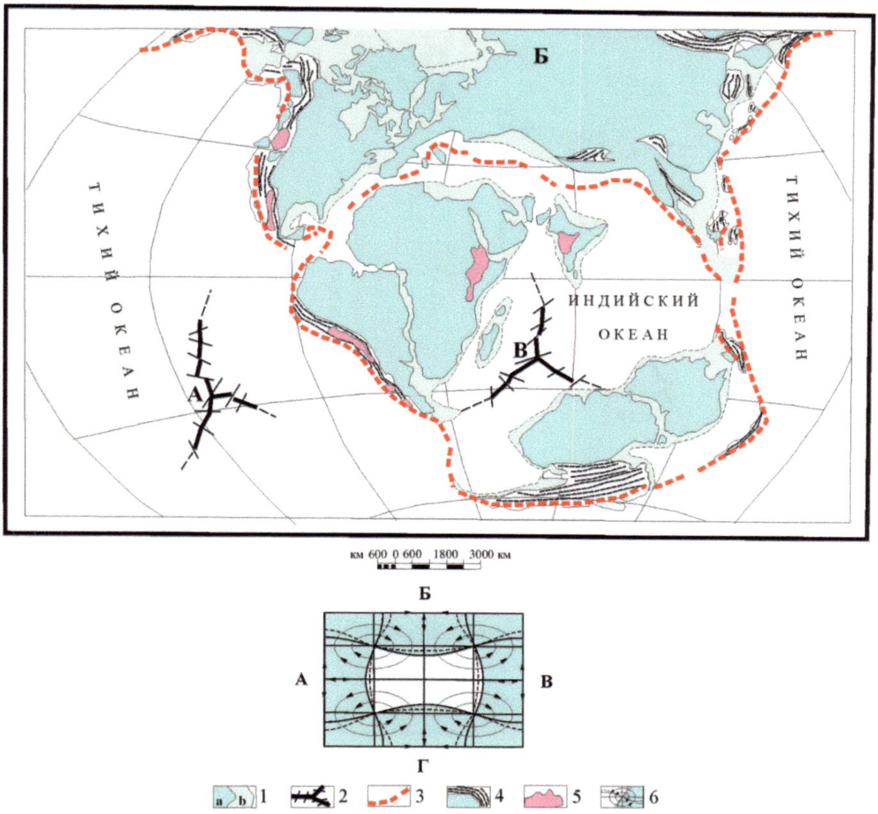

Fig. 4.18 Global paleogeodynamic scheme-reconstruction, the end of the Early—the beginning of the Late Cretaceous (99 Ma). Polyconic projection of the Central Institute of geodesy, cartography and aerial laser scanning, version of the Great Soviet Encyclopaedia. 1—inland crust (**a**) and shelf zones (**b**); 2—boundaries of the central zones and directions of spreading of the Earth's crust; 3—subduction zones of the oceanic crust (including overthrusts and underthrusts); 4—Late Cimmerian (Austrian, etc.) folded accretionary complexes, zones of tectonic melange development and intensive deformations; 5—zones of Early Cretaceous volcanism; 6—displacements and streamlines in the Earth's gravitational instability structures of the second order

Obviously, it was these processes that determined the final disintegration of the Hyperborean platform into separate large blocks of the continental crust, and their rift induced spread caused intensive Mesozoic orogeny processes in the framework of the Amerasian basin. The structure of the Earth's crust of this oceanic basin differs from others in the presence of significant areas of the granite layer on the spaces of depressions and underwater ridges. The rises of Mendeleev-Alpha and Lomonosov, as well as individual deeply caved blocks of cratons, are relic continental structures. At the same time the depressions of the Arctic Ocean with their edges cross the structural elements of the continents surrounding them: the edge of the Arctic belt of America, the Mesozoic structures of the north of Siberia and the extremity of the Atlantic belt.

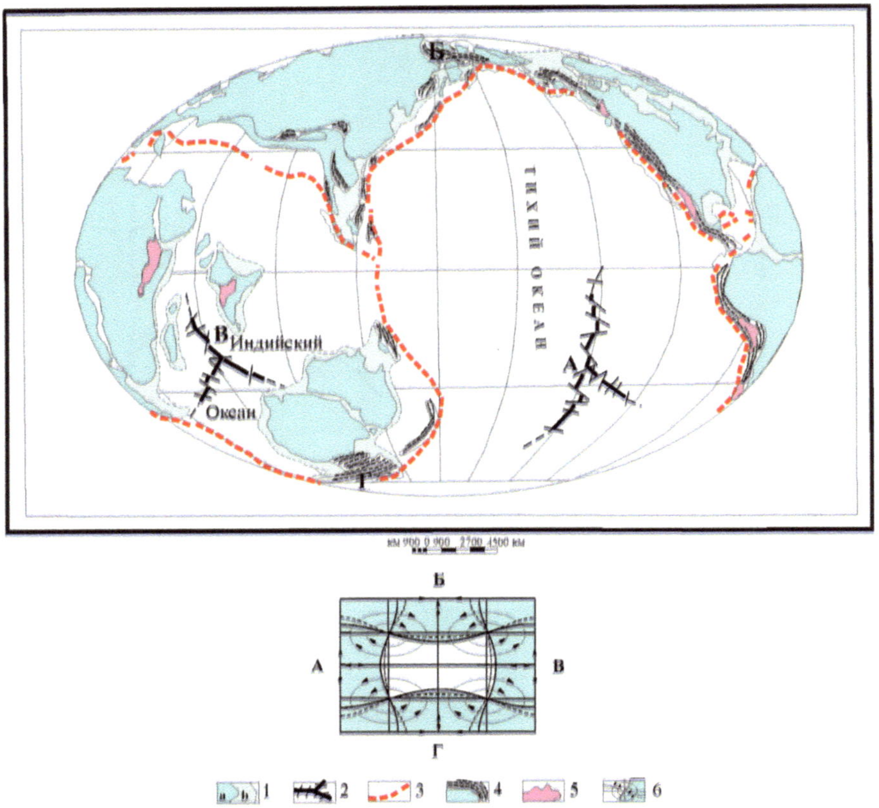

Fig. 4.19 Global paleogeodynamic scheme-reconstruction, from the end of the Early—to the beginning of the Late Cretaceous (99 Ma). In a pseudocylindrical elliptic Mollweide homalographic projection. Explanations are given in Fig. 4.18

Thus, with the course of geological time, the dimensions of newly originating generations of gravitational instability structures increase. The role of plate tectonic processes in the upper mantle and the lithosphere increases, and global structures of the first-order Earth's gravitational instability of the G-hexagon type in the eastern hemisphere and the L-hexagon type in the west increasingly manifest themselves.

4.11 Late Cretaceous–Present (99–0 Ma)

A qualitative change in the global geodynamic situation, associated with the activation of the global structures of the first-order Earth's gravitational instability (tenth generation), occurred in the Apto-Albian time, including the Cenomanian occasionally (Figs. 4.20 and 4.21). Neotectonic restructuring affected the alteration

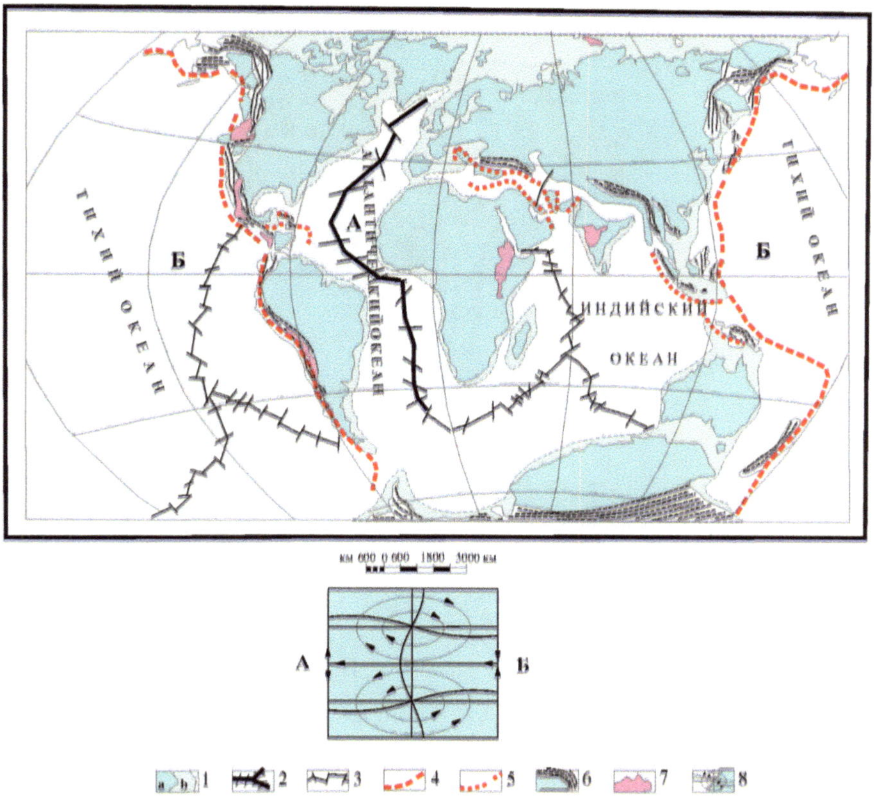

Fig. 4.20 Global paleogeodynamic scheme-reconstruction, boundary between the Mesozoic and the Cenozoic (65 Ma). Polyconic projection of the Central Institute of geodesy, cartography and aerial laser scanning, version of the Great Soviet Encyclopaedia. 1—inland crust (a) and shelf zones (b); 2—boundaries of the central zones and directions of spreading of the Earth's crust, referred to the structures of the Earth's gravitational instability of the first order of the L-hexagon type; 3—the same, of the G-hexagon type; 4—subduction zones of the oceanic crust, referred to the structures of the Earth's gravitational instability of the first order; 5—the same, for lower orders; 6—folded accretionary complexes including continental fragments (microcontinents, median masses); 7—zones of intensive Upper Cretaceous and the Tertiary volcanism; 8—displacements and streamlines in the structures of the Earth's gravitational instability of the first order

in the structure of the Earth's crust and upper mantle, geophysical fields, endogenous regimes, the design of modern contours of the continents and oceans, the development of the main Atlantic part of the world system of mid-oceanic ridges, the formation of marginal seas, deepest ocean depressions and trenches (Nikolaev 1988). This mark is associated with an increase in the overall amplitude of the Earth's relief, as well as with the deformation of the geoid morphology and its adaptation to the new depth conditions of gravitational instability.

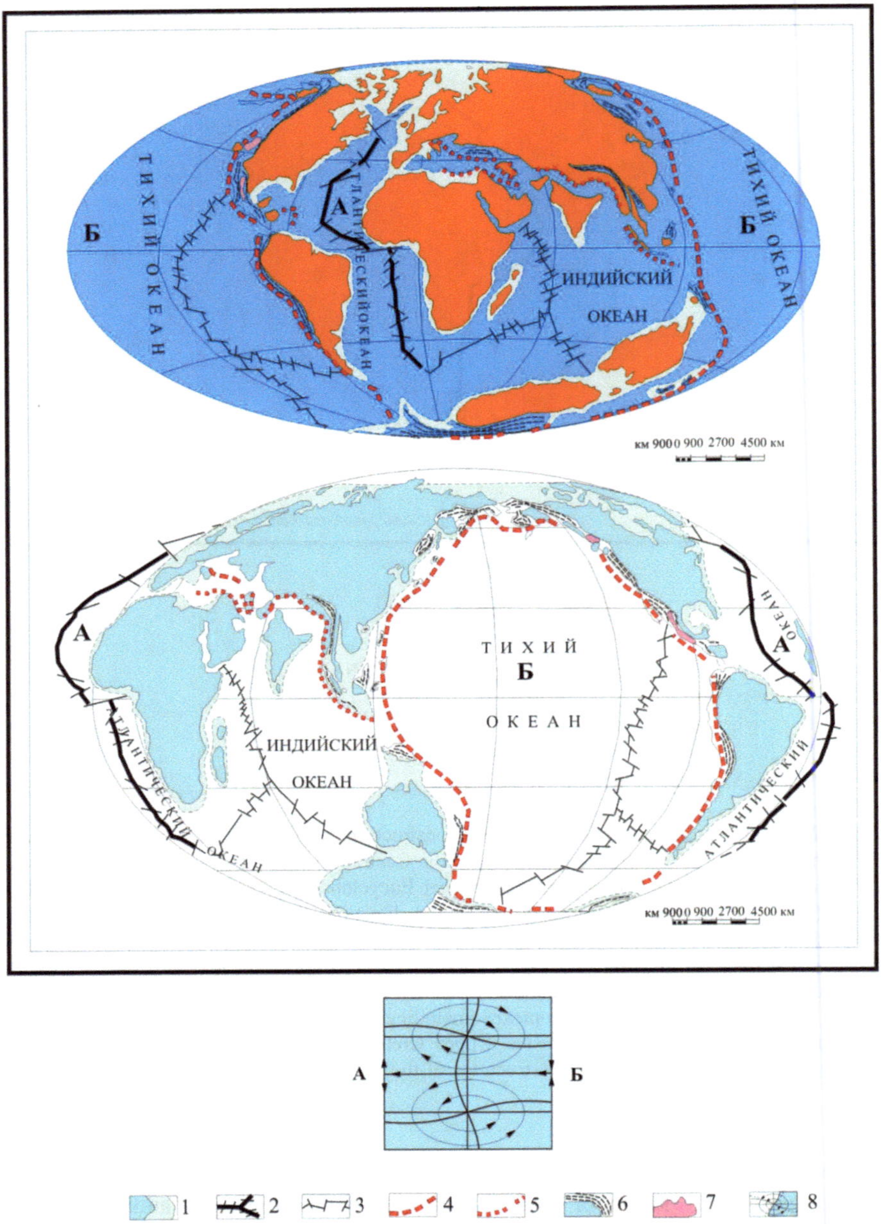

Fig. 4.21 Global paleogeodynamic scheme-reconstruction, the boundary between the Mesozoic and the Cenozoic (65 Ma.) (a)—in the arbitrary projection of Aitov-Gammer; (b)—in pseudo-cylindrical elliptic Mollweide homalographic projection. Explanations are given in Fig. 4.20

The analysis of the Earth's shape, gravitational field and seismic tomography data indicate that throughout this time the geodynamic situation in the western hemisphere is controlled by the global structure of the Earth's gravitational instability of L-hexagon type, while in the eastern hemisphere the gravitation instability results in inverse structures of the G-hexagon type (Figs. 4.20, 4.21 and 4.22). On the surface of the Earth, the boundaries of these global cellular structures are confined to the main planetary geodiscontinuities and the typomorphic features of their structure are reflected in geoid anomalies and geophysical fields, emphasised by the distribution pattern of the continental and oceanic crust.

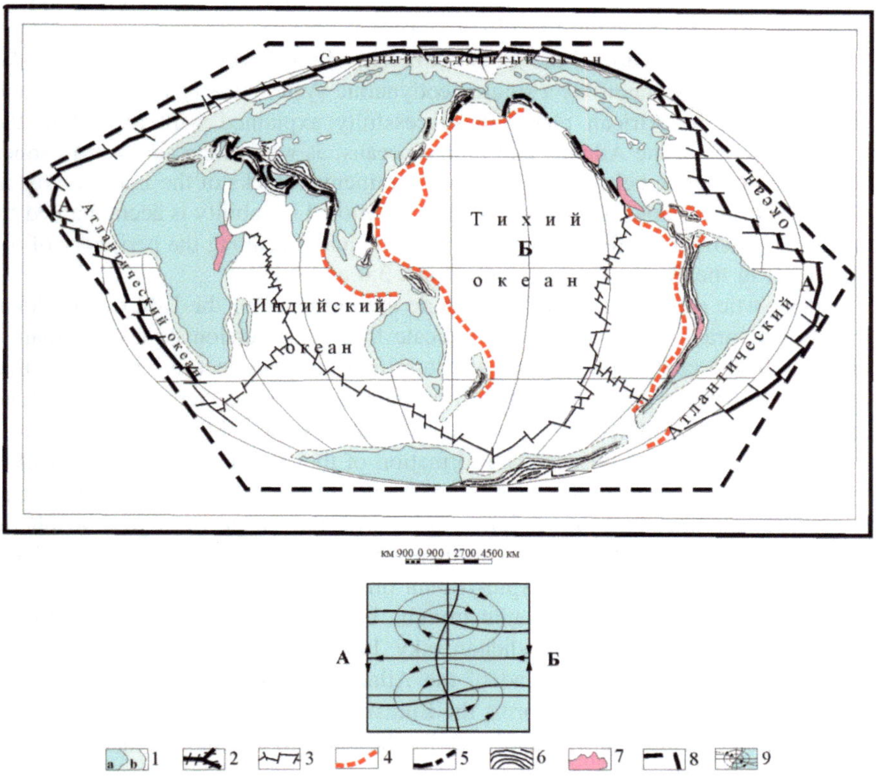

Fig. 4.22 Global paleogeodynamic scheme-reconstruction, present time. In the modified pseudocylindrical elliptic Mollweide homalographic projection. 1—inland crust (a) and shelf zones (b); 2—boundaries of the central zones and directions of spreading of the Earth's crust, referred to the structures of the Earth's gravitational instability of the first order; 3—the same, for lower orders; 4—subduction zones of the oceanic crust; 5—folded accretionary complexes; 6—main ophiolite belts, zones of tectonic melange development and intensive deformations; 7—zones of intensive Tertiary volcanism; 8—boundaries of the structures of the thermogravitational instability of the Earth of the first order; 9—displacements and streamlines in the Earth's gravitational instability structures of the first order

Seismic tomography data detailing the 3D picture of Earth's deep interior structure made it possible to establish that in the western hemisphere the central, arc-shaped region of the emersion of the deep mantle substrate within the L-hexagon structure of the gravitational instability covers both the Atlantic and Indian oceans, uniting them into a single Indo-Atlantic geodynamic system (Figs. 3.4 and 4.20). Within the boundaries of this geodynamic system, the plates increase the area and volume as they move from the Mid-Atlantic Ridge and the median ridges of the Indian Ocean, and the plates under the Pacific Ocean reduce the area and volume with the movement of the North and South America and Eurasia to the centre of the Pacific Ocean (Figs. 4.20, 4.21 and 4.22). The study of the heat flux of the Earth, the gravitational field and seismicity indicates that these movements cover not only the continents, but also the underlying mantle to depths of at least 400 km (Structural Geology… 1991).

Within the framework of the global structure of the Earth's gravitational instability (like the L-hexagon) as a single geodynamic system, the spatial position and kinematics of the African plate are successfully explained with respect to the spreading zones of the Atlantic and Indian oceans, as well as the subduction zones of the Pacific mobile belt, as the spread of continental blocks of the Earth's crust in the region of the Indo-Atlantic system of gravitational instability is accompanied by the development of the regime of active tectonic margins along the periphery of the continents in the Pacific ocean region (Figs. 3.4 and 4.20).

Palinspastic reconstructions that exclude the influence of the L-hexagon global structure of gravitational instability indicate that the emersion of a deep mantle substrate in the central part of this structure and the plate tectonic processes that conform to it in the upper horizons of the mantle and lithosphere during the Late Cretaceous era determined the formation of the depressions of the Atlantic and, to a large extent, the Indian oceans. The formation of the Atlantic is associated with a split in the Earth's crust in the zone of emergence of the deep mantle substrate and the subsequent spreading immersion coupled with the spread of large lithospheric plates including the contemporary continents in the process of development of this gravitational instability. In the Apto-Albian time, South America completely separated from Africa and the oceanic troughs began to develop actively between Europe and North America (Pushcharovsky 1992). The Maastrichtian-Paleocene time is associated with the spread of the Atlantic to the north, up to the Arctic Ocean, where its continuation forms the Eurasian basin. Here there is evidence of the formation of a modern structural plan for the east part of the South and partly for the Central Atlantic.

Tectonically, the depression of the Atlantic Ocean crosses the most diverse elements of neighbouring continental structures. At the same time, as Yu.M. Pushcharovsky notes (Pushcharovsky 1992, 2005), the Atlantic, along with the spreading mechanism of bottom emersion, is characterised by a collapse mechanism, which is reliably established within the wide zones of oceanic margins. Thus, deep-sea drilling in the part of the North Atlantic adjacent to Western Europe has shown the consistent emergence and deepening of the ocean basins that began as early as the Late Jurassic–Early Cretaceous with the formation of shallow-water lagoon epicontinental basins, the accumulation of sapropels in them, and the continuing intensive collapse in the post-Albian period and in the Oligocene–Miocene (Initial reports… 1969–1983, Vol. 12, 47, 48, 80; Rudich 1983).

Along the western margin of the South African block, the basement undergoes a flexure-like immersion, complicated by faults that crush it into a series of blocks. The surface of the continental body, composed of metamorphic rocks with ages ranging from Precambrian to Perm-Carboniferous, is submerged here to a depth of 5 km. The Upper Jurassic–Lower Cretaceous basaltic covers of the basement of the transition zone are comparable with the Lower Jurassic trappean basaltic covers of the Kaokofeld and Stromberg (basalts of Carrus) developed on land. Both these and others can be considered as traps, corresponding to the initial stage of the formation of the oceanic depression.

The tectonic character of the morphostructure of the northern and southern America's margins opposite to Western Europe and Africa is quite close. In both cases, we are dealing with the fragmentation in the Late Jurassic–Early Cretaceous of ancient platform blocks by a system of faults, the formation of step blocks, as well as the system of horsts and grabens. The fragmentation was accompanied by a stable emersion and a progressive collapse of the continental margin towards the ocean, as well as by the intense magmatism and the opening of its subsequent Late Cretaceous spreading. In general, a similar geochronological sequence of events is established for a number of oceanic margins and central zones of the Indian Ocean, although, as shown above, it is assumed that its genesis took place earlier.

Thus, the paleogeodynamic reconstructions of the formation of the Atlantic and Indian oceans are in good agreement with the conformal and reversed nature of the correlations between the surface and deep elements of the organisation of cellular structures of the Earth's gravitational instability. As might be expected in this case, the zones of emergence of the deep mantle substrate in the upper horizons of the mantle and the lithosphere are responsible for the spreading mechanism of the ocean floor emersion and its collapse.

In the eastern hemisphere, since the Apto-Albian time, the geodynamic situation was controlled by the G-hexagon global structure of the Earth's gravitational instability inverted with respect to the global structure considered above (see Figs. 3.4, 4.20 and 4.21). In the tectonic plan, the central region of the emersion is emphasised by the arc-shaped curved world system of islands in the west of the Pacific and the east of the Indian oceans.

The time of the formation of deep-sea trenches, island arcs and deep-sea depressions of the marginal seas is different (Tectonics of the continents… 1988). The age of these tectonic structures varies from the Late Cretaceous (Tasman Sea) and the Paleocene (the Aleutian arc and the trench accompanying it) to the Pliocene–Quaternary time (the Marian interfluvial downfold) and reflects the complex manifestation of tectonic movements occurring here mainly in the Cenozoic. These movements, being confined to the central, submerging part of the structure of the gravitational instability of the Earth such as the G-hexagons, are determined by the repeated interaction of the continental and oceanic lithosphere in the conditions of the neighbouring continents (Asia–Australia), on the one hand, and the oceanic regions (Pacific Ocean–Indian Ocean) on the other.

The marginal parts of the G-hexagon global structure of the Earth's gravitational instability are characterised by the extensive development of inland and oceanic rift systems. B. Heezen and D. Fornari showed that the most ancient, Jurassic–Cretaceous basalts are distributed in the western part of the Pacific Ocean, while the youngest, those of the Late Cenozoic, are confined to the East Pacific Rise (Heezen and Fornari 1976). Basalts of other regions, both in the Pacific and Indian oceans, have an intermediate age. In addition, a large number of relatively young (Late Mesozoic–Cenozoic) intracontinental rift zones in North America, Antarctica, West Africa, Eastern and Western Siberia are associated with the marginal part of this global structure of the Earth's gravitational instability. These inland rift zones cross the continents and their margins and cause the formation of large crustal downfolds (sedimentary basins or young plates) both within the continents themselves and under the seas of Antarctica and the Arctic.

In the abyssal of the Arctic Ocean, the closing of the two types of the global Earth's gravitational instability structures—G-hexagons and L-hexagons occurs (Figs. 3.3, 3.4, 4.20 and 4.22). With this zoning, the water area of the Arctic Ocean is divided into three tectonically isolated deep-water basins: Norwegian-Greenlandic and Eurasian, belonging to the Atlantic segment of the L-hexagon structure of the Earth's gravitational instability, and Amerasian, which is the marginal part of the Pacific segment of the G-hexagon structure of the Earth's gravitational instability. The Amerasian basin encompasses deep-water basins and submarine ridges lying between the Lomonosov rise and the continental margins of North America and northeast Asia. The Eurasian basin, located between the shelf of the Kara and Barents seas, on the one hand, and the submarine rise of Lomonosov, on the other, crosses the marginal part of the Pacific segment of the G-hexagon structure of the Earth's gravitational instability. At the same time, the Gakkel ridge, located in its centre, according to its morphological and geophysical characteristics, belongs to the number of classical mid-ocean ridges. Beginning at the Spitsbergen fault, the Gakkel Ridge passes through the near polar part of the basin in the direction of the Laptev Sea, crossing the marginal part of the G-hexagon global structure and preserving the morphostructural unity throughout 2000 km. At 80°N, in approximately 250 km from the edge of the continental slope, it flattens and peters out at a depth of more than 3000 m. However, the seismically active zone, confined to the axial rift valley, passes further through the shelf of the Laptev Sea to the northern spurs of the Verkhoyansk Ridge (Avetisov 1979; Tectonics of the Continents… 1988).

As it was mentioned before, it was very difficult to determine the boundaries of the considered global structures of the Earth's gravitational instability of the G-hexagon and L-hexagon types, since these objects consist of interconnected and interpenetrating fractal structures. Over 80% of the area and volume of the planet appear to be common for both structures; therefore, the dimensionality of these objects is fractional (Figs. 3.3, 3.4, 4.20, 4.21 and 4.22).

Such a character of the correlation between the Earth's global structures of the gravitational instability of the G-hexagon and L-hexagon types illustrates the schemes of displacements and streamlines showed in Figs. 3.4, 4.20, 4.21 and 4.22. In such a structure, each point is the limit (i.e., there are set points in any of its vicinities). In this regard, it is not always possible to draw the boundary of the structure in the form of a dividing line. Without taking into account the fractal dimensionality of the global structures of the Earth's gravitational instability, it is impossible to describe the kinematics of the plate tectonic processes covering the upper mantle and the lithosphere, and, consequently, to understand the reason for its division into the Pacific and Indo-Atlantic tectonic segments.

In addition, having a fractal dimensionality, the global structures of the Earth's gravitational instability repeat themselves at different scales, determining the predominant confinement of L-hexagon structures to the continents of the Indo-Atlantic segment and structures like G-hexagons to the Eurasian continents of the Pacific segment. Paleogeodynamic reconstructions show that this pattern in the spatial distribution of various types of structures of gravitational instability is noted already at an early stage of the geological history of the Earth (more than 3800 Ma ago) and is increasingly traced in all subsequent stages of its development, ultimately determining the features of the geodynamics of Laurasia and Gondwana (Figs. 4.1, 4.2, 4.3, 4.4, 4.5, 4.6, 4.7, 4.8, 4.9, 4.10, 4.11, 4.12, 4.13, 4.14, 4.15, 4.16, 4.17, 4.18, 4.19, 4.20, 4.21 and 4.22).

References

Aplonov S.V. Paleodynamics of the West Siberian Plate. Sov. geologiya. 1989; 7:27–36. (In Russ.).

Avetisov G.P. Seismicity and deep structure of the Earth's crust in the continental continuation of the Middle Arctic belt of earthquakes (the Laptev Sea and the Novosibirsk Islands). Published summary of the Candidate's dissertation. L.: LGU; 1979. (In Russ.).

Barbey P., Condert J., Morean B. et al. Petrogenesis and evolution of an Early Proterozoic orogenic belt: the Granulite belt of Lapland and Beloniorides (Fennoscandia). Bull. Geol. Soc. Finl. 1984; 56(1, 2):161–187.

Belov A.A. Tectonic development of the Alpine folded region in the Paleozoic. M.: Nauka; 1981. (In Russ.).

Borukaev Ch.B. Comparative tectonics of the shields of Laurentia and Gondwana. Russian geology and geophysics. 1979; 3:13–24. (In Russ.).

Borukaev Ch.B. Precambrian structure and plate tectonics. Novosibirsk: Nauka; 1985. (In Russ.).

Bozhko N.A. Geodynamic inversion in the polar systems of the northern and southern hemispheres of the Earth. Moscow University Geology Bulletin. 1992; 1:27–38. (In Russ.).

Bozhko N.A. Geodynamic inversion in polar systems of the northern and southern hemispheres of the Earth. Bulletin of Moscow State University. Series 4, Geol. 1992; 1: pp. 27-38. (In Russ.).

Bray W.C. A periodic reaction in homogeneous solution and its reaction to catalysis. J. Amer. Chem. Soc. 1921; 43(6):1262–1267.

Brooks M. Some aspects of the Paleogene evolution of Western Britain in the context of an underlying mantle hot spot. J. of Geology. 1973; 81(1):81–88.

Brun J.P. The clasterridge pattern of mantled gneiss domes in Eastern Finland: evidence for largescale gravitational instability of the Proterozoic crust. Earth and Planet. Sci. Lett. 1980; 47: 441–449.

Dixon J. M. Structural Geology and Plate Tectonics. M.: Mir; 1991. (In Russ.).

Dziewonski A.M., Woodhouse J.H. Threedimensional Earth structure and mantle convection. Abstracts. 1989; 1(28th IGC):427–428.

Early history of the Earth. Ed. B. Windley. M.: Mir; 1980. (In Russ.).

Friend C.R.L. The Origin of the Closepet granites and the implications for the crustal evolution of Southern Karnataha. J. Geol. Soc. India. 1984; 25(2):73–84.

Glukhovskii M.Z. Geological basement evolution of the ancient platformes. M.: Nedra; 1990. (In Russ.).

Glukhovskii M.Z., Moralev V.M., Kuz'min M.I. Hot belt of the early Earth and its evolution. Geotectonics. 1994; 3:3–15. (In Russ.).

Grachev A.F., Fedorovskii V.S. Greenstone Precambrian belts: rift zones or island arcs? Geotectonics. 1980; 5:3–24. (In Russ.).

Heezen B.C., Fornari D.J. Geological map of the Pacific Ocean. Eos. Trans. Amer. Geophys. Union. 1976; 57(4):264.

Hynes A. Stability of oceanic tectonosphere—a model for Early Proterozoic intercratonic orogeny. Earth and Planet. Sci. Lett. 1982; 61:333–345.

Khain V.E. Paleozoic stage of the development of the earth's crust. Proceedings of Higher Schools. Geology and exploration. 1989a; 9:3–15. (In Russ.).

Khain V.E. Stratification of the Earth and multilevel convection as the basis of a truly global geodynamic model. DAN SSSR. 1989b; 308(6):1437–1440. (In Russ.).

Khain V.E. Regional tectonics. Extra-Alpine Asia and Australia. M.: Nedra; 1979. (In Russ.).

Khain V.E., Bozhko N.A. Historical Geotectonics, the Precambrian. M.: Nedra; 1988. (In Russ.).

Kinsman D.J. Salt floors to geosynclines. Nature. 1975; 255(5507):375–378.

Kontinen A. An Early Proterozoic ophiolite, the Jormua maficultramafic complex, Northeastern Finland. Precambrian Res. 1987; 35:313–342.

Kumazawa M., Maruyama S. Whole Earth tectonics. J. Geol. Soc. Japan. 1994; 100(1) 81–102.

Larson R.L., Olson P. Mantle plumes control magnetic reversal frequency. Earth and Planet. Sci. Lett. 1991; 3:437–447.

Laz'ko E.M., Sivoronov A.A., Bobrov A.B. The problem of the tonalite layer in the granite-greenstone areas. Izv. AN SSSR. Ser. Geol; 1982. (In Russ.).

Maruyama S., Kumazawa M., Kawakami S. Towards a new paradigm on the Earth's dynamics. J. Geol. Soc. Japan. 1994; 100(1):1–3.

Milanovsky E.E., Rifting evolution in geological history. Tectonophysics. 1987; 143 (1–3): 103–118.

Morgan J.S. The periodic evolution of carbon monoxide. J. Chem. Soc. London, Trans. 1916; 109:274–283.

Nikolaev N.I. Neotectonics and geodynamics of the lithosphere. M.: Nedra; 1988. (In Russ.).

Paleotectonic maps of Gondwana. Ed. N.A. Bozhko, V.E. Khain. M.; 1987. (In Russ.).

Picard C., Lamothe D., Piboul M., Oliver R. Magmatic and geotectonic evolution of a Proterozoic oceanic basin system: the Cape Smithn Thrust-Fold Belt (New Quebec). Prec. Res. 1990; 47:223–249.

Piper J.D.A. The Precambrian palaeomagnetic records for the Proterozoic Supercontinent. Earth Planet. Sci. Lett. 1982; 55(1):61–89.

Pogrebitskii Yu.E. Geodynamic system of the Arctic Ocean and its structural evolution. Sov. geologiya. 1976; 12:3–12. (In Russ.).

Pushcharovskii Yu.M. Selected works. Tectonics of the Earth. Etudes in 2 volumes. Vol. 1: Tectonics and geodynamics. Geol. in-t. M.: Nauka; 2005. (In Russ.).

Pushcharovskii Yu.M. Tectonics of the oceans and nonlinear geodynamics. Dokl. RAN. 1992; 324 (1):157–161. (In Russ.).

Ronkina Z.Z., Vishnevskaya T.N. Glaucophane in the sedimentary strata of the Western part of the Soviet Arctic. Sov. geologiya. 1982; 2:90–93. (In Russ.).

Rudich E.M. Moving continents and the evolution of the ocean bed. M.: Nedra; 1983. (In Russ.).
Saleeby J.B. Accretionary tectonics of the North American. Earth Planet. Sci. Lett. 1983; 15:45–73.
Schwerdtner W.M. Salt stocks as natural analogues of Archaean gneiss diapers. Geol. Rundshau. 1982; 71:370–379.
Schwerdtner W.M., Sutcliffe R.H., Troeng B. Patterns of total strain within the crustal regions of immature diapers. Canad. J. Earth Sci. 1978; 15:1437–1447.
The Siberian platform. The geological structure of the USSR and the regularities of the location of minerals. V. 4. Ed. N.S. Malich, V.L. Masaitis, V.S. Surkov. L.: Nedra; 1987. (In Russ.).
Sivoronov A.A. Formations and origin of the Lower Precambrian greenstone complexes of the East European Platform. Published summary of the Doctor's dissertation. M.; 1988. (In Russ.).
Stratigraphic code of Russia. 3rd editon. Ed. A.I. Zhamoida. SPb: VSEGEI; 2006. (In Russ.).
Structural Geology and Plate Tectonics. In three volumes. V. 1-2. / Edited by K. Seifert. M.: Mir; 1991. (In Russ.).
Surkov V.S., Zhero O.G. The basement and development of the platform cover of the West Siberian plate. M.: Nedra; 1981. (In Russ.).
Talbot C.J. Thermal convection below the solidus in a mantled gneiss dome, Fungwi Reserve, Rhodesia. J Geol Soc 1971; 127(4):377–410
Tectonics of Continents and Oceans: Explanatory Note to the International Tectonic map of the World. 1:15 000 000. M.: Nauka; 1988. (In Russ.).
Turcotte D., Schubert J. Geodynamics: geological applications of the physics of continuous media. M.: Mir; 1985. (In Russ.).
Veevers J.J. Middle-late triassic (230 Ma) singularity in the stratigraphic and magmatic history of the Pangean heat anomaly. Geology. 1989; 17:784–787.
Vogt P.R. Evidence for global synchronism in mantle plume convection and possible significance for geology. Nature. 1972; 20:338–342.
Vogt P.R. Global magmatic episodes: New evidence and implications for the steadystate midoceanic ridge. Geology. 1979; 7:93–98.
Vogt P.R. Plumes, subaxial pure flow and topography along the midoceanic ridge. Earth and Planet. Sci. Lett. 1976; 29:209–325.
Windley B. The evolving continents. London; 1985.
Zhai M.G., Vang R.V., Zhoi W.J., Lu J. Geochemistry and evolution of the Qinguyan Archaean granitogreenstone terrain. Prec. Res. 1985; 27:37–62.
Zonenshain L.P., Kuz'min M.I. Deep geodynamics of the Earth. Russian geology and geophysics. 1993; 4:3–12. (In Russ.).

Chapter 5
Fractal Hierarchies and the Historical Irreversibility of Geodynamic Processes

The invariance of the revealed regularity in the spatial distribution of different-type hierarchies of the Earth's gravitational instability structures throughout its geological history, along with the data discussed above on the distribution of mantle plumes and the sialic protocrust with respect to global structures such as G-hexagons in the eastern hemisphere and the L-hexagons in the western, confirms the emergence of these global structures, which determined the fundamental asymmetry of the Earth and divided it into the Indo-Atlantic and Pacific tectonic segments during the early Archean, simultaneously with the isolation of the liquid core, i.e. more than 3500 Ma. At the same time, paleogeodynamic reconstructions show that generational structures emerge during each subsequent stage of the Earth's evolution, along with disordered instabilities. Moreover, these increasingly massive generational structures are presented predominantly by G-hexagon type in the eastern hemisphere and the L-hexagon type in the western. With the consideration of the relationship between the size and the stratification depth of the structures of gravitational instability, this means that with the course of geological time, the increasingly deep layers of the Earth's crust and mantle became gravitationally unstable with respect to the underlying shells. From a physical point of view, the increase in the thickness of gravitationally unstable layers over time can easily be explained by the nonequilibrium cooling of the Earth through its surface, i.e., by external sinks of heat. With the increase in the size of the Earth's gravitational instability structures, the associated tectonic regimes also change, ranging from fine-cellular intraplate (greenstone and granulite, protoaulacogenic and protogeosynclinal) to marginal-plate and global riftogenic regimes. In the terms of geodynamics, the horizontal component in the movement of the Earth's crust becomes especially critical when the magnitude of its gravitational instability increases. The scale of its destruction increases, and a gradation from small-plate and intraplate tectonics to the tectonics of lithospheric plates can be observed.

Paleogeodynamic reconstructions have also discovered that periodic geochemical crises and geodynamic catastrophes associated with the emergence of new generations of Earth's gravitational instability structures against the background of

O. V. Petrov, *The Earth's Dissipative Structures*, Springer Geophysics,
https://doi.org/10.1007/978-3-319-93614-7_5

Table 5.1 Generations of the structures of gravitational instability of the earth

Order of generation	Order of the structure of gravitational instability	Geochronological time (Ma ago)			Diameter (km)	Area (km^2)	Tectonic regimes	Structure type of gravitational instability
		Deposition	Decline	Development time				
First–third	8–10	Undivided generations: Middle Lopian (more than 3500–3150) And two Late Lopian (3150–2500)		More than 1000	Small cells, diameter of tens and hundreds	$(1.0–1.5) \times 10^4$	Greenstone belts: isometric, cellular (Rhodesian and North Asian types), linear extended (North Canadian type)	G-hexagons, L-hexagons and rolls
Fourth	7	Early Proterozoic (2500)	Late Proterozoic (1650)	850	$(1.25–1.5) \times 10^3$ and $(1.0–2.0) \times 10^2$	$(1.0–1.5) \times 10^6$ and $(1.0–1.5) \times 10^4$	Intracratonic ensimatic geosynclinal and protoaulacogenic. Ensialic geosynclinal and nongeosynclinal basement processing (TTP). Intracontinental basins with the crust of transitional and oceanic types, with the development of late greenstone belts	G-hexagons and L-hexagons
Fifth	6	Early Riphean (1650)	Late Riphean (1030)	620	$(2.0–2.5) \times 10^3$	$(3.0–4.0) \times 10^6$	Intracratonic ensimatic and ensialic geosynclinal, aulacogenic, intraplate depressions, of the nongeosynclinal tectono-thermal basement processing (TTP). Accretionary marginal plate	

(continued)

Table 5.1 (continued)

Order of generation	Order of the structure of gravitational instability	Geochronological time (Ma ago)			Diameter (km)	Area (km^2)	Tectonic regimes	Structure type of gravitational instability
		Deposition	Decline	Development time				
Sixth	5	Late Riphean (1030)	Middle Cambrian (500)	530	$(4.0–6.0) \times 10^3$	$(1.25–2.0) \times 10^7$	Intracratonic ensimatic and ensialic geosynclinal, aulacogenic, of the nongeosynclinal tectono-thermal basement processing (TTP). At the level of the middle of the Late Riphean—rift induced and marginal plate	G-hexagons and L-hexagons
Seventh	4	Late Cambrian (500)	Early Devonian (416)	84	8.5×10^3	$(6.0–7.5) \times 10^7$	Global rift induced and marginal plate accretionary. Rift induced and intraplate	G-hexagons and L-hexagons
Eighth	3	Early Devonian (416)	Middle Jurassic (161)	255	1.4×10^4	8.5×10^7		
Ninth	2	Late Jurassic (161)	Late Cretaceous (99)	62	2.4×10^4	1.7×10^7		
Tenth	1	Late Cretaceous (99)	Present time	–	3.4×10^4	5.1×10^8		G-hexagons

smooth evolutionary structural changes over hundreds of millions of years have been observed in geological history. The periodicity of such "geodynamic crises" covers different time intervals, from 850 million years in the Early and Late Proterozoic to 62 million years in the Mesozoic (Table 5.1). Qualitative reconstructions of the Earth's gravitational instability structures occurred during relatively few, significantly time-limited phases of tectogenesis, which had a global spread. Each subsequent generation of the structures of the Earth's gravitational instability emerges catastrophically and is not determined by the evolution of the preceding one, but rather takes place as an independent episode. At the same time, in the intervals between different phases, structural dislocations (for example, folding, rifting, etc.) takes long timespans, sometimes comparable to the duration of the existence of one or another generation of the structures of the Earth's gravitational instability (the geodynamic cycle). These data on the duration of the folding process once again underline the inconsistency of Stille's conclusions about orogenic (fold-forming) phases, which in his view were presented as short-term episodes (1964).

The accepted cyclicity in geodynamic evolution of the Earth is manifested in the repetition of very similar oscillatory motions, block movements, folding, horizontal movements, volcanism and sedimentation. However, with the increase of the structural size in each cycle, all these processes take place in a different environment or on a different basis. Therefore, the cycles do not completely reproduce those having taken place previously. The boundaries between the main geodynamic cycles are determined by the moments of the emergence of a new structural generation of gravitational instability of the Earth. These moments cause the transition of diverse regions into tectonic regimes different from those preceding them (Table 5.1).

Thus, the regularity in the predominant spatial distribution of different types of hierarchies of gravitational instability structures in the eastern and western hemispheres was revealed during the geological history. The information about the character of the distribution of the sialic protocrust and mantle plumes as regards the global structures of G-hexagon and L-hexagon gravitational instability was obtained. All these data indicate the emergence of these global structures, which, in turn, reflect the main heterogeneity of the Earth and determine its fundamental asymmetry (its partition into Indo-Atlantic and Pacific tectonic segments) in the early Archean, immediately following the isolation of the liquid core. The further existence of fractal hierarchies of these global structures for more than 3500 million years turned out to be closely related to the thermal history of the Earth and is determined both by internal sources of energy and matter and by external heat sinks (nonequilibrium cooling of the Earth through its surface). Such particularities are typical for the so-called open nonequilibrium thermodynamic systems, which, unlike closed systems, can exchange energy, matter or information with the environment. The appearance of these nonlinear terms (sources and sinks of heat) drastically changes the behaviour of the system under consideration and forces us to use such nonequilibrium thermodynamic concepts as "fractal dimensions", "chaos" and "strange attractors" in describing the global phenomena of the gravitational instability of the Earth.

The cooling of the lithosphere and mantle through the surface of the Earth with time leads to a nonequilibrium increase in the density of lithospheric and mantle rocks at increasingly depths. This increase, in turn, causes defects in their gravitational stability (Rayleigh-Taylor instability) with respect to the underlying shells and global structures of gravitational instability of the Earth manifested as a G-hexagons in the eastern hemisphere and L-hexagons in the western hemisphere. Therefore, the emergence of each new generation of structures of the Earth's gravitational instability is preceded by numerous irregular chaotic fluctuations of gravitational instabilities that make a sort of scan or test of the common "background" and activate the global structures of gravitational instability. Due to the density inversion, each perturbation will increase with time; therefore, it is possible to calculate the fractional increase in the amplitudes of these wavelengths using mathematical analysis (Danes 1964; Selig 1965; Biot and Ode 1965; Ramberg 1968, 1972; Fletcher 1972). The system is dominated by perturbations that grow at maximum speed and give a rise to the predominant high-amplitude perturbations, the characteristic size of which is referred to as the dominant wavelength. As was already mentioned earlier, the dominant wavelength and the amplitude increment of the predominant perturbations depend primarily on the thickness of the upper, gravitationally unstable horizons. In other words, an increase in the relative thickness of the upper horizon causes an increase in the dominant wavelength and growth rate of the perturbation (Biot and Ode 1965; Ramberg 1968). In this case, the multi-ordinal structures of the Earth's gravitational instability emerge during geological history as a result of the selection of the dominant wavelengths commensurate with the thickness of the overlying nonequilibrium cooled, gravitationally unstable layers. Throughout the whole of geological history, the type of gravitational instability structures is controlled during the selection of dominant wavelengths by global structures of the Earth's gravitational instability such as G-hexagons in the eastern hemisphere and L-hexagons in the western hemisphere. Thus, the process of selecting the dominant wavelengths takes place in constant interaction with global structures of the Earth's gravitational instability. This is a very powerful process in terms of allowing us to understand the essence of the emergence of the fractal dimension of the Earth's global gravitational instability structures and the reason why these structures repeat themselves across different scales, causing the predominant confinedness of the respective multi-ordinal structures to the eastern and western hemispheres.

In addition, a similar dependence of the selection of the dominant wavelengths, which are commensurable to the thickness of the overlapping gravitationally unstable layers, was also characteristic of mantle plumes. According to the global seismic tomography data (Maruyamo et al. 1994; Dziewonski and Woodhouse 1989), the superplume corresponding to the areas of the emergence of the lower mantle within the boundaries of the global structures of the Earth's gravitational instability is split into several smaller plumes in the upper mantle, which, in turn, divide into even smaller ones at the base of the plate, and so on. Such a consistent split of the plumes leads to the emergence of fractal clusters.

Thus, the fractal dimension of the global gravitational instability structures of the Earth turns out to be closely related to the probabilistic processes of selecting the dominant wavelengths, which are commensurable with the thickness of the overlapping gravitationally-unstable horizons. At the same time, the key moment in describing the global structures of the Earth's gravitational instability is the results of studying their most recent fractal hierarchies and paleogeodynamic reconstructions. These reconstructions show that, beginning at 3800 million years ago, at each subsequent stage of the Earth's evolution, along with irregular instabilities, increasingly large size generations of G-hexagons in the eastern hemisphere and L-hexagons in the western hemisphere prevail, i.e., some structural regularity or self-organisation emerges in the system. This regularity is often referred to in terms of dissipative structures to emphasise the role played in its formation by the dissipative processes associated with energy dissipation (Nicolis and Prigozhin 1979, 1990; Prigozhin and Stengers 1986; Kurdyumov et al. 1989).

The formation of the dissipative structures of the Earth is energetically very productive. On the one hand, they act to transform the energy additionally supplied to the system due to the nonequilibrium superadiabatic cooling of the overlying layers and the differentiation of the outer liquid core. On the other hand, they sharply increase the heat transfer of the environment and ensure the practical uniformity of heat flow on its surface. Upon that dissipative structures present themselves as a convective form of heat transfer and manifest irregularly or chaotically at highly diverse depths.

Strictly speaking, the fractal hierarchies of the global structures of the Earth's gravitational instability considered in this paper are in some way the organised processes of its global gravitational instability. These processes are associated with both internal sources of energy and matter (the energy of decay of radioactive isotopes, gravitational energy released during the formation and growth of the Earth's core, the internal energy of tidal interaction with the moon, the energy of chemical-density, crystallisation, thermodiffusion and other process) and external heat sinks (the nonequilibrium cooling of the Earth through its surface). And the very word "structure" with respect to the processes mentioned above should be quoted, since in terms of fractal sets such objects cannot be either points or curves or surfaces, or in general a topological manifold. Such objects, often referred to as strange (or chaotic) attractors, appear to be a natural model of the emergence of chaotic behaviour resembling turbulence. And the very presence of a strange attractor is the criterion for the existence of an irregular chaotic regime in the geodynamic system under consideration.

The first example of such objects was discovered in 1963 by the American meteorologist Edward Lorenz, for which he was awarded the Nobel Prize. His model provides a simplified description of the thermal convection in the Earth's atmosphere and leads to an attractor endowed with all the basic features of turbulent convection. This phenomenon, common to the terrestrial atmosphere, produces certain difficulties in the area of weather forecasting. Moreover, it follows that the weather and the climate are inherently unpredictable, since they have a basic property of chaotic dynamics—an essential dependence on the initial conditions.

The work of Lorenz remained unnoticed until the 1970s; however, since then, a veritable explosion has taken place in the field of chaotic dynamics, leading to the emergence of new results confirming the relevance of his work. Ideas simultaneously came from the basis of many disciplines, of which the contributions of pure mathematics, a wide range of computer modelling studies, theoretical and experimental physics and chemistry, as well as population biology should be noted. These ideas led to interesting "prototypical" equations and further to descriptions of chaotic behaviour, as well as a number of "scenarios" for achieving a chaotic attractor. And now, as many studies have shown, the attempt to combine dynamics and thermodynamics in the Earth sciences can provide us with a new and amazing example of the existence of a strange (or chaotic) attractor over a geological evolutionary timescale. This strange attractor can be seen to determine the development of processes linked to the gravitational instability of the Earth, connected both with internal sources of energy and matter, as well as with external heat sinks.

The nature of this chaotic attractor can be understood by a simple experiment with Bénard cells, which in this case can serve as a prototype of the phenomena of thermogravitational instability. As Nicolis and Prigogine (1990) showed, on the one hand, the experiment with Bénard cells is characterised by an ideal reproducibility, since under the same conditions, the exceeding of a certain critical value of the temperature difference ΔT_C always leads to the appearance of a convection pattern. On the other hand, as can be seen from Fig. 5.1, the substance is structured into cells with alternately right- and left-handed rotation.

Once established, the direction of rotation is subsequently saved. And yet, as Nicolis and Prigogine (1990) note, no matter how we try to reproduce the experimental conditions, after reaching the critical threshold value ΔT, two qualitatively different situations are possible. As soon as ΔT slightly exceeds ΔT_C, a cellular structure of gravitational instability emerges. Thus, this phenomenon is subject to strict determinism. On the contrary, the direction of rotation in the cells is unpredictable and uncontrollable. Only a random event, represented by various perturbations that dominate at the beginning of the formation of these cells, decides what will be the movement in the given cell: right or left-handed (Fig. 5.2).

Since the formation of cells of gravitational instability in the dissipative structures of the Earth occurs irregularly at entirely varied depth levels, the number of macroscopic spatial and temporal scales is so great that the behaviour of the system seems completely chaotic. Such a regime leads to an attractor endowed with all the basic features of turbulent convection.

Thus, we can see that far from being at equilibrium, i.e. with a significant constraint, the system can adapt itself to its environment in several different ways. That means that several different solutions are possible with the same values of the parameters and only a random event will determine which of these solutions will be implemented. The fact that only one was chosen from many possible variants, gives the system a "historical dimension", a kind of "memory" about the past event that

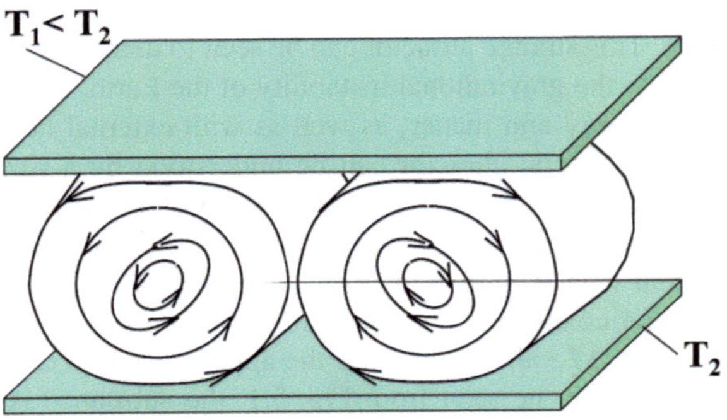

Fig. 5.1 Two images of convection (Bénard) cells. The emphasis is on the opposite directions of movement inside two neighbour cells (Nicolis and Prigogine 1990)

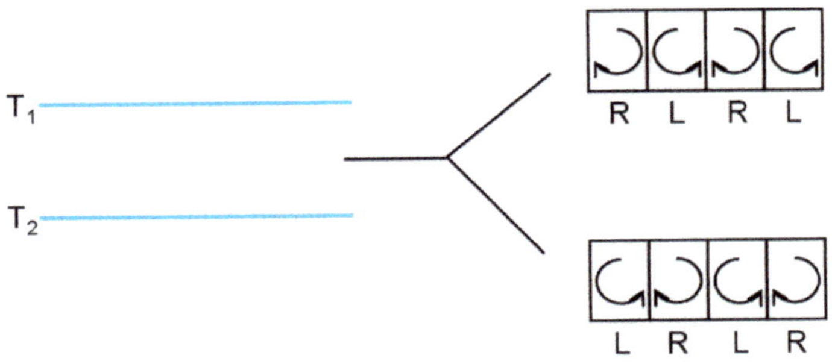

Fig. 5.2 The multiplicity of solutions above the cutoff of the thermal convection instability. Depending on specific conditions this spatial region may be a part of a cell with a left or right rotation orientation (Nicolis and Prigogine 1990)

occurred at a critical moment and influenced subsequent evolutionary processes (Nicolis and Prigozhin 1990).

Summarising the above, both with respect to the Bénard experiment, and to the global phenomena of the Earth's gravitational instability, a graph known as the bifurcation diagram can be obtained (Fig. 5.3). The only solution "α-stable", corresponding to the thermodynamic branch, loses its stability at λ_c. With this value of the control parameter, new branches of the solution appear (b_1, b_2), which in this case are stable.

With the small values of the control parameter λ, only one solution is possible, corresponding to the quiescent state in the Bénard experiment or global phenomena of the Earth's gravitational instability. This solution is a direct extrapolation of thermodynamic equilibrium and, similarly to equilibrium, is characterised by an important property—asymptotic stability. This follows from the fact that, due to the density layering of the Earth and the presence of discontinuous density boundaries, the system is capable of extinguishing internal fluctuations or external perturbations at a certain depth level in this region. For this reason, we shall call such a branch of states a thermodynamic branch (Nicolis and Prigozhin 1990). When passing through the critical value of the control parameter λ_c (Fig. 5.3), the states on this branch become unstable. The reason is that fluctuations or irregular gravitational perturbations are no longer extinguished at a certain depth level, but excite global structures of the Earth's gravitational instability. Acting like an amplifier, the system deviates from the stationary state and passes to a new regime in the case of the Bénard experiment, corresponding to the state of stationary convection. And, in the case of global gravitational instability, the system passes to the appearance of a

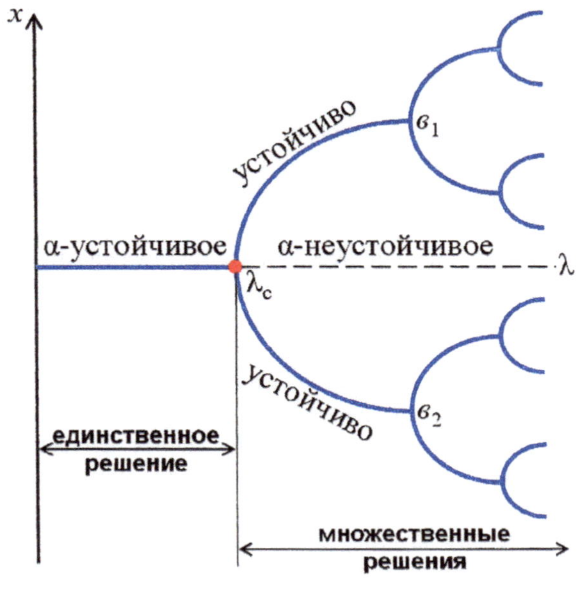

Fig. 5.3 Bifurcation diagram showing the effect of alteration of the control parameter λ on the state variable χ. Explanations are given in the text

new generation of dissipative structures of the Earth commensurate with the thicknesses of the gravitationally unstable horizons.

Both these regimes merge at $\lambda = \lambda_c$ and differ at $\lambda > \lambda_c$, resulting in the phenomenon referred to as bifurcation (Nicolis and Prigozhin 1990). It is easy to understand the reasons why this phenomenon should be associated with catastrophic changes. In fact, at the crucial point of the transition, the system must make a critical choice (in the neighbourhood of $\lambda = \lambda_c$), which in Bénard's problem is associated with the appearance of right- and left-handed cells in a certain spatial region (Fig. 5.3, b_1 or b_2). Whereas in the phenomena of gravitational instability, it is associated also with the emergence of various types of structures of G-hexagons, L-hexagons or Rolls, as well as with the probabilistic selection of the dominant wavelengths commensurate with the thicknesses of gravitationally unstable horizons. There is nothing in the initial parameters of the system that would allow the observer to determine in advance which state will be selected. Only a random event will determine it through the dynamics of the fluctuations. These instability fluctuations become especially interesting near the bifurcation point, where long-wavelength spatial correlations emerge, i.e., statistically reproducible ratios between the remote parts of the system. Such systems seem to "fluctuate" before choosing one of several paths of evolution. Here a striking analogy arises between the instability of nonequilibrium origin and the phase transitions (Nicolis and Prigozhin 1979). These two phenomena are closely connected in the Earth's interior, since its internal structure assumes multiple phase transitions of both the first and second order.

Paleogeodynamic reconstructions have also discovered that periodic "crises" and "geodynamic (thermogravitational) catastrophes" associated with the emergence of new generations of dissipative structures against the background of smooth evolutionary structural and temperature changes over hundreds of millions of years have been observed in geological history. The periodicity of such "thermogravitational crises" covers different time intervals, from 850 million years in the Early and Late Proterozoic to 62 million years in the Mesozoic (Table 5.1). Analysis of paleogeodynamic reconstructions indicates that qualitative rearrangements of dissipative structures occurred in relatively few and narrowly limited in time phases, which had a global spread. Moreover, each subsequent generation of the dissipative structures emerges catastrophically and is not determined by the evolution of the preceding one, but takes place as an independent episode. This behaviour refers to nothing more than self-organisation and a "forgetting" of the details of the initial data. It points to the fact of the spontaneous emergence of order in the system, the formation of localised structures, as well as of the possibility of identifying several basic order parameters, to which all the others are adjusted.

Therefore, the emergence of each new generation of dissipative structures is preceded by numerous irregular chaotic fluctuations of gravitational instabilities that make a sort of scan or test of the common "background". As it was noted above, up to a certain depth level these fluctuations are extinguished as a result of the density layering of the Earth. However, with the further non-equilibrium cooling of the Earth, these fluctuations increase at deeper levels, activating the global structures of gravitational instability of the G-hexagon type in the eastern hemisphere and of the L-hexagon type in the western hemisphere. These global structures of gravitational instability remain dominant throughout the geological history of the Earth. And all irregular chaotic gravitational instabilities (of local, regional and supra-regional levels) associated with both internal sources of energy and matter, as well as with external heat sinks, will tend to this attractive set. At the same time, both mantle plumes and heterochronous dissipative structures of the Earth emerge during geological history as a result of the selection of the dominant wavelengths, commensurate with the thicknesses of the overlying, nonequilibrium cooled and gravitationally unstable horizons. Moreover, the type of gravitational instability structures and the character of the distribution of mantle plumes are controlled during the selection process of the dominant wavelengths by global structures such as G-hexagons in the eastern hemisphere and L-hexagons in the western hemisphere. This or that fluctuation, stabilised in accordance with the thickness of the gravitationally unstable horizon and corresponding to the critical values of the control parameter λ_c, transforms the system into a kind of historical object in the sense that its further evolution will depend on this critical choice. The second law of thermodynamics, the law of increasing entropy, fulfils here the role of the function of the selection rule: the restriction of the initial conditions of the geodynamic processes that propagate at subsequent instants in accordance with the laws of dynamics.

In addition, the model of probabilistic selection of the processes of structuration as a result of the selection of the dominant wavelengths, commensurate with the thickness of the overlying nonequilibrium cooled, gravitationally unstable horizons under the conditions of a decreasing heat flux, leads to the generation of dissipative structures increasingly large in size at each subsequent historical stage. This allows us to take into account, on a single physical basis, the entire variety of tectonic regimes that successively follow each other during geological history. What is more, it is due to the presence of various types of thermogravitational instability structures that particular aspects of the geodynamics of Laurasia and Gondwana are explained. In this sense, the dissipative structures of the Earth appear as historical categories. Therefore, many endogenous tectonic regimes associated with them are evidentially manifested only in the past and inherently cannot be detected in the contemporary state of the lithosphere. As evidenced by paleogeodynamic reconstructions, the emergence of the generations of dissipative structures of increasingly large dimensions at each subsequent historical stage increases the significance of horizontal tectonic movements of the lithosphere. With an increase in the size of the structures, the tectonic regimes associated with them also change, ranging from fine-cellular greenstone and granulite, protoaulacogenic and protogeosynclinal to global riftogenic and marginal-plate types (Table 5.1). In this connection, the tectonics of lithospheric plates (or plate tectonics), at least as it is represented nowadays, began to manifest itself only in the late Mesozoic. Thereby, the discovery of the dissipative structures of the Earth forms the theoretical basis for the development of a large variety of paleogeodynamic settings that determined the tectonic development of the Earth, and points to the limited approaches and the impossibility of understanding Earth's tectonic history solely from the standpoint of modern plate tectonics.

Global gravitational instability also provides a good explanation for what causes the emergence of the Earth's magnetic field and the chaotic changes in its orientation over the course of geological history. In order to understand the physical essence of the geodynamic processes occurring in the liquid core, leading to a chaotic change in the orientation of the Earth's magnetic field, let us return to the fractal dimension of global structures and Earth's gravitational instability. On the one hand, it is obvious that the very existence of a global gravitational instability of the G-hexagon type in the eastern hemisphere and of the L-hexagon type in the western core-mantle boundary will cause forced convection in the liquid core (Fig. 3.3). On the other hand, when each point within the structure is the limiting one and it is not possible to draw the boundary of the structure in the form of a dividing line, the fractal (fractional) dimension of the global gravitational instability of the Earth will obviously provide a chaotic change of directions of convective motions in the liquid and dynamic core with the course of geological time. Additionally, this process results in the chaotic change in the orientations of the Earth's magnetic field.

In order to simulate this surprising natural phenomenon, Rikitake proposed in 1958 an original physical model, now known as the Rikitake Dynamo. This represents two dynamo discs connected so that the current from each disk passes through the coil of the other. The same torque acts on the shafts. Disks can be considered in this model as representing two large vortices in the core of the Earth. It turns out that in this simple model of two rotating disks with interacting magnetic fields, chaotic regimes of polarity reversal of the magnetic field also exist. These correspond to strange attractors in one system of three ordinary differential equations (Ershov et al. 1989). An example of a strange attractor in the Rikitake dynamo model is shown in Fig. 5.4, with its species projection resembling two glued bands.

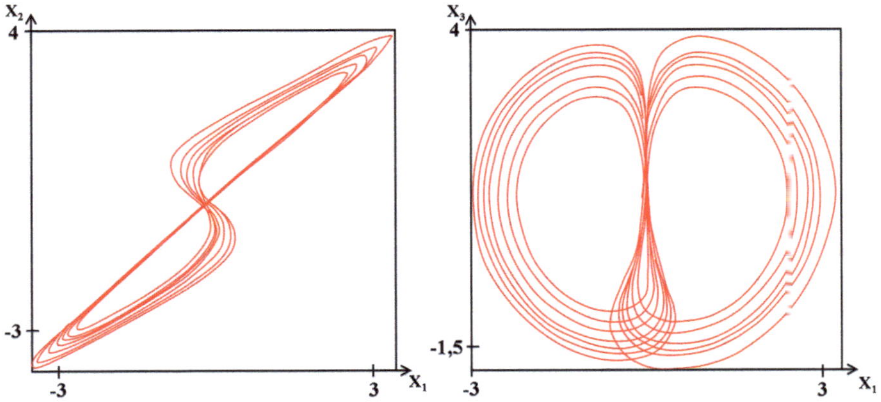

Fig. 5.4 Two projections of a strange attractor in a system of equations describing the Rikitake Dynamo

If we follow the point characterising the state of the system while it "runs" along the attractor on a display screen, then we see that it randomly falls on the left or right band.

However, the strangeness of the chaotic attractors is not only confined to its exotic appearance. Like the objects they represent, the attractors also have remarkable geometric and dynamic properties—a Cantor or fractal structure that repeats itself on a smaller scale. If we dissect the attractor bands in Fig. 5.4 with a certain surface, then with a successive decrease in the scale of the examination it turns out that each of the bands splits into two, then again in two, and so on. The strange attractor in this section is constructed in the same way as the Cantor set. An extraordinary geometric object constructed at the beginning of the century to serve as an argument in abstract mathematical reasoning, turned out to be typical for the behaviour of the most of the natural systems.

Thus, we can state that the unification of dynamics and thermodynamics in the Earth sciences as a result of the discovery of its global gravitational instability structures, which have a fractal dimension, solves the problem of truly global coverage of irregular chaotic, often simultaneously acting and seemingly completely independent geodynamic phenomena. Forced convections in an external liquid core with a chaotic change in the orientations of convective motions are of this type; consequently, the orientation of the Earth's magnetic field corresponds to the fundamental asymmetry of the Earth, which divided it back in the early Katarchean into the Indo-Atlantic and Pacific tectonic segments. Here we may also discuss the distribution of mantle plumes and their projections on the Earth's surface; the hot belt and hot spots of the Earth at an early stage of its geological history (Protopangeia), which provide modern indicators of the advective heat and mass transfer processes; the formation of multi-ordinal and polytypic hierarchies of dissipative structures of the Earth, determining during its geological history a great variety of paleogeodynamic settings and endogenous tectonic regimes, as well as influencing the anomaly rearrangements of the geoid and the relief of the Earth's surface. At the same time, the energy side of the evolution of global fractal structures of gravitational instability responsible for the geodynamic excitation of the Earth's interior is determined by the secular cooling of the Earth. This cooling is, in turn, a consequence, on the one hand, of nonequilibrium cooling through its surface; on the other hand, it is associated with a gradual depletion of radioactive elements, the depth differentiation slowdown of the outer liquid core, as well as with a decrease in the intensity of the lunar tides, and so on. Secular cooling of the Earth is a directional and irreversible thermodynamic process; therefore, its geological history bears lines of direction and irreversibility. However, the controlling effect of global fractal structures of gravitational instability, or cellular standing inner gravity waves of the Earth, can be seen at all historical stages of its geological evolution. Possessing a fractal dimension based on well-studied processes of decomposition of wave vibrational motions into modes, these global structures consist of a natural chaotic (strange) attractor. Therefore, all irregular chaotic instabilities inside the Earth, including the lithosphere, mantle and core, tend to this attracting set during the geological history. At first glance it seems incredible and paradoxical that global fractal structures of the Earth's gravitational instability can control such a variety of irregular chaotic geodynamic phenomena. However, this is the consequence of the revolutionary discovery of the manifestations of the wave properties of matter for these new geodynamic objects, which determined the transition from the time-reversible Newtonian description of the geological structuring processes to a probabilistic description based on their wave nature.

References

Danes Z.F. Mathematical formulation of saltdome dynamics. Geophysics. 1964; 29:414–424.
Selig F. A theoretical prediction of salt dome patterns. Geophysics. 1965; 30:633–643
Biot M.A., Ode H. Theory of gravity instability with variable overburden and compaction. Geophysics. 1965; 30: 213–227.
Ramberg H. Instability of layered systems in the field of gravity, parts 1 and 2. Phys. Earth and Planet. Inter. 1968; 1:427–474.
Ramberg H. Theoretical models of density stratification and diapirism in the Earth. J. Geophys. Res. 1972; 77:877–889.
Fletcher R.C. Application of a mathematical model to the emplacement of mantled gneiss domes. Am. J. Sci. 1972; 272:197–216.
Dziewonski A.M., Woodhouse J.H. Threedimensional Earth structure and mantle convection. Abstracts. 1989; 1(28th IGC):427–428.
Nikolis G., Prigogine I. Samoorganizatsiya v neravnovesnykh sistemakh. Ot dissipativnykh struktur k uporyadochennosti cherez fluktuatsii. M.: Mir; 1979 [Nikolis G., Prigogine I. Self-organisation in nonequilibrium systems. From dissipative structures to ordering through fluctuations. M.: Mir; 1979 (In Russ.)].
Prigogine I.R., Stengers I. Poryadok iz khaosa: novyi dialog cheloveka s prirodoi. Obshch. red. V. I. Arshinova, Yu.L. Klimontovicha i Yu.V. Sachkova. M.: Progress; 1986 [Prigogine I.R., Stengers I. Order out of chaos: Man's new dialogue with nature. Ed. V.I. Arshinov, Yu.L. Klimontovich and Yu.V. Sachkov. M.: Progress; 1986 (in Russ.)].
Nikolis G., Prigogine I. Poznanie slozhnogo. M.: Mir; 1990 [Nikolis G., Prigogine I. Exploring Complexity. M.: Mir; 1990 (In Russ.)].
Kurdyumov S.P., Malinetskii G.G., Potapov A.B. Sinergetika - novye napravleniya. M.: Znanie. 1989; 11 [Kurdyumov S.P., Malinetskii G.G., Potapov A.B. Synergetics - new directions. M.: Znanie. 1989; 11 (In Russ.)].
Ershov S.V., Malinvetskij G.G., Ruzmaikin A.A. A generalized twodisk dynamo model. Geophys. and Astrophys. Fluid. Dyn. 1989; 47:251–277.
Maruyama S., Kumazawa M., Kawakami S. Towards a new paradigm on the Earth's dynamics. J. Geol. Soc. Japan. 1994; 100(1):1–3.
Stille H. Izbrannye trudy. M.: Mir; 1964. [Stille H. Selected works. M.: Mir; 1964 (In Russ.)].

Chapter 6
Applied Aspects of Dissipative Structuring Theory to the Nonequilibrium Geological Environment

The manifestation of the wave properties of matter in the processes of spontaneous structuring of density-unstable masses, originally described as a natural phenomenon characteristic of both the nonequilibrium density-unstable laboratory physical models, and the geological planetary layers and upper horizons of the Earth's crust, is the basis for new approaches to geological mapping and to the interpretation of geological and geological-geophysical data at all stages of their processing. One of the characteristic properties of cellular standing internal gravitational waves, which determines the wave character of the dissipative structuring mentioned above, is the presence of fractal dimensionality in regular geological structures that reveals itself on the Earth's surface as a system of lineaments, a global regmatic fault network and planetary fracturing. As is known, all rocks, regardless of their origin, age and location, are intersected by a dense network of fractures (diaclases), which, like the regmatic fault network, is naturally oriented with respect to terrestrial coordinates. The lineaments marking the geological fracturing form a hexagonal system with splitting into pairs NW–SE and NE–SW, WNW–ESW and NNW–SSE and transverse to them directions (Khain 1971; Odessky 2005; Anokhin 2006). We consider fractal theory and fractal geometry, as applied to these lineaments and microfractures, as an abstract mathematical reflection of the manifestation of the fundamental wave properties of matter—namely, as a result of the expanding the wave oscillatory process into modes and the formation of self-similar structures. The analysis of the regmatic fault and microfracture network, based on fractal theory, sheds light on the fundamental wave properties of density-unstable media: the formation of three-dimensional cellular (or hexagonal) standing internal gravity waves and their subsequent expansion into modes accompanied by the emergence of a hierarchy of geoblock divisibility of the lithosphere. The methods of fractal analysis, revealing the characteristic scales of self-similar geoblocks, enable us to solve the inverse problem of establishing deep levels of density instability of matter in the interior of our planet.

According to Mirlin (2001), the fractal discreteness of the lithosphere can be considered as an eligible term within the contemporary specialised geological

O. V. Petrov, *The Earth's Dissipative Structures*, Springer Geophysics,
https://doi.org/10.1007/978-3-319-93614-7_6

language, describing the universal natural phenomenon of the domination of the self-similar regular objects in the interior of the planet that possess the deterministic fractional dimensionality within their individual families. The fractal objects, like wave processes, are characterised by self-similarity, that is, the similarity of its parts to the whole, thus reflecting the fundamental wave properties of matter—the expansion of the wave oscillatory process into modes—within the framework of our representations. Since traditional geology, despite the development of a system of geodynamic models using fundamental physical laws alongside a mathematical apparatus, is still a descriptive science, an empirical approach remains very popular until nowadays for the study of this phenomenon. The latter comprises, as a rule, the study of the subordination of structural objects at various scales with the calculation of fractal dimensionality, taking the form of a coefficient that determines their spatial hierarchy. The importance of this approach from the perspective of new concepts is determined by a number of classification problems that solve the manifestations problems of the wave properties of matter in nature.

As we noted above, one of the unexpected examples of fractals is the coastline of continents and islands, which prompted Mandelbrot (1977) to think about the fractal geometry of nature. The fractal nature of these objects is indicated by their geometric commonality: a lozenge-shaped or similar form consisting of separate segments separated by disjunctive families or individual large faults. When measuring the length of a coastline L, the complex intersected coastline is replaced by a broken line consisting of units having a length of at least ε. For ordinary smooth curves with decreasing ε, the value $L(\varepsilon)$ tends to a certain amount, which comprises the length of this curve. However, for the British coast, it turned out that L depends on ε (in the interval from 10 to 1000 km) according to the power law $L(\varepsilon) \approx \varepsilon^{1-D}$, $D \approx 1.3$. On the basis of this methodology, the dimensionalities of the coastlines were calculated for the German coast $D \approx 1.12$, Portugal $D \approx 1.12$, Australia $D \approx 1.13$ and South Africa $D \approx 1.04$. A similar dependence determines the length of the other coasts as well, the dismemberment of the relief of the ocean and river beds, the path traversed by the particle under Brownian motion, and so on. As a result, along the continent—ocean boundary, a characteristic step-like pattern with different values of "steps" or different displacement amplitudes along the fault planes can now be observed. The multi-scale structures of the pull-apart crests of the mid-ocean ridges, delimited by transform faults, also proved to be related by identical dependencies (Mirlin 2001).

Similar dimensional dependencies and self-similarity effects of shapes are established for the set of terrains of the northeast of Russia $D \approx 1.377$ and the western part of Alaska $D \approx 1.220$ (Vadkovsky et al. 2002). In the opinion of these authors, the presence of a power law of the distribution of areas and perimeters of many terrains and the calculated fractal dimensionality indicate the absence of any identified characteristic sizes of geological bodies in the range from the first kilometres to thousands of kilometres, that is, from individual geoblocks to entire continents. Thus, a fairly narrow range of values of fractal dimensionality for such different objects of the Earth's crust provides grounds for assuming the unity of the wave mechanisms that underlie their interaction—the merging and separation of continents.

The first major synthesis of works connected with the application of fractals in studying the relief of the ocean floor with the use of wave functions was undertaken by Malinverno (1995). His main postulate concerns the waveform of the Earth's surface relief forms, namely, that this surface can be regarded as a superposition of two components—a deterministic long wave and random short waves. In the latter case, in particular, this relates to the possibility of describing the anisotropy of the morphology of the system of abyssal hills on the basis of statistical methods and fractal representations, as well as the construction of a model for their formation along with a determination of the dominant of tectonic or volcanic processes.

In the study of volcanoes, fractal analysis was used to parameterise the periodicity of eruptions and quiet periods of activity of hot spot type volcanoes such as Mauna Loa and Kilauea in the Hawaiian Islands as well as Piton de la Fournaise on the island of Réunion. Attention is drawn to the close connection between the periodic oscillations and wave processes, with the latter consisting in the propagation of oscillations in space; that is, if fractality is a characteristic property of a wave, then in some form or another it must be contained in the oscillations that form the wave. Indeed, two primary modes of activity were also detected in this problem, having short and long periods respectively. Obviously, different regimes of volcanic eruptions can be interpreted as the signs of different dynamics of processes in the magmatic chamber (Dubois and Cheminée 1991).

And yet the most obvious example of the manifestation of the multi-ordinal fragmentation of the lithosphere is a discrete, often co-ordinated and branching system of geological faults and fractures. The conclusion about the possibility of analysing their spatial hierarchy on the basis of fractal images is suggested by the geometric similarity of the fractal branching structure and the tree-like structure of the disjunctive system. In particular, during the tectonic division of the Amur region, a significant complication and condensation of a fault family playing the role of the boundaries of multi-ordinal structures formed due to repeated activations was observed, which led to significant progress in understanding the Baikal, Stanovoi and Mongolo-Okhotsk lineament zones. Disjunctives of the Amur region, represented by data on 3844 faults, are approximated by lineaments of finite length and deterministic strike azimuth (Sherman et al. 2001). A similar fractal analysis of the fault network (more than 3000 faults) was carried out in the work of Semyon I. Sherman and Gladkov (1999) for the Baikal rift zone. In this sense, the faults have the properties of a "coastline map", whose degree of intersection depends on the scale of the map, and increases with increasing image detail, as was discussed in one of the first works on fractal geometry in nature (Mandelbrot 1977). This generalisation effect, manifested in the decay of objects of a planetary scale into families of higher-order formations related to the first by strike azimuth, geometry and the expected depth of their occurrence, is widely known in aero- and planetary geology.

One such example is described by Gzovsky (1971) as applied to the study of the structure of disjunctives (Fig. 6.1). It is noted that, on survey geological maps, on

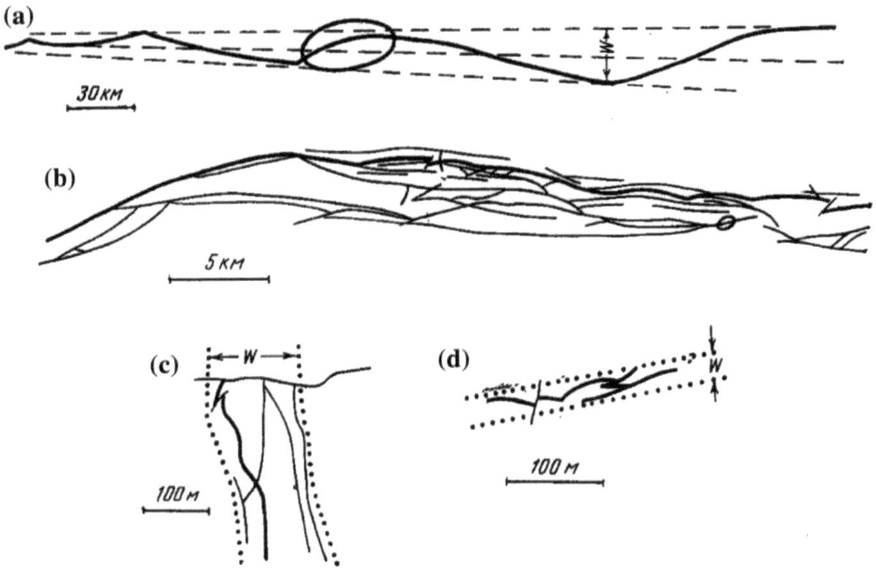

Fig. 6.1 The structure of the geological boundary for various details of its consideration, according to Gzovsky (1971) **a** disjunctive on the overview map; **b** on the detailed map (the area, circled on **a**); **c** on the vertical section of the mine (the area circled on **b**); **d** on the plan of the mine horizon (the area circled on **b**). Explanations are given in the text

which some discontinuity may have the form of a straight line, when the scale is increased by an order of magnitude, the discontinuity is represented by one line of complex shape combining smooth curves with sharp breaks and a peculiar wavelet of the discontinuity along its strike emerges. This author proposes to quantify this wavelet according to the length of the half-wave l (the average distance between the points of maximum deviations from the midline) and its span w (the average deviation amplitude of the fault plane from its midline). Qualitatively marked waves, as M. V. Gzovsky points out, differ in their form—sinusoidal, zigzag and garland-like (alternating arcs with sharp breaks, reminiscent of the cycloid). The knowledge of the fracture wavelet and its quantitative indices allows the disjunctive to be traced along the strike and to the depth. In particular, the wavelet and its span w act as seismic hazard criteria in the construction sites, since they determine the stress concentration points at a distance between 0.5 and 1.0 w from the fracture line. A further increase in the scale of research using the example of the Main Karatau fracture of the Baijansai region indicated that a single wave-like surface of rock destruction is transformed into a fracture zone, characterised by the complex branching, intersections and merges. This complication of the disjunctive structure as the scale increases is a consequence of several stages of their development, on each of which a given part of the Earth's crust was subjected to a deterministic load system.

During a discrete stage, the loads are uneven; therefore, the growth of discontinuities occurs in the form of multiple impulses, referred to in geology as movements. In this case, one particular movement itself consists of five stages:

- the emergence of a network of small preparatory fractures;
- combining a part of the fractures into a single wave-like fracture surface and the appearance of associated shearing surfaces in conditions of the main alteration in the stress state of the surrounding rocks;
- the main displacement along a relatively large fracture surface, causing the formation of associated minor fractures associated with friction between the wings of the fracture;
- prolonged slip along the fracture;
- healing of the cavities formed with a mineral substance along with the restoration of continuity of the geological environment and strengthening of heterogeneity of the structure of this section of the Earth's crust.

Despite the fact that Gzovsky did not explicitly use the terms "self-similarity" and "fractal", nevertheless, the combination of wave-like local fracture surfaces in disjunctives of lower order up to a regional fracture surface that also has a wave-like form is clearly referred to. It is curious here that a combination of wave properties and fractal self-similarity in the fault system is observed. In addition, these properties, as the above review shows, are combined in a wide range of other geological objects—for example, the day relief surface forms, the ocean floor and the relief of the roof of the acoustic basement. This conclusion reinforces the mathematical justification for fractal self-similarity being seen as a characteristic property of cellular internal gravity waves formed in density-unstable media, which in this case is represented by a geological environment. Thus, there is an assumption that the elements of the structure of the geological environment can be reconstructed on the basis of wave and fractal approximations by indirect indications, for example, those given by relief forms or geophysical observations.

6.1 Fractal Nature of Relief Structure, Geological Formations, Geophysical Fields and Elements of Their Wave Representation

Since waves are inherently genetically related to the oscillatory process (its propagation in some extended medium), we are forced to state that the wave nature of the development of geological processes was also indicated by ancient Greek scientists who observed the movements of the shoreline. The same regularity was noted by Mikhal V. Lomonosov; it was emphasised by geologists and geographers of the 19th and 20th centuries: F. Yu. Levinson-Lessing, A. P. Karpinsky, E. Rekmo, S. Bubnov, Ch. Shukhert, G. Shtille, V. V. Belousov, M. M. Tetyaev, A. N. Mazarovich, V. I. Popov, V. E. Khain, N. I. Nikolaev, E. E. Milanovsky,

N. S. Shatsky, V. V. Piotrovsky and many others. Nikolaev (1949), summarising a vast amount of material on the study of the latest and modern tectonic movements of the Earth's crust, came to the conclusion that they are all rhythmic and clearly oscillatory in nature. He understood rhythmicity as a certain repetition of movements; in his opinion, it consisted in "the repetition of oscillations manifesting in time in the form of a full wave: subsidence is necessarily replaced by uplifts, which are like … standing waves …".

In geological history, all these rhythms are superimposed one on another over time. This overlap, proved by the works of many researchers, Nikolaev (1949) depicted graphically (Fig. 6.2). Considering this drawing within the framework of the absolute geochronological scale, it can be seen that those movements referred to as the most recent fit on the scale of the layout, essentially speaking, at one point. However, in terms of amplitude and period, the most recent movements are as diverse as the more ancient movements (Fig. 6.3).

These rhythms in turn are divided into rhythms having a shorter period. For the late and postglacial periods in the St. Petersburg area, fluctuations with a period of the order of 2000–3000 years are revealed. Movements having similar periodicities are established for the south of the European part of Russia. Finally, the consideration of the data obtained by applying instrumental methods indicates rhythms having even shorter periods (Fig. 6.4). These can be evaluated over the years and decades, noting the presence of propagation of both standing and running waves.

In developing the ideas and methods for revealing and studying the new wave structures of the Earth, their three-dimensional wave nature, providing a genetic connection between the surface forms and geological (tectonic) structures, as well as the deep representation of these structures in geophysical fields, acquires a special significance. Despite the fact that, in practical geology, the data on the

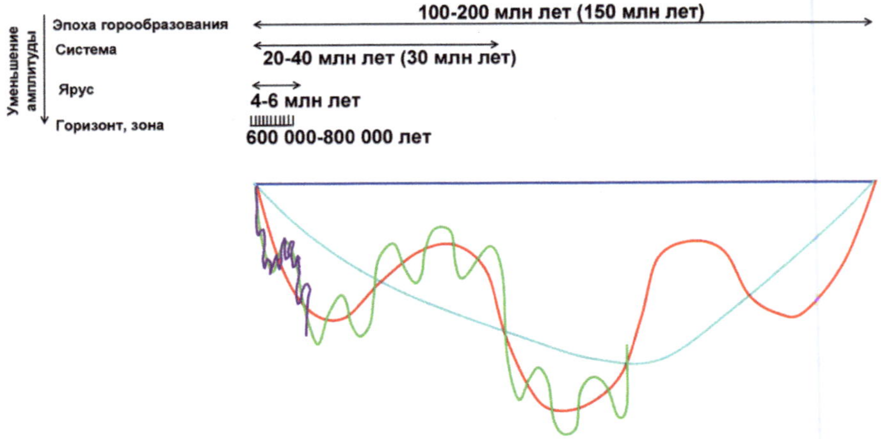

Fig. 6.2 Scheme of eurhythmy of superimposed oscillations of the Earth's crust, according to Nikolaev (1949)

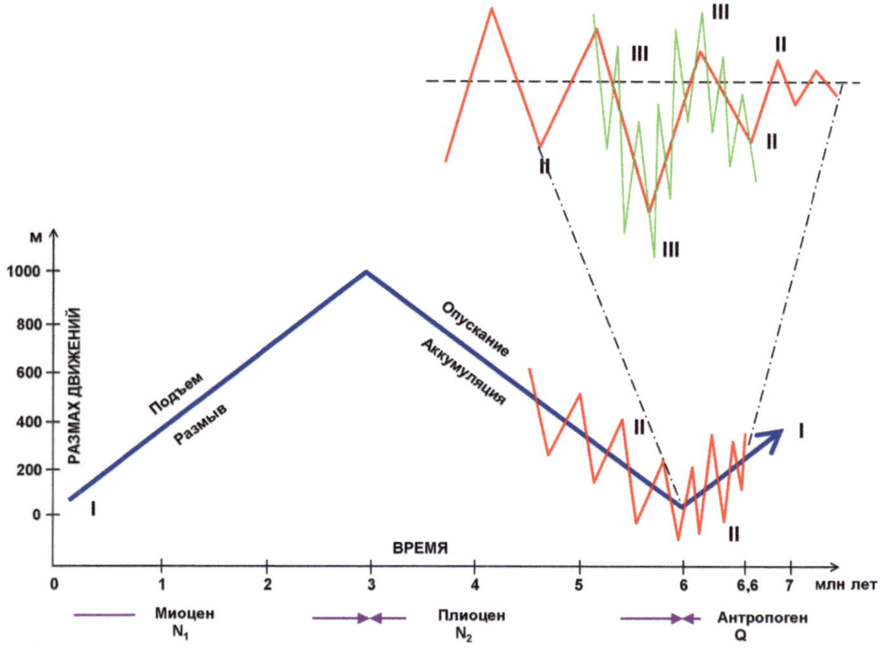

Fig. 6.3 Graph of vibrational tectonic motions with an indication of superposition of the motions of different amplitudes and periods (of different orders—I, II, III) within the European platform, according to Nikolaev (1949] {Legends on Fig. 6.3, r to l, t to b: span of motion, uplift, downthrow, washing-out, accumulation, time, Miocene, Pleiocene, Athropogene, Ma}

nature of the relief were more or less always used for the analysis of geological structures (Gerasimov 1946; Meshcheryakov 1957; Filosofov 1963; Piotrovsky 1964, and others), the true nature of their interrelation becomes obvious only from the standpoint of generalising the features of the manifestation of the wave properties of matter in the processes of spontaneous structuring of density unstable masses. In this case, the consideration of three-dimensional volume elements of the Earth's crust, will obviously not be sufficiently complete without an analysis of their external orographic expression. Curiously enough, many outstanding Russian geologists and geomorphologists have approached similar representations already at the end of the 50s–60s of the last century. At that time, a special branch of knowledge began to form, having tectonically-conditioned, primarily large-scale relief elements as its subject, and studying the relationship between surface forms and geological structures. For the development of this branch of knowledge, the works of Gerasimov (1946) and Meshcheryakov (1957) had particular significance since they introduced the concept of morphostructures into geology and geomorphology and indicated the main types of morphostructures inherent in the image of the Earth. According to Gerasimov (1946), morphostructures are large-scale and at the same time perfectly integrated orographic and structural-geological formations

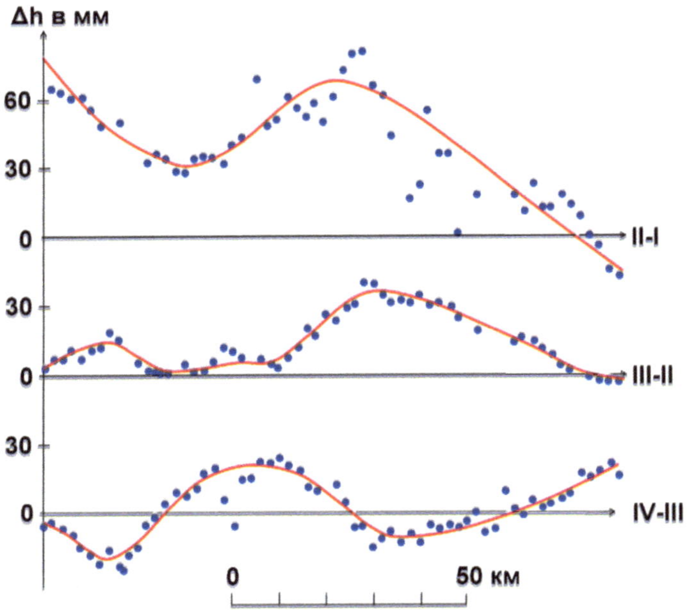

Fig. 6.4 Alteration of the altitude level line from Takasaki to Simosuvi; leveling was produced four times (*I–IV*). According to Nikolaev (1949). *I*—1888; *II*—1925; *III*—1926; *IV*—1932. Curves *II–I* and *III–II* have a common slope to the right, curves *IV–III*—the opposite slope. Point of maximum uplift during the time period can display the maximum uplift or the maximum depression during the following time period, according to Tsuboi (1983)

that are clearly visible on all contemporary composite orographic, hypsometric and geological maps. During the 1960s, Piotrovsky (1964), in placing the relief forms and tectonic structures according to their morphometric indicators, established that the ratios of these indicators are regularly repeated, and the forms themselves are grouped in a general morphometric series consisting of eighteen orders. Within these orders, including the smallest forms of relief such as sand ripples and the tiniest folds developed in plastic clays, shales, etc., with an invariable relationship between length, width and development along the vertical, expressed in the ratio of one to three and to ten, are located the most common forms of relief and tectonic structures of the Earth.

The regularity of the transition from the forms of one order to the next, expressed in an increase of three times, is particularly visible in a relief having a clearly wave origin (sand ripples, barchanic ridges, etc.). Figure 6.5 indicates a scheme of sand ripples formed from local material on a horizontal surface while maintaining a common ground level. The diagram indicates that if we take as the initial two adjacent forms—positive and negative—for one "wave", then the following order will be obtained by merging three such "waves".

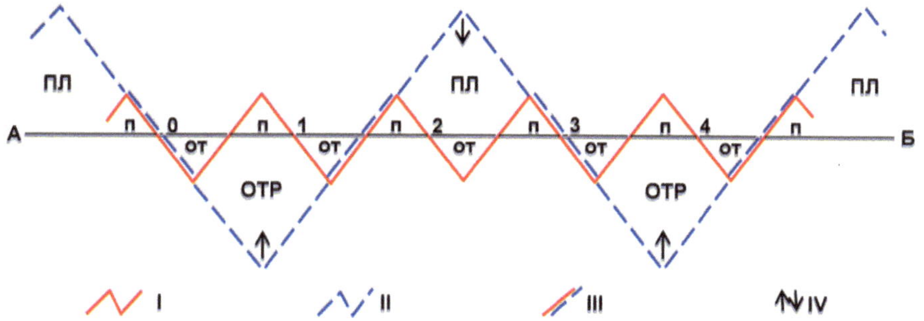

Fig. 6.5 Scheme of addition of "waves" of two adjacent orders. 0, 1, 2, 3, 4—nodes of "waves"; *I*—small "waves"; *II*—large "waves"; *III*—common surfaces; *IV*—in the movement direction—according to Piotrovsky (1964). {Legends on Fig. 6.5, l to r: A, PL, p, ot, p, OTR, ot, p, ot, PL, p, ot, p, OTR, ot, p, PL, B}

Correctly supposing that any form of relief and any tectonic structure have "roots", Piotrovsky (1964) analysed the material at his disposal at that time for locating the centres (hypocentres) of earthquakes (Fig. 6.6). In the graphs constructed on the basis of these materials, a surprising identity with the graphs of dimensionalities of morphometric and tectonic structures constructed by him was found. Their comparison confirmed the assumption that the morphometric series of surface structures breaks down into several orders, and the "roots" of these structures are located in the Earth's crust not in a disorderly manner, but in layers, i.e., in accordance with the well-known from fluid dynamics regularity for standing internal gravitational waves: larger structures are characterised by greater depth of penetration into density-unstable environments. Thus, the morphometric, geological and seismic data obtained before the 1970s made it possible to conclude that a general morphometric series of relief forms and tectonic structural forms is a digital expression of the general (oscillatory, and hence wave) regularity traced in nature.

Under the framework of actively developing digital GIS-technology, these previously disaggregated data are systematised in the form of thematic regional maps. More than thirty years of application in applied computer science has determined the emergence of a wide range of computational algorithms for describing the structure of some of these maps, which have the ultimate formalisation for geology. As a rule, they include maps of geophysical fields. The technique based on these algorithms is oriented mostly towards recalculating the amplitude-frequency characteristics of these fields into their characteristics, not measured by instrumental methods: the strike azimuths of the anomalies, entropy, and so on. Prolonged approbation of the method, many elements of which originate in the cameral methods of manual processing of field material (for example, sliding window averaging, calculation of the upper edge of the geological body using the points of inflection of the anomaly, visual division into spatial variability of the field), allows this technique to be properly referred to as classical. Below are some results of its use with reference to digital maps of the anomalous magnetic and

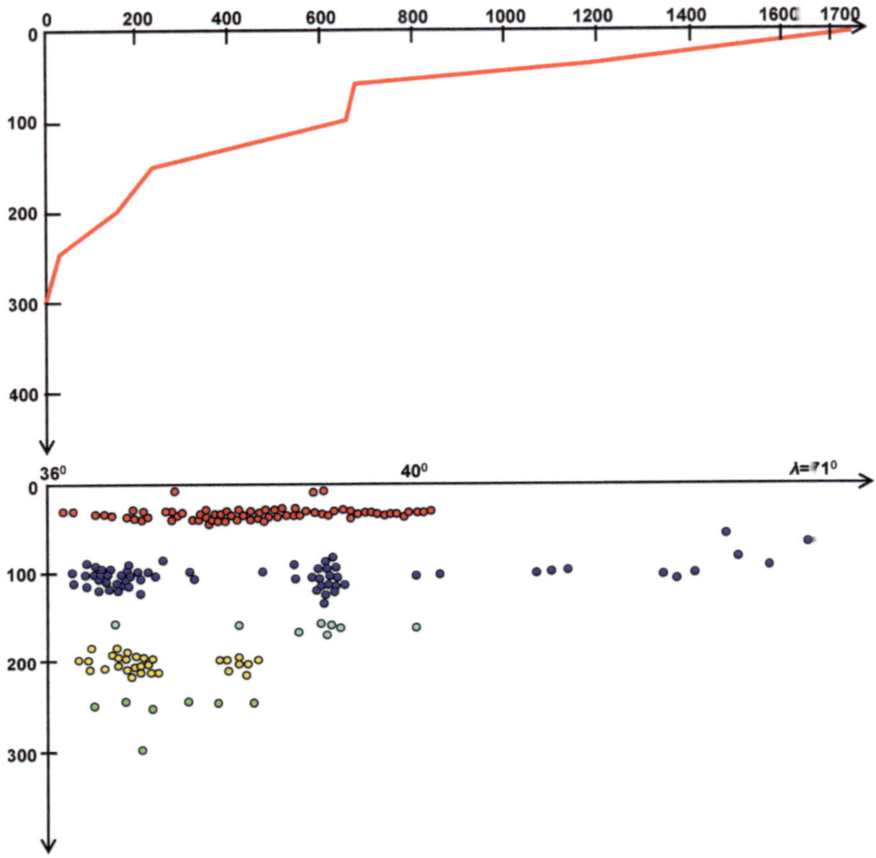

Fig. 6.6 Location of earthquake foci in the mountain regions of Central Asia (Piotrovsky 1964). {Legends on Fig. 6.6, t to b: Number of earthquakes, Depth, km.}

gravimetric fields of Russia to identify the main structural features of their task area and, in particular, for a visual representation of the fractal discreteness of the lithosphere. The mentioned maps were created in recent years in VSEGEI under the general editorship of the Main Editorial Board for Geological Mapping (O. V. Petrov—Chairman, A. F. Morozov, A. V. Lipilin, T. P. Litvinova, F. V. Myasnikov, Yu. M. Erinchek, V. D. Kaminsky) and were the result of a generalisation of aeromagnetic and gravimetric survey work performed in Russia over the past 50 years.

At a glance at the map of the anomalous magnetic field of Russia and the adjacent water areas (Fig. 6.7a), a combination of chaos and orderliness in the distribution of individual multi-ordinal anomalies strikingly coincides with the hierarchy of selected morphostructural blocks from the local supra-regional order to the fifteenth local order (Table 3.1). Just like the morphostructural blocks of the

L-hexagons, G-hexagons and Rolls types themselves, the anomalous magnetic field is characterised by similar morphostructural features (Fig. 6.7d). In the spectrum of the anomalous magnetic field (AMF) there is a regional component with the wavelengths of 100–500 km and local with a wide range of frequencies. AMF is characterised by structural and quantitative characteristics. Structural features that reflect the morphology of the field include strike, shape, linear dimensions, extent and differentiation of anomalies, quantitative intensity (amplitude), order of anomalies (intensity ranking) and their gradient.

According to the configuration of isolines, linear anomalies are distinguished, with outlines elongated in plan, with a clearly expressed predominance of the major axis; nonlinear, subdivided into isometric and irregular, having a complex configuration. Anomalies are grouped into systems, forming anomalous regions, zones, bands, sections, etc., subdivided according to their principal, typical features within the prevailing linear, mosaic, and annular anomaly types. The horizontal AMF gradient in Russia reaches tens or hundreds nanotesla (nT) per kilometre in contrast to a very low normal field gradient. The amplitude of the anomalies basically does not exceed 10% of the intensity of the normal field, except for the local anomalies

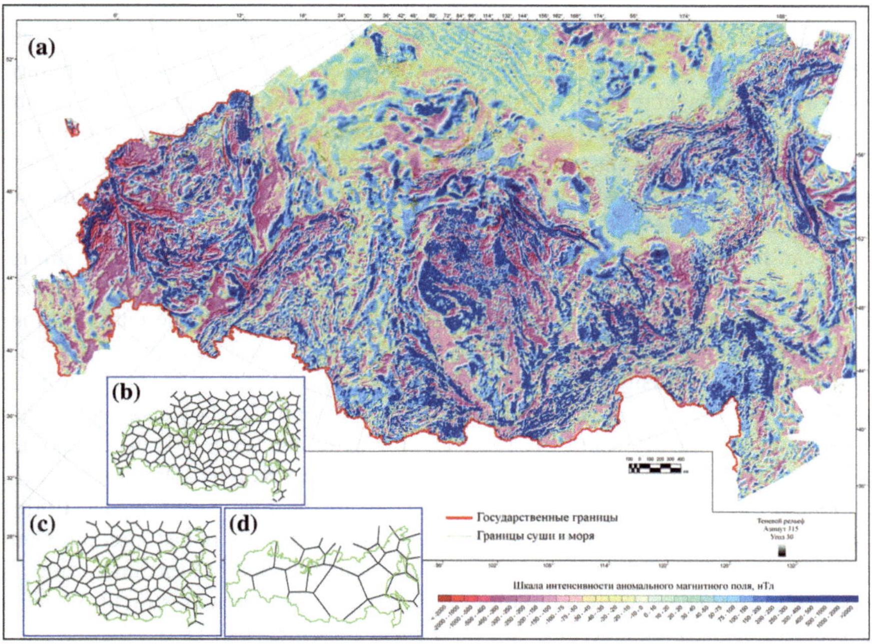

Fig. 6.7 Wave representation of the map of the anomalous magnetic field of Russia scale 1:10,000,000 (**a**) in the form of decomposition into cellular structures of different order: λ = 300–400 km (**b**), λ = 400–500 km (**c**) and λ = 900–1300 km (**d**). The wavelength λ is given according to an evaluation of the periodicity of cellular structures. {Legend on Fig. 6.7: Scale of intensity of anomalous magnetic field, nT}

over the iron ore deposits of the Kursk magnetic anomaly. The availability of a digital AMF database allows the field to be reliably broken down into different spectral components in contrast with the manual means for its division. To create a map of AMF transformations in the form of a regional component map, a digital cartographic model (DCM) of the anomalous magnetic field of scale 1:10,000,000 was used.

The map of the regional component of the anomalous magnetic field of Russia (Fig. 6.8) displays anomalies averaged with a radius of 150 km. This averaging radius gives an idea of anomalies with a wavelength set of 300 km or more. The structure of the regional anomalous magnetic field reveals the main structural-tectonic elements associated with the deep structure of the consolidated crust. Two ancient platforms (Russian and East Siberian) are distinguished by super-regional cellular structures, but with different spectral characteristics of the field. The Russian platform is characterised by low background values with separate linear and local positive and negative anomalies reaching ±150 nT. The East Siberian platform is structurally framed by positive regional anomalies having an intensity of up to ±200 nT. The Urals does not have a pronounced expression among the structure of the regional anomalous magnetic field. A cellular structure in the regional field is allocated to the Barents and Kara Seas. Deep faults of a planetary nature find a clear expression in the nature of the regional field.

The map of residual anomalies is obtained by subtraction from the initial anomalous magnetic field of the regional component calculated by T. P. Litvinova with a radius of averaging of 160 km (Fig. 6.9). In this case, it appears that the map of residual anomalies displays the distribution of magnetised formations mainly up to a depth of 2 km. In this case, the magnetised objects of the crystalline basement, emerging on the Earth's surface, also find a clear expression in the overall ensemble of residual anomalies.

It can be seen on the map that a high-intensity (from 10 to 1000 nT), strongly differentiated and predominantly positive field is distributed within the Russian and Siberian platforms, consisting of the Ural fold system. Almost the entire northern part of the territory is characterised by a slightly differentiated alternating field of low-to-average intensity (100–200 nT). The exception is the north of the Siberian platform. A field close to the normal is marked by the Verkhoyansk fold system. Two structures clearly stand out in it, topologically and geometrically resembling vortex structures (twist structures). One of these is the Kolyma-Omolon semi-ring anomaly; the other is located on the west coast of the East Siberian Sea. Both of them seem to originate from a complex system of intense linear, predominantly positive anomalies of the north-western strike, which gradually turns into a ring and semi-ring anomalies with a minimum in the middle. Such a structure of ring anomalies of the magnetic field is very typical for other regions of Russia.

A digital gravimetric map of scale 1:5,000,000 was created on the basis of an analogous gravimetric map of the USSR at a scale of 1:2,500,000 (edited by P. P. Stepanov M. A. Yanushevich; Moscow, VNIIGeofizika (1999). The map is compiled according to the Bouguer reduction (σ = 2.67 g/cm^3). A correction for the relief is introduced, r = 200 km. The normal Helmert's transformation of 1901–1909

(a)

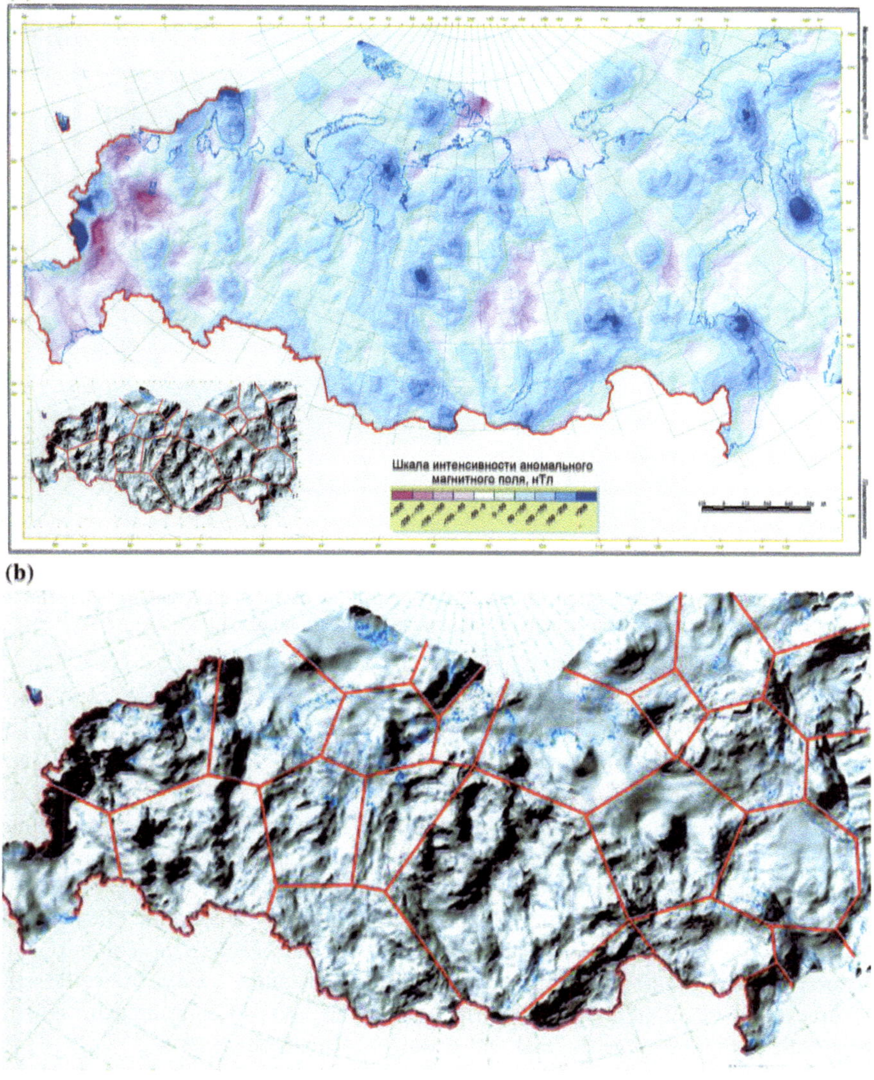

(b)

Fig. 6.8 Map of the regional component of the anomalous magnetic field of Russia (**a**) with a set of wavelengths l = 300 km and more (**b**) (the wavelength is given on the basis of the evaluation of frequency composition of the transformant). Explanations are given in the text

with the amendment of 14 mGal, 1971 system, digital model (Fig. 6.10a) was created in the VSEGEI by ArcInfo 8.2 software package. The editorial board of the digital mapping project: O. V. Petrov (chairman), A. F. Morozov, A. V. Lipilin, T. P. Litvinova, F. V. Myasnikov, Yu. M. Erinchek, V. D. Kaminsky. Authors of the

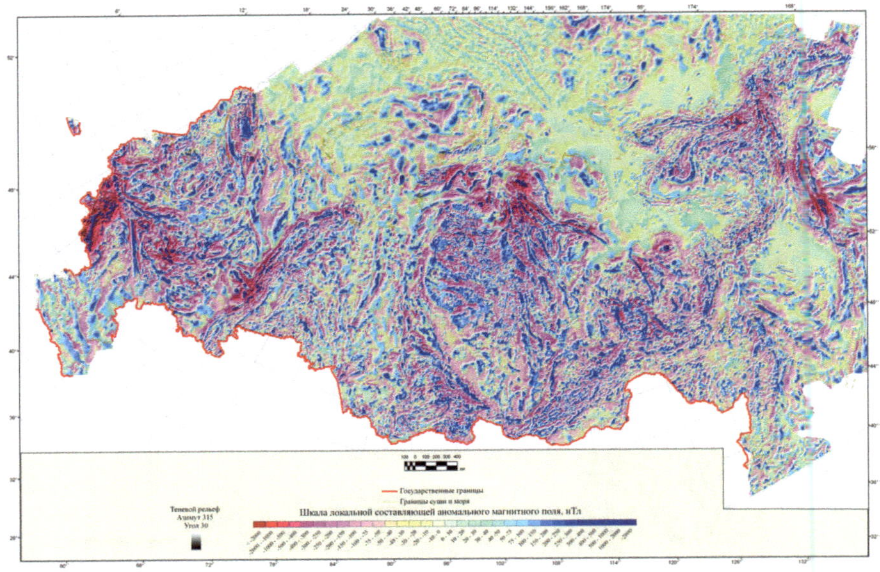

Fig. 6.9 Map of residual anomalies of anomalous magnetic field with a set of wavelengths λ = 5–15 km (result of evaluation of the frequency composition of the transformant). {Legend on Fig. 6.9: Scale of the local component of anomalous magnetic field, nT}

digital model: T. P. Litvinova (the head of the work and the editor), A. A. Trunin, O. B. Negrov (VSEGEI), B. V. Vaskovsky, Yu. V. Kangropol, L. B. Gutman, D. S. Zelenetsky, L. A. Tselisheva (VIRG—Rudgeofizika); A. I. Atakov (Sevmorgeo).

The map of the gravitational field is used to study the deep structure of the Earth's crust, tectonic and petrographic zoning of large regions, geological mapping of closed areas, search for oil and gas deposits, prospecting and exploration of solid minerals: coal, ores and non-metallic raw materials.

As long experience in interpreting gravimetric data has shown, the effectiveness of gravity prospecting is determined by the physico-geological conditions of occurrence and the physical properties of the object under study, as well as the accuracy and detail of the gravimetric survey, the level of interference, and the study of the region by geological and geophysical methods.

In the formation of observed gravitational anomalies, an important role is played primarily by large, nearly vertical, lateral density inhomogeneities located not only in the Earth's crust, but also under its base layer, including deep in the mantle of the Earth; it is precisely the sum of all these lateral density inhomogeneities that causes the formation of the observed initial gravity anomalies. It is assumed that such near-vertical density inhomogeneities are formed during the removal of the deep heat and matter of the Earth, which occurs unevenly throughout its surface and is confined to its more permeable zones in the form of global mobile belts of the Earth (Nikolsky 1996).

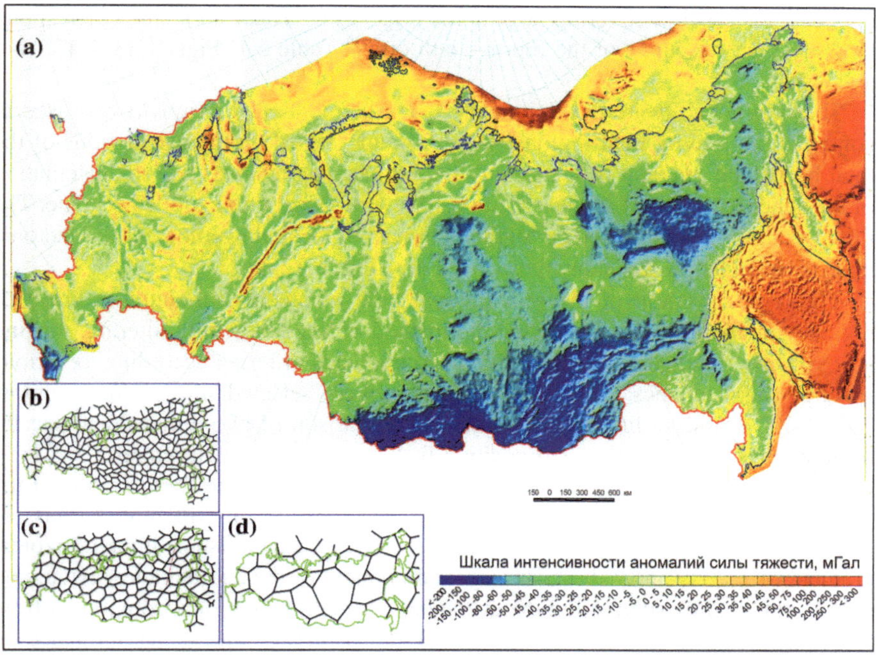

Fig. 6.10 Wave representation of the **a** gravimetric map of Russia in the form of decomposition into cellular structures (standing waves) of different orders: **b** λ = 300–400 km, **c** λ = 400–600 km and **d** λ = 700–1100 km. The wavelength λ is given by an evaluation of the periodicity of cellular structures

In the practice of geological interpretation of the results of gravity prospecting, various methods of division and transformation of gravitational anomalies are widely used. The need for such operations is due to the fact that the observed picture of gravitational anomalies is caused by the whole set of anomalous masses, by all interfaces of the density, and reflects the effects of many geological causes. The most important of these are the thickness, composition and structural features of sedimentary formations, the surface relief and petrographic heterogeneity of the crystalline basement, as well as the deep structure of the crust. The division of the field into regional and local components makes it possible to approximately extract the components having a different geological nature from the observed field. It is assumed that the features of the geological structure, which sharply differ in their scale and nature, create changes in the anomalous field with different correlations, i.e., with varying degrees of alteration.

Regional anomalies of the gravitational field are obtained by averaging the initial data with a radius of 250 km. Due to the fact that the force of attraction is inversely proportional to the square of the distance, the effect of deep-lying masses on the surface is very weak. This effect is clearly visible on the map of regional anomalies, which characterise the distribution of deep-lying large masses that give rise to noticeable anomalies covering large areas (Fig. 6.11). These anomalies correspond to dissipative structures of the fourth–sixth order (Table 3.1, Figs. 1.43–1.47, 1.49 and 1.50).

The original software development (A. I. Atakov, T. P. Litvinova) was used to construct the gravity field regionalisation scheme, based on the calculations of the total horizontal gradient of the gravity field, lineament analysis and the calculation of the general directions of the structure strike, integrated into a unified scheme. The final version is a synthesised cartographic image of the main geological and tectonic objects, reflected in the anomalies of gravity (Fig. 6.12a).

The regionalisation of the gravitational field along the horizontal gradient of regional anomalies was made using the method of frequency field filtering, making it possible to divide the field into several components differing according to various spectral characteristics. The filter parameters were selected in a sliding window 200 × 200 km according to the results of comparison of two-dimensional spectra of anomalies over negative and positive first-order basement structures.

The map illustrates the distribution of the lineaments of the Earth's crust of the seventh and eighth orders (Table 3.1; Figs. 1.51 and 1.52). Different regions are characterised by a specific set of spectral-frequency characteristics of the horizontal gradient of gravity anomalies. The local component of the anomalous gravitational field is obtained by subtracting the averaged values of the regional component from the initial field (Fig. 6.13). The residual (local) field reflects the influence of

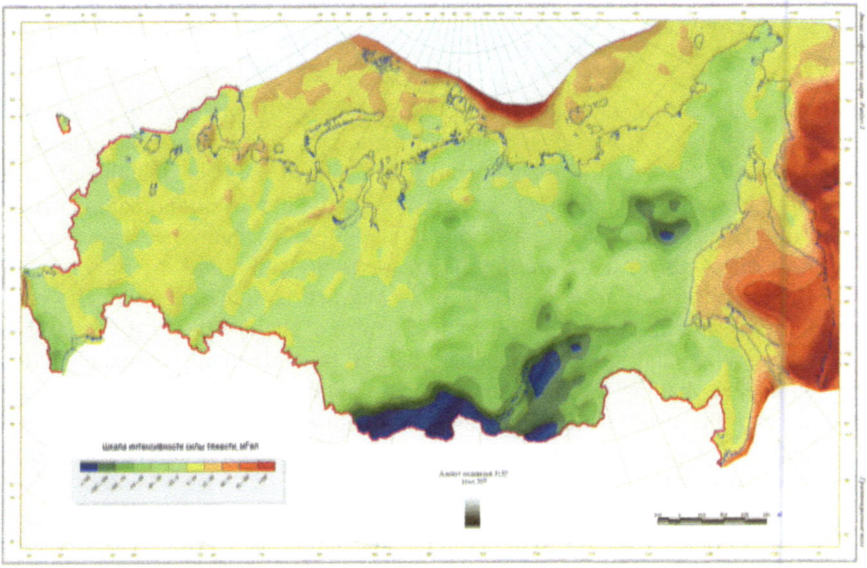

Fig. 6.11 Map of the regional component of the gravitational field of Russia. {Legend on Fig. 6.11: Scale of the intensity of gravity, mGal}

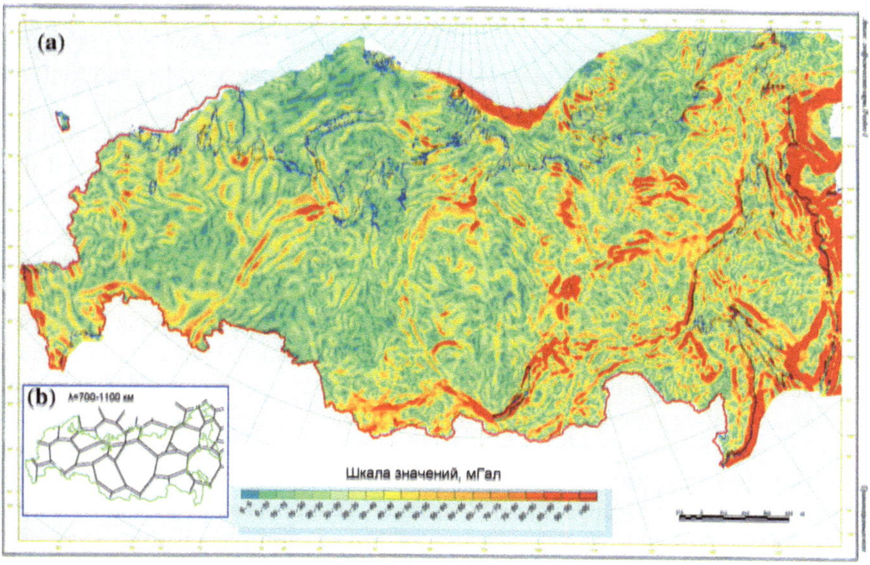

Fig. 6.12 **a** Map of the classification of Russia's gravitational field along the horizontal gradient of regional anomalies and **b** its cellular interpretation. {Legend on Fig. 6.12: Value scale, mGal}

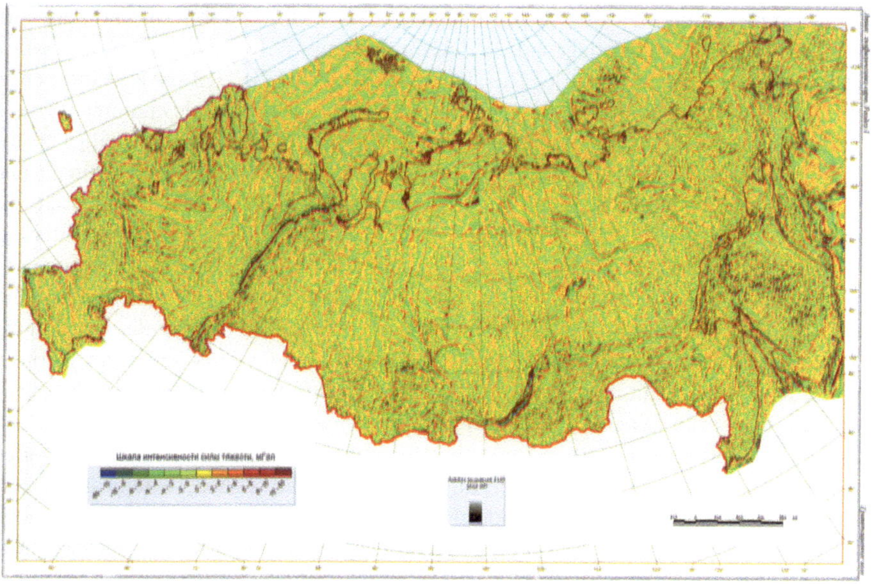

Fig. 6.13 Map of the local component of the gravitational field of Russia. {Legend: Scale of gravitational intensity, mGal}

anomalous masses close to the surface. Local linear and cellular structures of twelfth-fifteenth orders are clearly traced by local anomalies (Table 3.1, Figs. 1.59–1.65). Conditional sources of observed anomalies are located in the depth interval of 0–5 km.

It should be noted that, against the background of the above structural features in the maps of the initial anomalous values of the gravitational magnetic field and their transformants, a commonality is observed at the level of reflection of the main structural units of platform and folded regions and their boundaries on the territory of Russia. Primary processing of these maps by the original modified method of lineament-spectral analysis (Petrov et al. 2006) demonstrates that this quite obvious structural commonality is manifested in the form of a scheme of cellular polygonal formations, which, according to the hydrodynamic analogy, are associated with the reflection of internal standing gravitational waves. Thus, it is the wave process that determines the commonality of heterogeneous geophysical and gravimagnetic fields: it would be logical to expect the manifestation of this same commonality in other, less formalised geological materials. As the latter, we considered a digital model of the day relief, a cosmo-photographic map, a map of Quaternary sediments and a geological map of pre-Quaternary complexes of the territory of Russia, also compiled in VSEGEI in recent years under the general editorship of the Editorial Board for Geological Mapping.

Omitting the detailed description of these materials, as in the case of maps of geophysical fields, we pay attention to their fundamental heterogeneity in the sense of the nature of the spatial "signals" fixed on the maps. The digital relief model reflects the distribution of the excesses of different signs relative to the level surface (for example, Krasovsky's triaxial ellipsoid). The distribution of anomalies of relative excesses on regional maps is determined mainly by isostatic compensation processes. The cosmic image of Russia was formed in VSEGEI by Kirsanov et al. (2005) on the basis of spectrozonal space imagery, where each channel forms a halftone image of the relief surface. The distribution of halftone anomalies in the range from 0 to 256 gradations of brightness is associated with the intrinsic and reflected infrared radiation of the Earth, as well as with various terrain characteristics, including near-surface rock complexes specifying the degree of scattering, reflection, and absorption of the incident light. The map of Quaternary deposits of the territory of Russia contains information on the structure of a loose sedimentary cover dating from approximately 1670 thousand years to the present days within the so-called glacial and extra-glacial areas. The spatial signal here is formed by a colour gamut, where, according to legend, the colour itself reflects the genesis of deposition (alluvial, lacustrine, glacial, aeolian, etc.); the degree of colour saturation is the age of these deposits from the Pliocene to the Holocene. The exception is the undivided Quaternary complex, having orange-pink colours in the legend, which also determine the genesis of the rocks. In addition, the map contains a sufficiently detailed image of the hydrosystem and the designation of the boundaries of the maximum spread of the cover glaciations. The geological map of Russia reflects the distribution in three-dimensional space of different ages of sedimentary complexes, intrusive massifs and tectonic contacts. Based on the geochronological scale, the geological map, like the map of Quaternary sediments, forms a spatial signal, the characteristics of which are determined by colour division, where the colour itself marks the periods and sub-periods, and the colour saturation represents the individual epochs. Undoubtedly, the history of the formation of the main geostructural units of Russia determined the generality of the above spatial signals in the sense that each of the platform- and folded-regions has its own set of structural-material complexes and, correspondingly, a specific landscape. Another question is how this commonality manifests itself on the maps discussed above. The visual primary analysis of the relief indicates a sharp change in its character (the strike azimuths of regional anomalies of relative excesses and their actual values) during the transition from one geostructural unit to another. The change in relief combined with changes in the degree of covering of the territory by the soil and vegetation cover, the geometry of the hydrosystem, as well as near-surface structural-material complexes allows the boundaries of platform and folded regions to be unambiguously traced along the cosmic image of Russia. Both in relief and in the satellite image, the generalised perception of the contours of these regions is close to the perception of cellular polygonal structures. They are also distinguished in the system of colour objects on a geological map and a map of Quaternary sediments. Moreover, if, in the analysis of the structure of the relief and satellite image, the criterion for mapping the boundary of a single cell consisted of the traditional lineament representation, then, in the case of a colour spatial signal, this criterion would be an estimate of its spatial stationarity (the periodicity of colour objects, their geometry and area). As a consequence, we can assume the manifestation of the wave processes underlying the spontaneous structuring of a nonequilibrium geological environment is characteristic thereof in all direct and indirect measurements.

According to the ideas about the wave nature of the processes of spontaneous structuring of density unstable masses—and in addition to the morphometric methods of Filosofov used in the third chapter—a detailed analysis and interpretation of the modern digital geological and geophysical materials considered above was carried out. The basic idea of creating an algorithm for cellular wave (fractal) interpretation is based on the notion that any spatial (temporal) signal contains a discrete series of periodic components (the interval of permissible periods or frequencies is not filled continuously but with passes). The expansion procedure is typically performed in the frequency plane, for passing into which a direct Fourier transform is needed for transforming the original signal defined in the object (spatial or temporal) plane into a complex characteristic.

As applied to the satellite image of Russia (Fig. 6.14), an algorithm of the lineament tracing (linearised landscape elements, including its heterogeneous forms) of narrow-band frequency filtration was used to simulate the effect of generalisation and the subsequent unification of lineaments of a certain rank (length) into the cellular structures forming the framework, inseparable within the

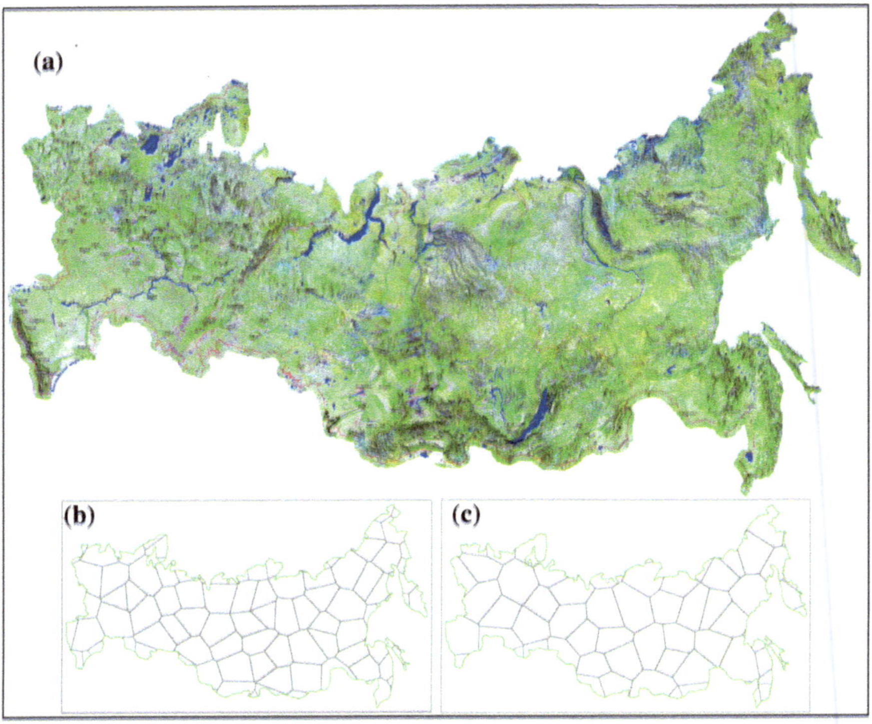

Fig. 6.14 **a** Wave representation of the structure of the cosmic image of Russia in the form of decomposition into cellular formations of different orders: **b** λ = 500–700 km and **c** λ = 600–900 km

polygon. The cellular interpretation of gravimagnetic fields was based on the same approach for mapping lineament structures; however, the latter were understood in terms of straightened gradient zones, sections of linearly extended anomalies, linearly extended chains of isometric anomalies and linearised boundaries of zones of spatial stationarity (Fig. 6.15).

The lineament in the case of the day relief differs from the lineament in the case of the satellite photograph in a way that the influence of the change in the forms of vegetation, soil cover and other elements of the landscape, not directly attributable to deep geological objects, is not taken into account. In addition, the satellite photograph has a much larger spatial resolution, which determines the significant difference in the result of its interpretation from the interpretation of the discrete digital relief model (Fig. 6.16).

Since the actual lineament representation is applicable to any map, for the geological map and the map of quaternary formations, the lineaments were drawn along the straightened sections of the structural-material complexes and the rectified disjunctive grid (Figs. 6.17 and 6.18).

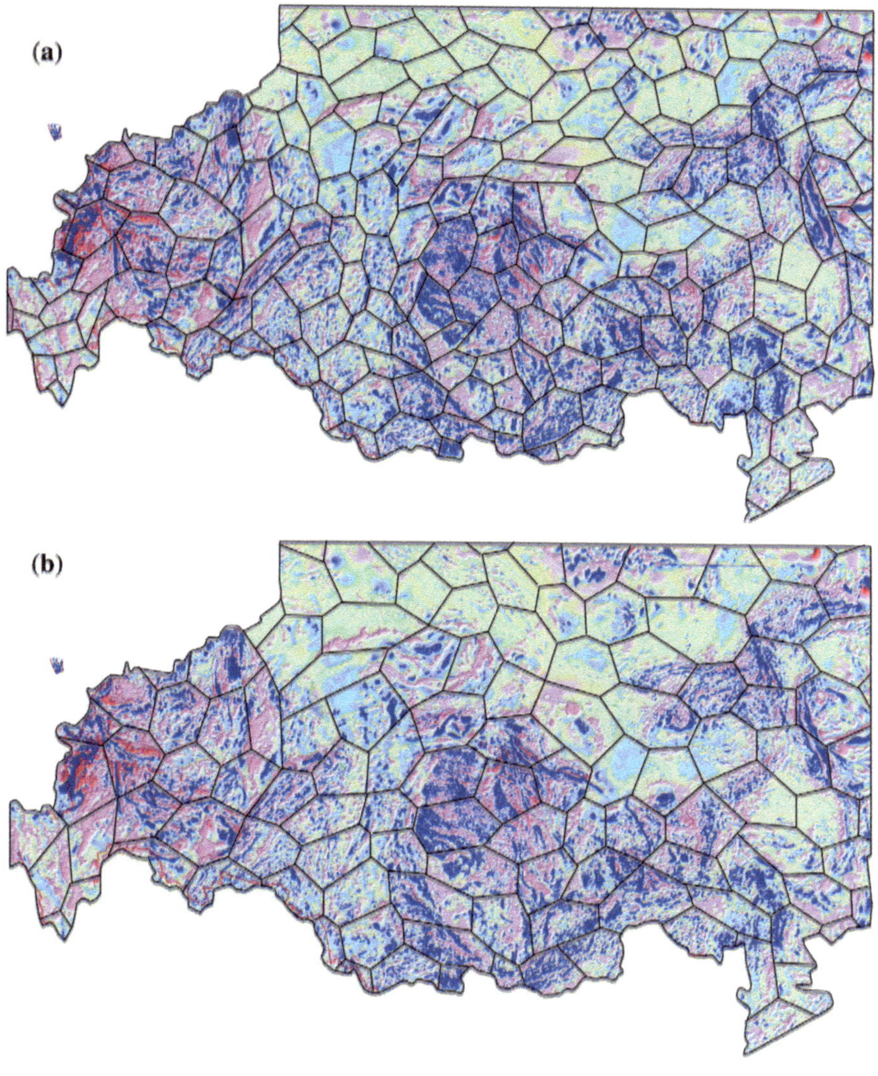

Fig. 6.15 Wave representation of the map of the anomalous magnetic field of Russia in the form of decompositions into cellular formations of different orders: **a** λ = 300–400 km and **b** λ = 400–500 km

As the primary quantitative indicator of the self-similarity effect in the cellular expansion of heterogeneous geological and geophysical fields, we have chosen the fractal dimensionality *D*, which, according to the traditional definition, is only the tangent of the slope in the graph of linear approximation of the actual data. In our

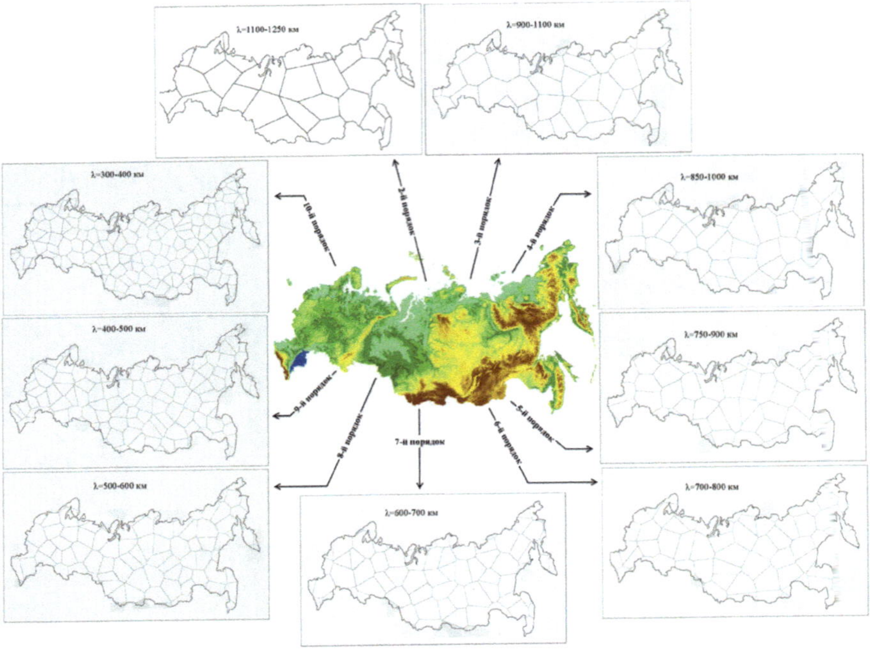

Fig. 6.16 Wave representation of the structure of the topographical relief in the form of decomposition into families of cellular formations of different orders in the analysis of quasi-independent Fourier spectrum harmonics {Legends on Fig. 6.16 (clockwise from top left): km, km, km, km, km, km, km, km, km, (centre) 10th order, 2nd order, 3rd order, 4th order, 5th order, 6th order, 7th order, 8th order, 9th order}

case, the latter are represented by the dependency of the number N—area polygonal objects that fit within the investigated polygon (territory of Russia), from the average length of this wave λ. Since logarithms appear in the general formula for the fractal dimensionality, in our case we also consider an approximation of the form $\log(N) = D \cdot \log(\lambda) + C$, where C is some arbitrary constant that defines the intersection graph with the ordinate axis at a single wavelength. We calculated the fractal dimensionality for anomalous gravitational and magnetic fields, as well as for the digital model of the relief. As can be seen from Figs. 6.19, 6.20 and 6.21, in all listed cases the value of D falls within a rather narrow interval of values (1.56–1.84), which allows the manifestation of one type of fractal self-similarity to be assumed in all experimental data characterising a density nonequilibrium geological environment.

The final materials, represented by the multi-ordinal cellular structure of the mentioned maps, cannot be unequivocally comparable, for example, with traditional lineament schemes. If, in the latter, the mapping of reliable structures is based on such criteria as image stability, the effect of inheritance at different levels of generalisation, spatial correlation (the unification of the integral image from the set

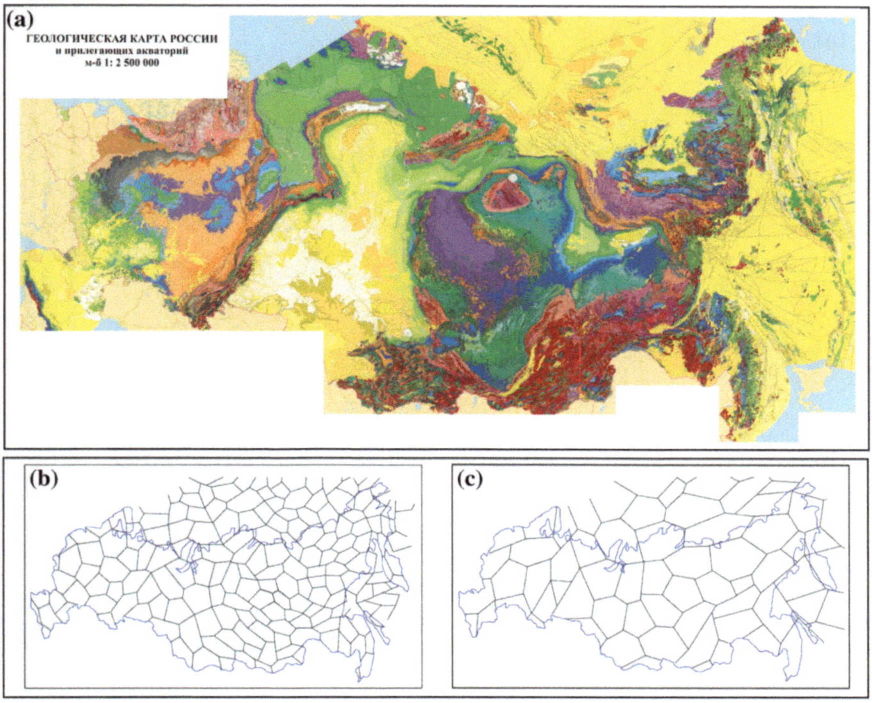

Fig. 6.17 Wave representation of the **a** geological map of Russia in the form of decomposition into cellular structures of different orders: **b** λ = 250–700 km and **c** λ = 600–1200 km

of single-oriented elementary lineaments) and deep correlation of the density unstable structures, then in cellular interpretation another criterion is added, all the allocated structures, according to the hydrodynamic analogy, are convex. This position can substantially burden the deciphering result, since it assumes only the effect of the multi-depth standing inner gravity waves, and not the total surface effect of the running, shearing, torsional and other waves, which determines a number of notable deviations of the cellular structural boundaries from the tectonic structural boundaries known in classical geotectonics.

The overlapping of the result of cellular deciphering onto the map of the initial magnetic field, from which a clear connection is prominent between the boundaries of the polygonal formations and the linearised anomalous and gradient zones ΔT, is indicated in Fig. 6.15. For all the above materials, cellular interpretation of close orders (usually 10- and 9-th) is performed.

The cellular deciphering of the 2nd order, which distinguishes the boundaries of the platform and folded structures of Russia, is indicated in Fig. 6.15. An expansion of cellular structures onto several orders (from the 10th to the 2nd), illustrating the effect of inheritance (fractal self-similarity) of the transition from order to order

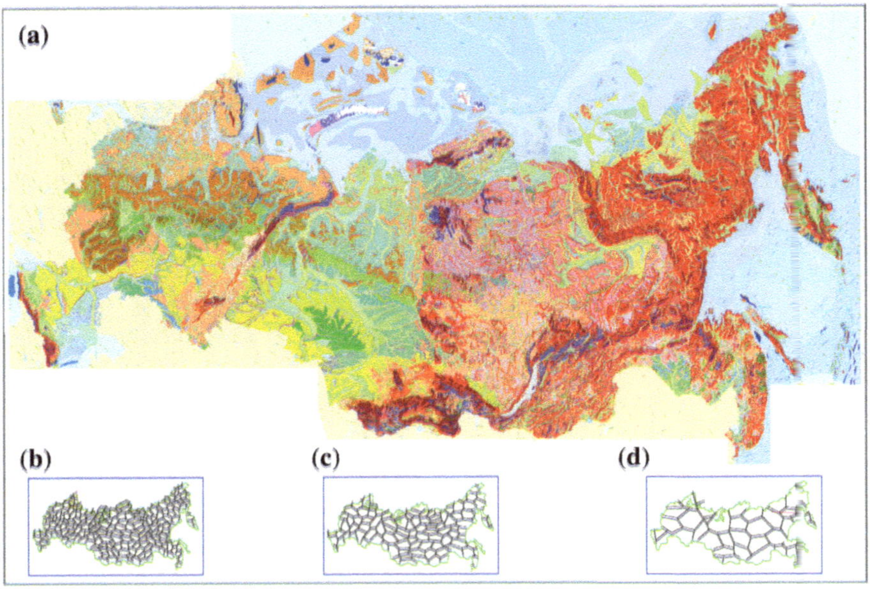

Fig. 6.18 Wave representation of the map of the Quaternary formations of Russia **a** in the form of decomposition into cellular formations of different orders: **b** λ = 250–450 km, **c** λ = 450–800 km and **d** λ = 790–1400

through a constant scale factor (fractal dimensionality), is given for the relief (Fig. 6.16).

Thus, our definition of the dissipative structures of the Earth forces us to recognise that the lithosphere, as well as its relief, is a complex and at the same time unified natural formation possessing a fractal dimensionality and being subordinate in its development to the general wave patterns of its natural development. In the lithosphere and its relief, the fundamental laws of manifestation of the wave properties of matter are fixed in the processes of spontaneous structuring of density unstable masses, which require the formation of new approaches in the construction and analysis of geological, tectonic, geophysical, geomorphological and other special maps during delineating and typifying geological, tectonic and morphometric structures of various orders, as well as in the development of terminology—the language of their description during the collection and processing of geological, geophysical and morphometric data, as well as for other scientific and practical purposes. For example, the commonality of heterogeneous geological and geophysical materials, manifested at the level of their cellular spatial structuring, opens the possibility of their joint quantitative interpretation, taking into account the hydrodynamic analogy determining the functional relationship of the linear dimensions of a single cell with the thickness of a density nonequilibrium layer. Further, we will consider this approach in relation to the quantitative interpretation of potential fields in the conditions of a minimum of a priori information.

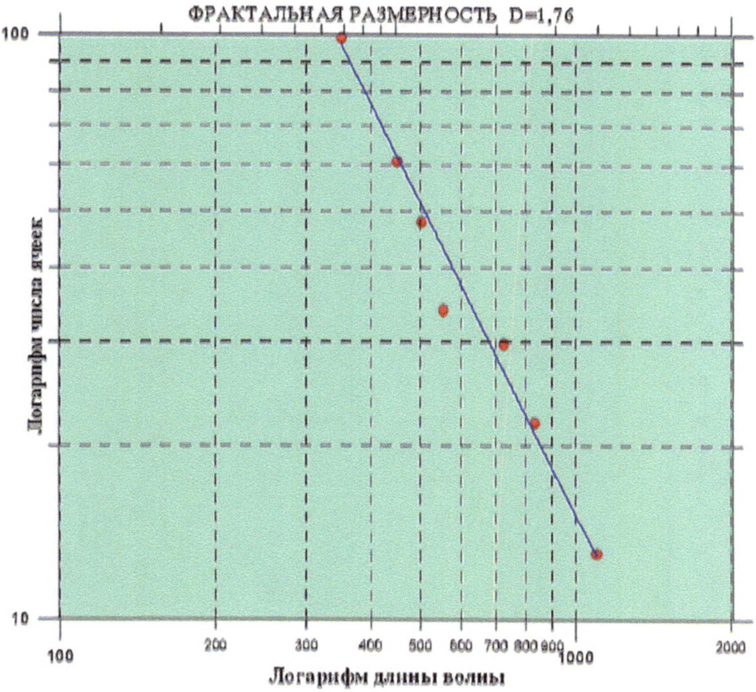

Fig. 6.19 Calculation of the fractal dimensionality of the anomalous magnetic field. Red points indicate the actual data (the correspondence between the number of the cells N and their average wavelength λ); the blue line reflects the linear dependence $\ln(N) = -1.76 \cdot \ln(\lambda) + 14.89$. {Legends on Fig. 6.19 (a, t to b, r to l): Fractal dimensionality, Cell number logarithm, Wavelength logarithm}

6.2 Fractal Nature of Geological Section Structure and Its Wave Stratification

In the first part of the book, it was indicated that dissipative structuring occupies the entire volume of a nonequilibrium layer, manifesting on its surface in the form of cellular fragmentation, and in the section—as a spontaneous stratification. Since the latter is due to an alteration of the dimensions of the convective cells—or, in other words, the lengths of the internal gravitational waves—it is possible to introduce the image of a volumetric cellular fragmentation of the dissipative system. In this case, the dominant cell geometry from the planetary to the microscopic scale turns out to be hexagonal. Simplifying the geometric description of the volumetric wave process, let us represent it as a consequence of the merging of several cells of the previous order. Continuing this process of fragmentation, we obtain a simplified geometric description of the spontaneous structuring of the Earth as a dissipative system (Fig. 6.22).

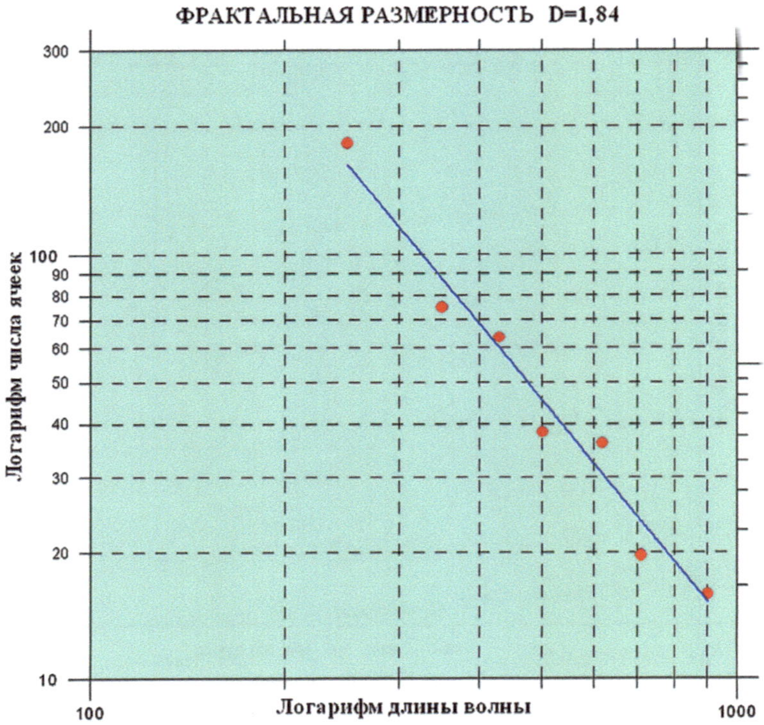

Fig. 6.20 Calculation of the fractal dimensionality of the gravimetric field. {Legends on Fig. 6.20 (a, t to b, r to l): Fractal dimensionality, Cell number logarithm, Wavelength logarithm}

The presented process of cellular wave fragmentation of the Earth's volume as a result of dissipative processes ideally continues both on the surface and in a section infinitely, from macroscopic to microscopic scales, causing the emergence of a regularly oriented regmatic fault network and planetary rock microfracturing. The faults, lineaments and fractures extend mainly latitudinally and meridionally, forming an orthogonal system and subordinated diagonal NW–SE and NE–SW azimuths, and, as Khain and Koronovskiy (2007) noted, a splitting of the diagonal lineaments into pairs WNW–ESE, NNW–SSE and transverse directions to them can be observed. This strict conformity to the sides of the hexagon and the well-studied regular orientation of faults and fractures has recently received convincing additional confirmation in the works of Odessky (2005) and Anokhin (2006). The specifics of the resulting structures will be determined by the geometry of the media and their rheology. Considering the model represented in Fig. 6.22, it can be seen that, despite its conventionality, the obvious correlations of embedded hexagonal cells of the first-to- fourth order are in evidence with a scale image of certain layers of the Earth and their discontinuities. For example, a fourth-order central hexagon marks the inner core. The central hexagon of the third order, twice as large as the

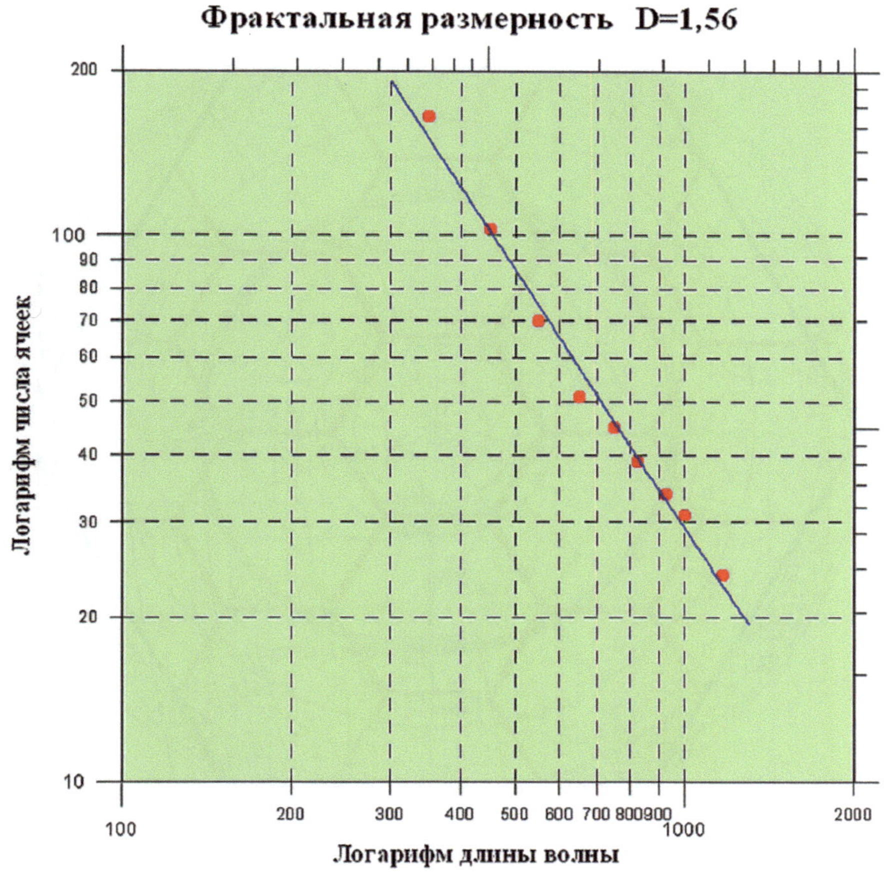

Fig. 6.21 Calculation of the fractal dimensionality dimensionality of the topographical relief. Red points denote the actual data (correspondence between the number of cells N and their average wavelength λ); blue line reflects linear dependence $\ln(N) = -1.56 \cdot \ln(\lambda) + 14.16$. {Legends on Fig. 6.21 (a, t to b, r to l): Fractal dimensionality, Cell number logarithm, Wavelength logarithm}

first-order hexagon, is located entirely within the outer core. The central hexagon of the second order is located near the discontinuity of the lower and upper mantle. Hexagonal cells of different orders form a regularly oriented regmatic fault network and planetary fracturing on the Earth's land surface, which manifests itself in space images as a hierarchy of lineaments.

Following the above presentation on the correlation between the fractal and wave properties of dissipative structures, let us estimate the fractal dimensionality D of this model using the deterministic approach:

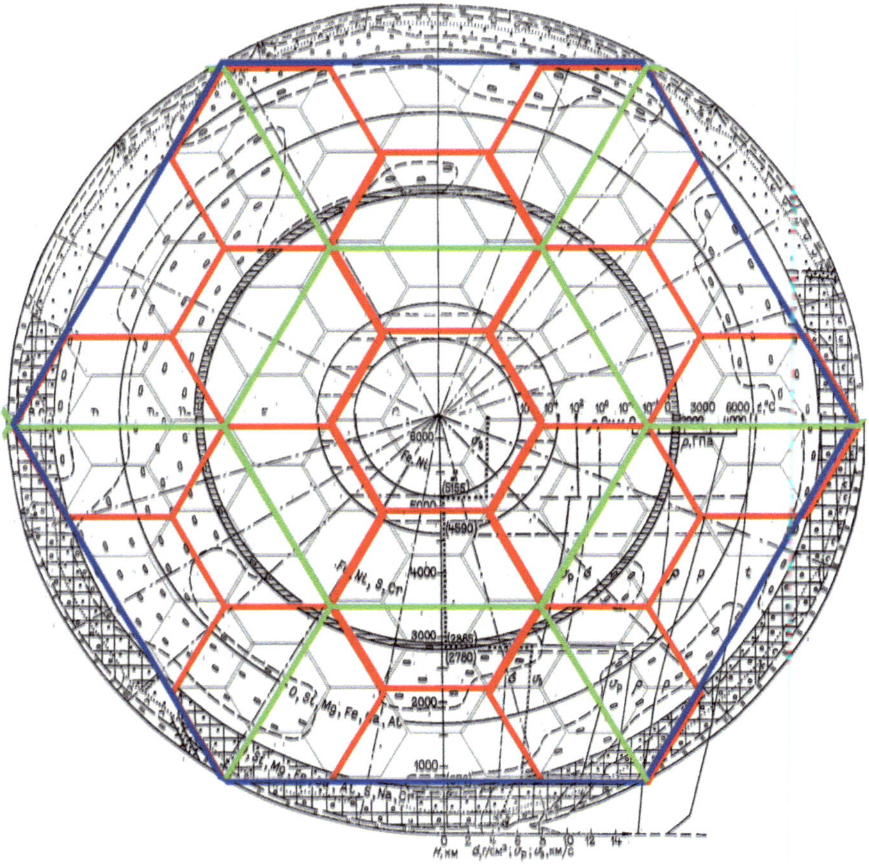

Fig. 6.22 Two-dimensional simplified model of cellular wave segmentation of the Earth Blue hexagon—cell of the first order; green—second order; red—third order; black—fourth order

$$D = \log(N)/\log(1/r), \tag{6.1}$$

where *N*—the number of microscopic objects of the same dimension that make up one macroscopic object, 1/*r*—the scale factor showing how many times the dimension of microscopic objects is less than macroscopic. In the first approach, the fractal is estimated by analogy with the structure of the Koch snowflake fractal (Fig. 6.23).

Here the half of the hexagonal cell is a self-similar object, each side of which is divided into three parts, half the dimension of the original side:

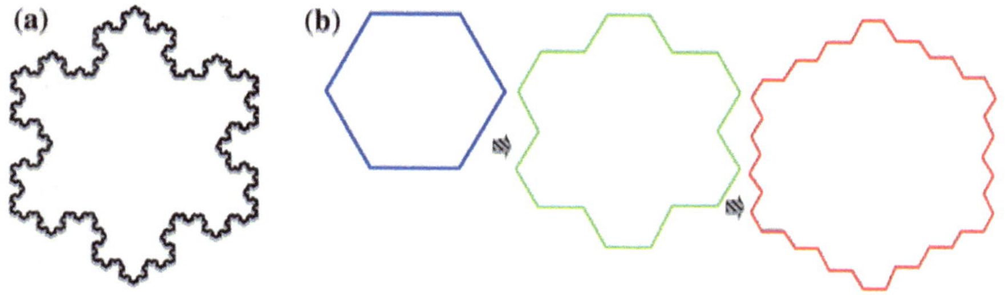

Fig. 6.23 Fractal structures: **a** classic "Koch's snowflake"; **b** fractal, implemented in the wave segmenting model in Fig. 6.22

$$D = \log(3)/\log(2) = 1.5850. \quad (6.2)$$

In the second approach, the fractal is estimated by analogy with the "Sierpinski carpet" (Fig. 6.24): the hexagonal cell of the n-th order is divided into four hexagonal cells n + 1-th order, whose dimensions are half the cell dimensions of the n-th order:

$$D = \log(4)/\log(2) = 2 \quad (6.3)$$

To compare the obtained values of D with more rigorous model estimates, we consider the fractal dimensionalities within the Busse model. Its essence lies in the fact that the velocity distribution on the surface of a nonequilibrium spherical layer

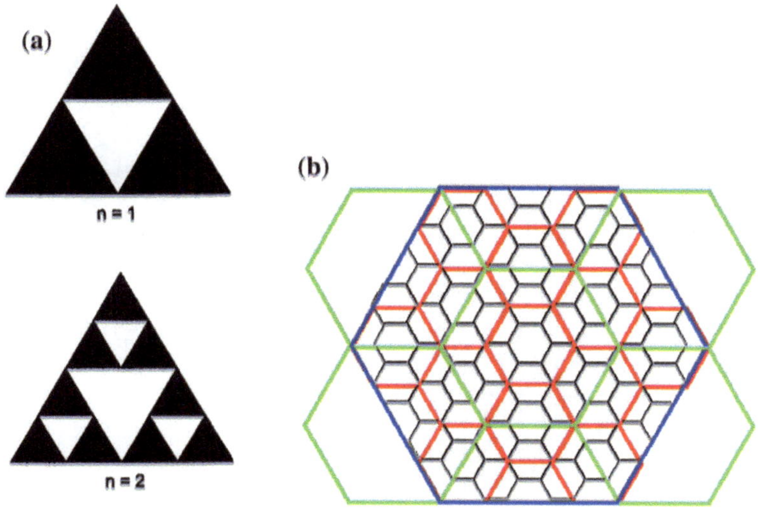

Fig. 6.24 Fractal structures: **a** the formation of the "Serpinskii carpet"; **b** formation by analogy with a cellular hexagonal structure

is expressed in terms of Legendre polynomials and, thus, is functionally related to the vibrational eigenmodes of this layer. As can be seen from Fig. 6.25, geometrically indicated velocity distribution is displayed as isolines, and the transition from the region of one sign to the region of another sign of this or that velocity component means a transition from the region of the stream emersion to the region of their immersion. The regular alternation of these regions along the surface of a density nonequilibrium spherical layer as well as their deterministic geometry allow us to state that the Busse model reflects the process of spontaneous cellular structuring of a nonequilibrium medium, which, both in geometry and in generalised physical conditions, is similar to the individual geodynamic layers of our planet. Comparing the Busse model with the geometric model (Fig. 6.22), it can be seen that dissipative structuring, the simplest in geometrical terms, is inherent to the inner core, and that the differential dissipative structure itself, which is a superposition of dissipative structures of all orders, appears on the land surface. This general conclusion allows the more or less deep type of structuring in the Busse model to be qualitatively determined (Fig. 6.25).

Let us estimate the fractal dimensionality D of regular dissipative structures in the Busse model, using the method of cell counting: the object structure is represented by a point embedded in the observation space. This space is divided into cells with an edge r, and then the number of cells K, containing at least one point of

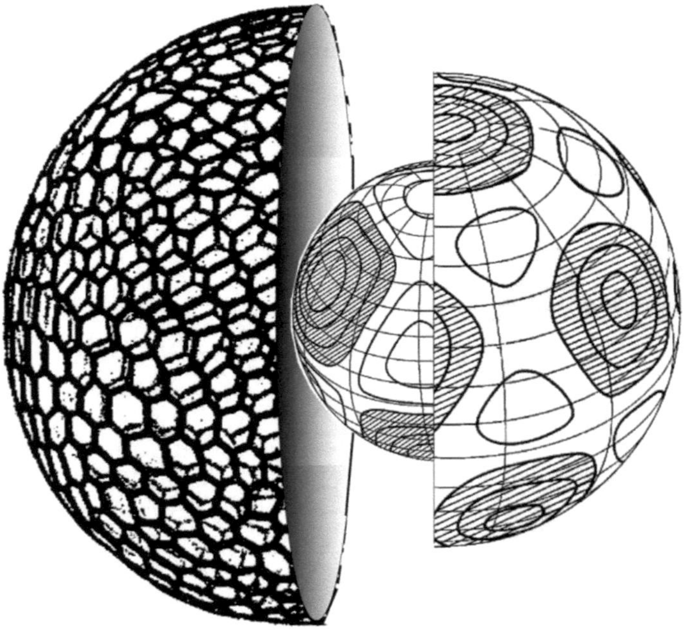

Fig. 6.25 Discrete wave dissipative structures in the model of a nonequilibrium spherical Busset layer

the set, is counted. The procedure is repeated many times for continuously decreasing r, with the approximation of the obtained dependency by the power equation

$$K = ar^{-D} \tag{6.4}$$

We noted above that the fractal dimensionality can be formally derived from the equation of any linear regression that determines the dependency of some geometric parameters of the scale invariant of the dissipative structure on other geometric parameters. For example, an estimate of the dependency of the area S of a hexagonal cell on its side length L or the length of its diagonal H gives logarithmic relations

$$\log(S) = 2.0071 \times \log(L) + 0.2554,$$
$$\log(S) = 1.7560 \times \log(H) - 0.1921,$$

i.e., as in the case of a deterministic algorithm, the fractal dimensionality D is close to 2.

Similar estimates for the Busse model have the form

$$\log(S) = 1.6186 \times \log(L) + 0.2554$$
$$\log(S) = 1.7560 \times \log(H) - 0.1921,$$

Here the value of the fractal dimensionality, first, falls within the range of values $D \in [1.5850; 2.0071]$, calculated above for the cellular model of dissipative wave structuring of the Earth, and, secondly, differs insignificantly from the fractal dimensionality D = 1.83–1.87, calculated for seismic tomography data. Since the hexagonal cell in the dissipative structure is uniquely associated with the Bénard cell, for which the discrete-wave nature is grounded, the mentioned correlations of the D values further confirm the concept of the discrete-wave mechanism of spontaneous (dissipative) structuring of the Earth.

Based on this concept, the analytical assessments of the structure of the geological section were carried out, considering one of the regional geotransects, in particular, Murmansk–Kyzyl (Petrov and Movchan 2003). The section was reconstructed from the set of interpretations of potential geophysical fields and seismic data, as well as the generalisation of earlier sections and materials of the longhole system. Such discontinuities as the lithospheric base layer, the Moho discontinuity, the roof of the granite-metamorphic layer, the bedding surfaces in the sedimentary cover, are not smooth and continuous everywhere, especially in the vicinity of suture zones and folded systems. Tracing each of the listed discontinuities, sustained in a section, and applying spline interpolation, we transform the geological-geophysical section into a pseudowave pattern (Fig. 6.26).

The visual analysis of the internal wave indicates that the wavelength increases with depth, from which direct analogies with Turner's experience are suggested

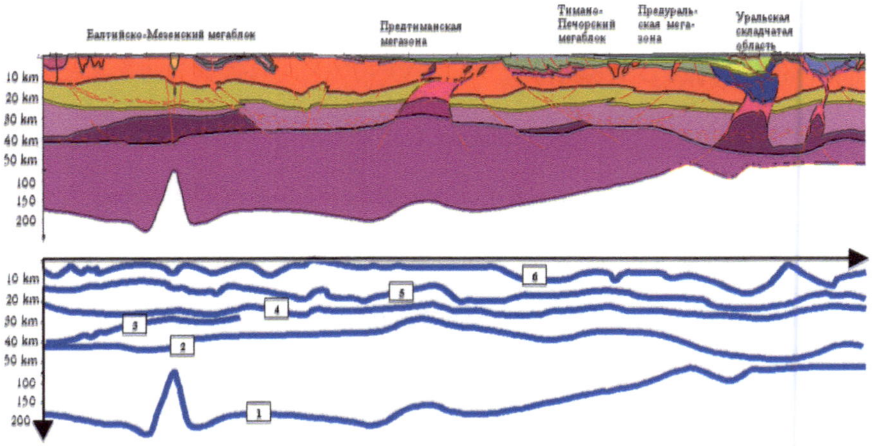

Fig. 6.26 Representation of a section of the northwestern fragment of the Murmansk-Kyzyl Geotransect, according to Egorov (2000), in the form of a system of wave surfaces 1—location of the lithosphere base layer; 2—the Moho discontinuity; 3—discontinuity of the ultrabasite-basite region of the crust-mantle mixture and the basite-granulite megalayer; 4—discontinuity of the lower crustal basite megalayer and the gneiss-basite-granulite megalayer; 5—discontinuity of the gneiss-basite-granulite megalayer and granite-gneiss layer; 6—sedimentary cover base of the volcanogenic sedimentary formations. {Legends on Fig. 6.26, l to r: Baltic-Mezensk megablock, Pre-Timan megazone, Timan-Pechora megablock, Pre-Ural megazone, Pre-Ural folded region}

(Fig. 6.27). The waveform contains unique records of oscillations at different distances from the free surface of the disturbed ocean. From the immersion it can be seen that, firstly, an inversion effect is observed when two neighbouring waves propagating in the horizontal or vertical directions oscillate in antiphase to each other and, secondly, that the relation between the lengths of the inter-deep waves obeys the dispersion law of the waves:

$$\omega \sim \frac{2\pi}{\lambda} gd, \tag{6.5}$$

where ω is the oscillation frequency; λ—the length of cellular internal gravity waves; *g*—the module of acceleration of free fall; *d* is the thickness of a density nonequilibrium layer. Assuming, by analogy with optics, the frequency ω as a function of only the primary source of disturbances and the magnitude of the constant, we have $\lambda \sim gd$: the length of internal gravitational waves increases as immersion takes place in a nonequilibrium medium.

In the geological section, a regular change in the length of the cellular internal gravity wave as a function of depth is observed in homogeneous regions, for example, within the East European platform framed by the Baltic crystalline shield and the Timan folded zone, or the Timano-Pechora plate, bounded by the Timan folded zone and the Urals. As soon as the section crosses the suture zone or folded

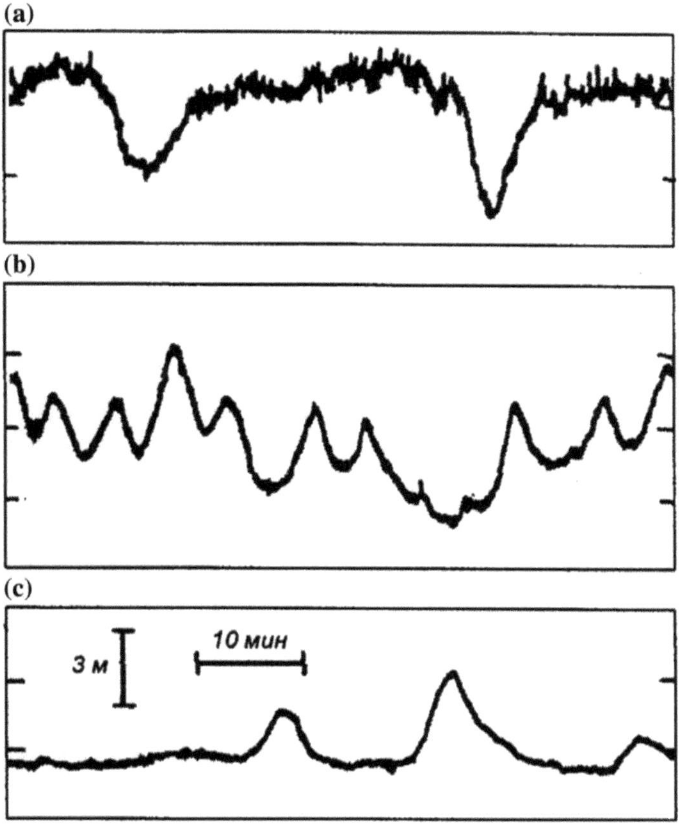

Fig. 6.27 Smoothed form records of the internal waves: **a** near the surface of the ocean; **b** at the intermediate depth; **c** near the bottom. According to Turner (1977). {Legends on Fig. 6.27c, l to r: 3 m, 10 min}

system, where the allocated wave surfaces are interrupted or experience sharp surges, this pattern is violated.

Taking into account the above considerations, this geological section is represented as a system of standing internal gravitational waves in a flat density nonequilibrium layer bounded laterally. According to Landau and Lifshitz (1988), in the two-dimensional problem for such a layer, whose free surface displacements can be described by harmonic functions, for example,

$$\Delta h = A\cos(2\pi x/\lambda - \omega t) + A\cos(2\pi x/\lambda + \omega t) \quad (6.6)$$

the dispersion law of standing gravitational waves acquires the form

$$\omega = \sqrt{\frac{\pi g(2n+1)}{L} \mathrm{th}\left(\frac{\pi d(2n+1)}{L}\right)}, \tag{6.7}$$

where L—the length of the nonequilibrium layer, here defined by the length of the geotransect segment. Standing internal gravitational wave, formed laterally or vertically in density nonequilibrium layer, has an integer number of half-waves:

$$k^2 = \pi^2 \left(\frac{m^2}{L_1^2} + \frac{n^2}{L_2^2} + \frac{p^2}{d^2}\right), \tag{6.8}$$

where $k = 2\pi/\lambda$ is the wave number; L_1 and L_2—the lateral dimensions of the nonequilibrium layer; d—its depth; m, n and p are integers. The last relation determines the effect of quantising the lengths of internal gravitational waves, or, in other words, a discrete (abrupt) change in the lengths of these waves in the volume of a nonequilibrium layer. Taking into account this effect, as well as approximating the change in density with depth by the exponential dependency (Boltzmann's law), we can obtain the change in the length of internal cellular gravitational waves as a function of depth:

$$\lambda(d) \sim \exp(C \cdot d) \cdot th\left(\frac{\pi d(2n+1)}{L}\right), \tag{6.9}$$

where C is some constant. For each of the six interfaces indicated in Fig. 6.26, the autocorrelation function is calculated, according to which the wavelength was determined in two ways:

- at the first intersection of the abscissa axis by the autocorrelation function graph; the change of this parameter with depth is reflected in Fig. 6.28a;
- by the average distance between the uniformly located extrema of the residual autocorrelation function (RAF); the dependency of this criterion on the depth is indicated in Fig. 6.28b.

The choice of the analytical dependency $\lambda(d)$ is made on the basis of the criterion χ^2 using the simplest dependencies (power polynomials, exponential, logarithmic) and indicates that the best approximation is achieved by the function

$$\lambda(d) = C_1 + C_2 \exp(C_3 d) th(C_4 d). \tag{6.10}$$

This relation confirms the effect of the dispersion law of standing gravitational waves on the planet's interior structuring and the change with the depth of lengths of these waves formed in a volume of density nonequilibrium medium.

Summarising the discrete-wave and fractal representations obtained for the Earth's crust, lithosphere and deeper geodynamic layers of the planet, let us consider the discrete-wave model of the deep geological section and try to generalise it

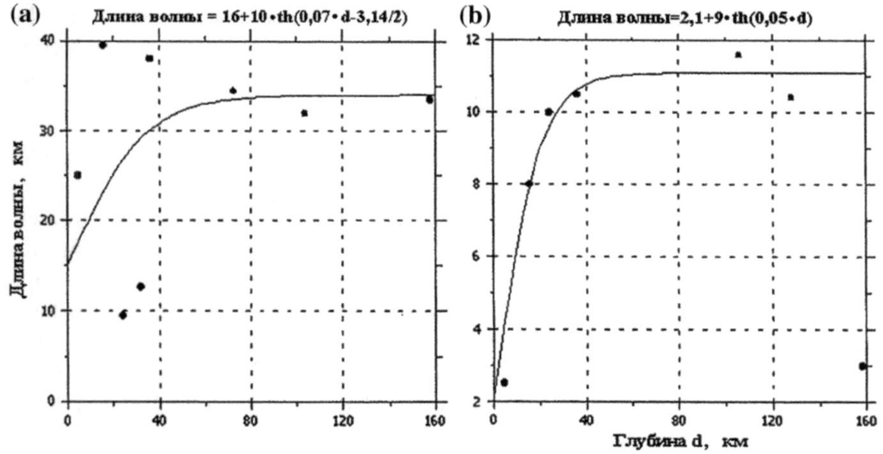

Fig. 6.28 Selection of the variation in the length of the gravitational wave with depth along the autocorrelation function **a** selection of the dependence $\lambda(d)$ with the definition of λ obtained from the first intersection of the ACF profile of the stratification surface with the abscissa axis (displacement axis); **b** the same as (**a**), but with λ being determined by the periodicity of the ACF extremes. {Legends on Fig. 6.28 **a** l to r: Wavelength km, Wavelength = 16 + 10 · th (0.07 · d − 3, 14/2); **b** t to b, Wavelength = 2.1 + 9 · th (0.05 · d).}

for a planetary scale. In the first case, the following provisions must be adopted as initial:

- model of the interrelation between convective and wave motions in the volume of a density nonequilibrium flat layer—here, interrelation refers to the deterministic geometric relationship of the streamlines (lines of particle circulation along closed trajectories during convection) and displacement lines (wave profiles describing particle oscillations caused by convective motion;
- one of the main corollaries of Bénard's experiments is the functional interrelation of the dimensions of convective cells, hence, the dimensions of cellular dissipative structures, i.e., the length of cellular internal gravitational waves, and the thickness of the density nonequilibrium layer. For example, the area of the Bénard's hexagonal cell S and the thickness of the nonequilibrium layer d are related as $S = (2.7–3.8)d^2$, and in the case of the cell diagonal length L the experiment has indicated the validity of the dependency in the form of $L = 2\sqrt{2}d$;
- the previous point is valid for thin density nonequilibrium layers that occur in a physical experiment. In the case of a section, and, even more, on the scale of the entire planet, we can speak about a layer whose vertical dimensions are commensurable with the lateral ones. Such a layer, as shown by Turner's experiment (Fig. 6.27), due to abrupt change with the depth of the length of standing internal gravitational waves, tends to spontaneous stratification;

– it should be emphasised by a special line that in the geological section wave-like discontinuities are considered to be standing waves, since the speed of tectonic movements can be regarded as significant only on a geological time scale, while on the scale of an individual human life it is negligible.

Finally, the model of the geological section is formed by a family of standing internal gravitational waves, the length of which is commensurate with the thickness of a separate density nonequilibrium layer, by analogy with the Bénard cells. Generally, the ultimate depth of the reconstructed section is selected, starting from the average depth of occurrence of one of the acoustic discontinuities; for example, the Moho discontinuity, the roof of granite-gneiss, granulite-basite and other layers can be chosen as such a discontinuity for deep sections. The average length of the lateral wave formed at the deepest section of the discontinuity is defined as the side of an equilateral hexagonal cell. Assuming that the layer initially was in the equilibrium state, we can apply the Boltzmann law, according to which the distribution of particles in the volume of a nonequilibrium medium placed in an external force field is given by an exponential law (Fig. 6.29).

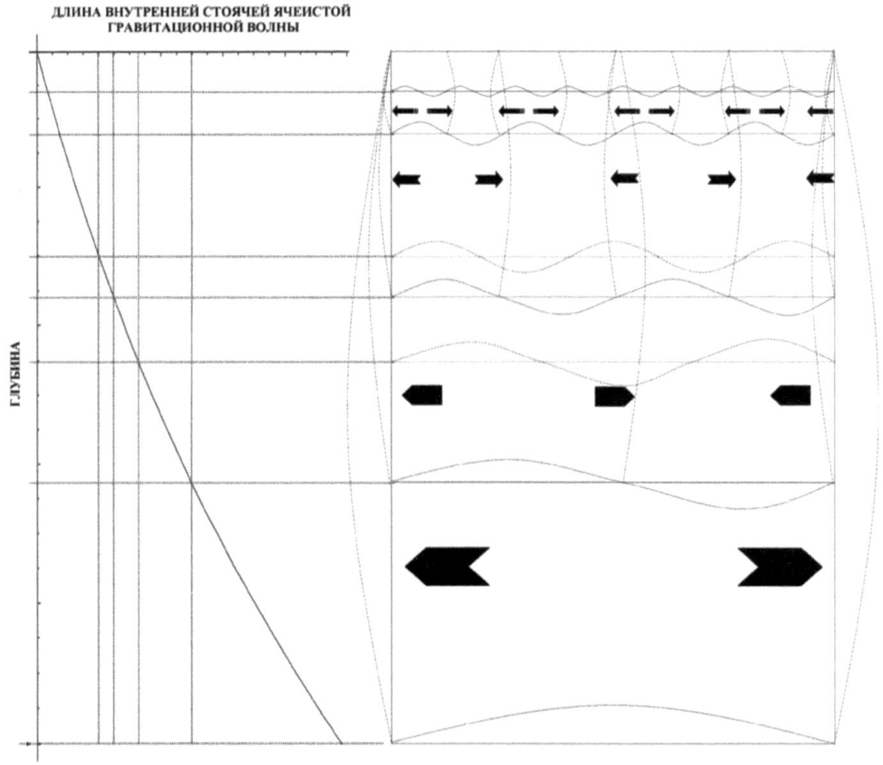

Fig. 6.29 Wave model of the geological section structured by the family of volumetric cellular internal gravitational waves {Legends on Fig. 6.29, t to b: Wavelength of internal standing cellular gravitational wave, Depth}

In this case, for the reconstruction of the geological section, the Boltzmann law applies to the deterministic position of the deep levels, with respect to which wave-like discontinuities with deterministic wavelength are depicted. For example, if the ultimate section depth is taken to be unity, then the maximum thickness of the lowermost nonequilibrium layer in the section whose sole is characterised by the greatest length of the lateral standing gravitational wave is 0.375 of the indicated ultimate depth. The formation overlapping this lower layer should, according to the exponential law, be twice as thin, and thus the length of the internal gravitational wave on its base layer will be halved in comparison with the length of the gravitational wave depicted at the section base. Thus, as you move upwards along the section, the thickness of each n-th overlying layer is reduced by half in comparison with the thickness of n − 1-th overlapped layer, and the length of the internal gravitational wave formed on the sole of the n-th layer is reduced in n times in comparison with the length of the gravitational wave depicted at the base of the section. On the land surface, the length of the gravitational wave is calculated only on the basis of a superposition of multi-depth internal gravitational waves. The sign of the topographical relief forms (positive for elevations and negative for depressions) is determined in this model, proceeding from the principle of conformity: the near-surface structures with respect to the deep structures are inverted (the synform at the same depth corresponds to the elevation having the same lateral dimensions on the day surface, and vice versa). This principle corresponds to the phenomenon of multi-level isostatic compensation known from the gravity surveying, the special cases of which are analysed by the Pratt and Airy models.

In addition to the family of lateral internal gravity waves, a family of vertical cellular gravitational waves is formed in the geological section due to the three-dimensional character of the wave process developing in density unstable media. Half of their wavelength is numerically taken to be equal to the selected depth, for which the characteristics of lateral gravitational waves are calculated. In this case, the deterministic correlation between the signs of vertical and lateral gravitational internal waves, taken from the Prandtl's model indicated in Fig. 1.29.

In the generated wave model of the geological section (Fig. 6.29), it is possible to observe certain contradictions. Firstly, according to Prandtl's model, describing the structuring of a density nonequilibrium flat layer by a family of standing internal gravitational waves, this layer is stratified so that the conformity principle at the invariable length of the gravitational wave is compiled on the roof and base of the n-th formation. In our exponential model, the lengths of internal gravity waves on the roof and sole of each formation are significantly different. Secondly, as the experience of our studies shows, the exponential nature of the variation in the length of the gravitational wave as a function of depth in the system of bedding surfaces is a rather crude approximation. For example, the dependency through the hyperbolic tangent function was given above. As another example, we give some formulations based on the classical work on the fluid dynamics of Landau and Lifshitz (1988).

A homogeneous layer of finite depth h is taken as a model, and it is required to determine the propagation velocity of gravitational waves on the surface of a liquid.

As the boundary conditions, the zero normal component of the velocity vector of the liquid flow at the bottom of the layer is chosen:

$$v_z = \frac{\partial \varphi}{\partial z} = 0 \quad \text{at} \quad z = -h \tag{6.11}$$

where φ is the velocity vector field potential. Then the general solution of the system of differential equations of motion and continuity will have the following form in terms of this potential:

$$\begin{aligned} \varphi &= \cos(kx - \omega t)(A \cdot \exp(kz) + B \cdot \exp(-kz)) \\ &= A\cos(kx - \omega t)ch(k(z+h)), \end{aligned} \tag{6.12}$$

where A and B are some constants; $k = 2\pi/\lambda$ is the wave vector; z—the coordinate axis, coinciding with the axis of depths and oriented vertically upwards; x—the coordinate axis, lying entirely in the plane of the roof or the sole of the layer; ω—the vibrational frequency of particles of a nonequilibrium flat layer, determined from the wave dispersion law

$$\omega^2 = gk \cdot th(kh). \tag{6.13}$$

With these relations in mind, a general expression is derived for the propagation velocity of a gravitational wave along a selected horizontal axis:

$$U = \frac{\sqrt{g}}{2\sqrt{k \cdot \mathrm{th}(kh)}}\left(\mathrm{th}(kh) + \frac{kh}{\mathrm{ch}^2(kh)}\right). \tag{6.14}$$

In this general solution, three particular cases are possible: as we immerse in a density nonequilibrium flat layer, we pass from a layer whose thickness is small in comparison with the length of the gravitational wave λ ($kh \ll 1$, th $(kh) \approx kh$), to the layer whose thickness is commensurable with the length of the gravitational wave ($kh \geq 1$, th$(kh) \approx 0.72$), and in the extreme case to a layer whose thickness exceeds the length the gravitational wave ($kh \gg 1$, th$(kh) = kh$). Considering each of these cases separately with respect to the wave dispersion law written above, we obtain an abrupt or discrete increase of λ with a depth h. In particular, for the first case, assuming the frequency ω as a function of only the primary source of disturbances (by analogy with optics), we obtain

$$\lambda = C \cdot \sqrt{gh}, \tag{6.15}$$

where C is some constant. As can be seen from Fig. 6.30, the wavelength distribution over the depth of the density nonequilibrium liquid layer differs strongly from the exponential model depicted in Fig. 6.29.

A compromise model of the wave structuring of the geological section is indicated in Fig. 6.31. From the exponential approximation according to the Boltzmann

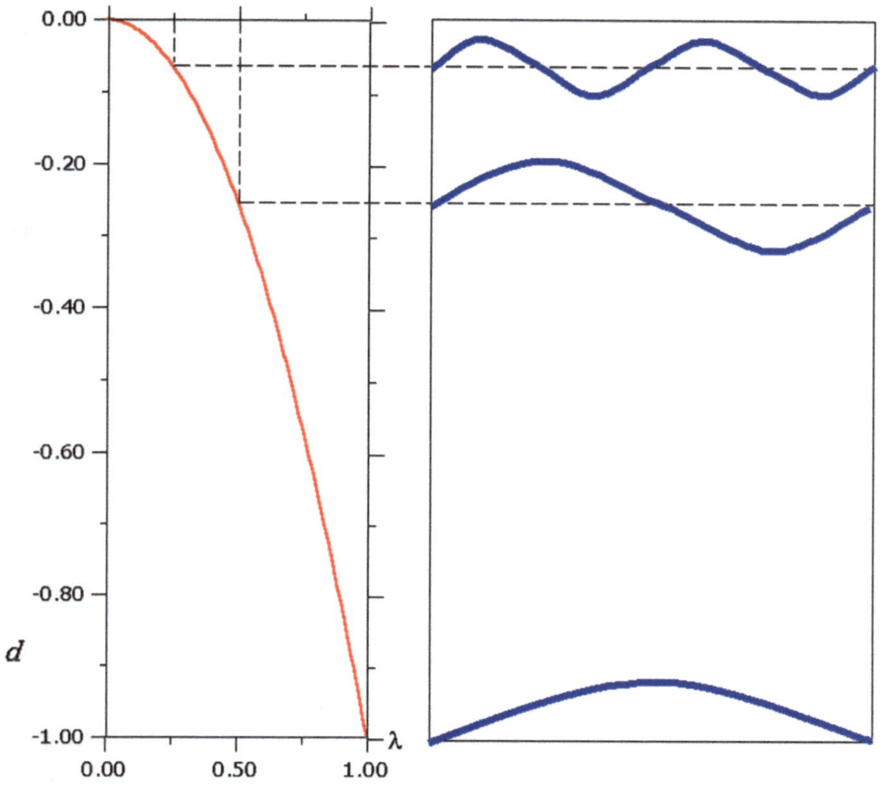

Fig. 6.30 Model of spontaneous stratification of the density nonequilibrium flat liquid layer according to the root length dependence of the internal gravitational waves on depth

law, the proportional change in the thickness of the formations remains during the movement vertically along the section: as displacement occurs from the bottom up, the thickness of the $(n - 1)$-th layer is two times larger than the thickness of the n-th layer. The gravitational wave in this model develops not only along the bedding surfaces, but also in the formation volume; as it rises from the bottom of the section up to the land surface, a more or less deterministic decrease in the length of the gravitational wave is assumed. We note that this assumption, which was repeatedly encountered in our constructions, corresponds to the empirical views of most geologists, that the deeper structures reflected in the topographical relief are larger and have larger spatial step in the case of the regular nature of these structures. Hence, returning to the description of the wave model of the section, we note that on the roof and the sole of an individual n-th formation, the gravitational wave develops in antiphase (in conformity) and does not change its length in an appreciable way. However, in the volume of the n-th formation, the length of the gravitational wave is halved, and a wave conformal to it develops on the roof of the

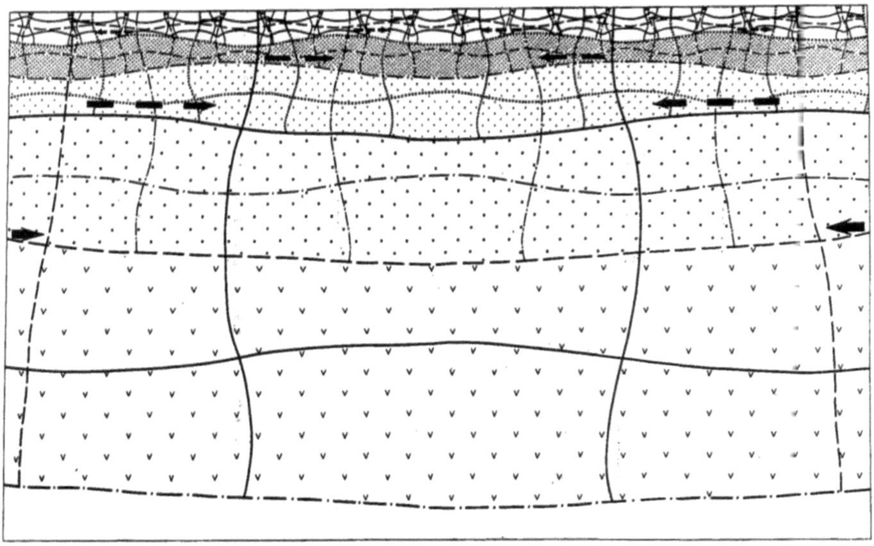

Fig. 6.31 Model of the development of volumetric internal standing gravitational waves, determining the dissipative structuring of the geological environment—in particular, its stratification

(n + 1)-th overlapping formation. In the volume (n + 1)-th formation, the length of the internal gravitational wave again decreases by half and the wave conformal to it is formed on the roof (n + 2)-th formation, and so on, up to the land surface. In view of the three-dimensional nature of the wave process noted above, along with lateral gravitational waves, vertical waves develop, the sign and length of which are connected deterministically with the sign and length of lateral waves. For example, in the case when a pair of conformally-coupled lateral gravitational waves forms an antinode (an increase in the formation thickness), vertical gravitational waves in the vicinity of the antinode reflect the reciprocal convergent displacements. This makes it possible to speak of the possibility of not only vertical, but also lateral migration of matter in a density nonequilibrium section. As the wavelength of the lateral gravitational waves decreases, the wavelength of the vertical gravitational waves also decreases in the same proportion.

In trying to compare Figs. 6.29, 6.30 and 6.31 with a specific geological section, it must be borne in mind that the wave images depicted in these figures are very conventional, primarily because they are depicted for a fixed (invariable or stationary in space) law of the dependency of the length of internal gravitational waves on their depth. In fact, one of the constants in such a law changes as a function of spatial coordinates in a geological section, and the models considered above will only apply to individual areas of this section. In other words, wave approximations are applicable only under the conditions of estimating the spatial stationarity of the geological section, performed either by direct observations, for example, by seismic-wave sections or by indirect observations, let us say on the basis of a description of the land relief forms or geophysical fields.

With respect to geophysical potential fields, for which the ratio between the linear (area) dimensions of the anomalies, their amplitude, on the one hand, and, on the other, the depth of occurrence and the average physical parameters of the anomalous source, is determined, this model is correct primarily for the interpretation of the gravitational field structure. Firstly, gravitational anomalies reflect the specificity of density anomalies in the geological environment; that is, they are directly related to the source of standing cellular internal gravitational waves. Secondly, the distribution of density anomalies in the geological environment, unlike anomalies of, for example, magnetisation or polarisability, has a continuous character everywhere, except for the land surface. We note here that continuity is one of the main properties of any wave process. In addition, the advantage of the gravitational field over other geophysical data, such as, for example, spectrozonal space images, is the relatively low cost of the former when gravity surveying large areas. The approbation of the idea of the wave representation of the geological section structure will be performed using the example of the interpretation of the gravitational field through the Bouguer reduction given along the AR-1 deep seismic profile. It extends from the northern slope of the Baltic anticlise to the South Barents depression (Fig. 6.32).

The initial geological section, considered as a reference for the interpretation of reconstructions along the gravitational field, is based on seismic materials: the decipherment of the sedimentary cover structures is based on the data of the seismic reflection method with the refinement of the velocity properties of terrigenous deposits from the data of seismic refraction method, and in the case of basement constructions and below, the significance of these seismic methods was the reverse.

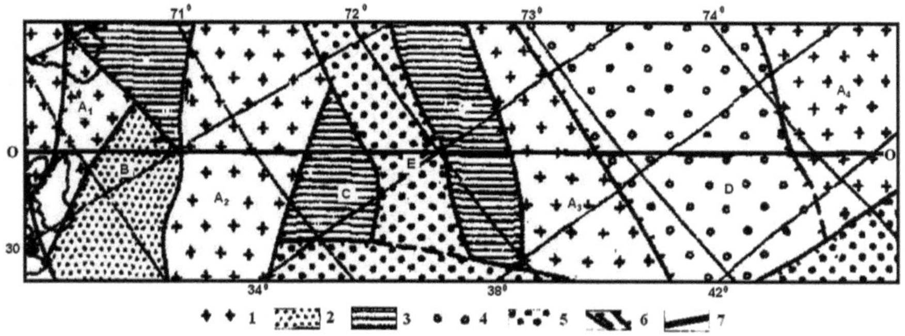

Fig. 6.32 Scheme of the location of tectonic structures occurring along the regional profile AR-1, indicated by the central line O. (1—blocks of the Karelian basement: A1—the northern slope of the Baltic Shield, partially overlapped with Riphean formations; A2—North Murmansk subsurface offset; A3—Fedinskii subsurface offset; A4—Fersman subsurface offset; 2—Riphean rifts: B—East Varanger; 3—Caledonian rifts: C—West Kola; 4—Middle Paleozoic aulacogens: D—Demidov; 5—Late Paleozoic Triassic rifts: E—Varanger graben; 6—regional faults and shifts; 7—the geotraverse line AR.1 and 100 km stations

The bottommost structural boundary, traced along the lateral in the section, was the Moho discontinuity, lying in the depth ranges from 35 to 40 km. At the same time, the minimum depth of its occurrence corresponds to the southern flank of the Demidov aulacogen, and the maximum—to the shelf continuation of the Baltic Shield structures. A paradox arises when considering the rise of the discontinuity M up to 35 km under the North-Murmansk buried offset of the Archaean basement, as well as the high-amplitude deflection of the M discontinuity under the northern wing of the Demidov aulacogen. The paradoxicality consists in the absence of explicit observance of the principle of isostatic compensation, or, according to the terms we have introduced, in the absence of conformity of the displacements on the roof and the sole of a density nonequilibrium layer. The Conrad discontinuity lying higher inside the consolidated crust repeats the main forms of the discontinuity relief M with a variation in the thickness of the lower crust in a relatively narrow range of 11–13 km.

Due to the riftogenic processing of the crust, the K discontinuity is not traced within the West Kola trough. The upper layer of the consolidated crust, also referred to as granitic, with a thickness of 10–24 km, can be traced throughout the section. The minimum values of this thickness correspond to the Riphean downfold in the region of the Kola-Kolguev monoclise, as well as the Demidov aulacogen. The maximum thickness of the granitic layer is established within the shelf continuation of the Baltic Shield. The overlying crystalline basement consists, according to seismic data, of two acoustic layers. The lower layer with a boundary velocity of 6300 m/s has a roof lying at an average depth of 10 km and is compared with the formation of the Archaean gneisses. The overlying metamorphic stratum with an average thickness of about 8 km is defined as the Lower Proterozoic Pechenga series. The shelf area, where the Lower Proterozoic stratum is widespread, is believed to belong to a large palaeodepression that adjoins the Murmansk Archaean block from the south and to a similar offset of the ancient North Murmansk basement from the north. Overlapping the roof of the crystalline basement, the seismic horizon significantly changes the depth of its roofing, varying along the profile from 0 to 12 km, and by indirect signs dates back to the Riphean by the analogy with the Rybachinsk series. To interpret the nature of the Riphean stratum, Khain's tectonic model of the incompleteness of the geosynclinal regime and its gradual change to platform is used. In the terrigenous formations, unconformably deposited on the Riphean deposits, three structural-material complexes are distinguished. The lowest of these, having a variegated material composition, covers the uppermost Cambrian, Ordovician and a significant part of the Silurian (Caledonides) ages. The middle, mostly terrigenous layer, includes the Upper Silurian and Carbonate Carboniferous-Lower Permian stratum (early Hercinides). The upper layer of the sedimentary cover is composed of terrigenous formations of Upper Permian, Triassic, Jurassic and Lower Cretaceous.

In the structural section reconstructed from the seismic picture (Fig. 6.33), the following features can be observed:

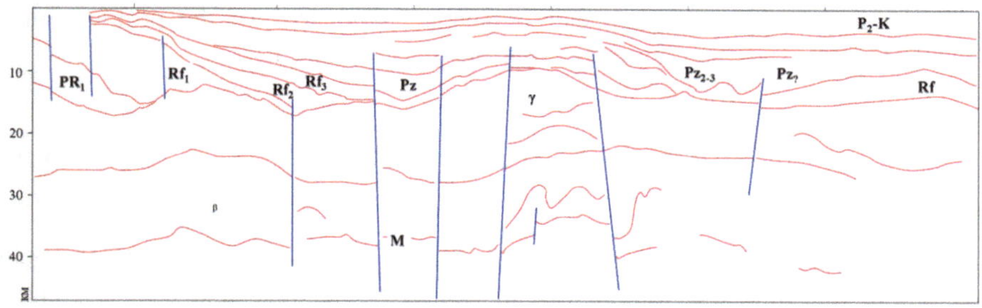

Fig. 6.33 Structural tectonic section along the profile of AR-1 (according to Verba and Atakov 1999). Red lines mark the stratification surfaces; blue lines mark the disjunctives reconstructed by the seismic-wave section

- all discontinuities comprise both long and short wave components (Fig. 6.33). Thus, the wave model can give a result only in the case of a statistically significant dominant of one of the wavelengths;
- the section includes vertically disposed disjunctives typical for wave patterns, reconstructed from the lateral tracing of individual horizons, as well as from their vertical displacements (Fig. 6.33). At the same time, the well-known physical-tectonic models deny the existence of steeply dipping faults that cut the entire Earth's crust, affirming the gradual smoothing of the occurrence of their planes as they approach from the land surface to the Moho discontinuity;
- lateral density distributions, known from seismic data within the limits of individual formations, form an integral picture explaining the rise of the M discontinuity in the region of the North Murmansk offset and its downfold in the vicinity of the Demidov aulacogen as isostatic compensation effects of density anomalous objects localised mainly below the sole of the Riphean formations. The alternations of increased and decreased values of the lateral density are characterised by different spatial steps for different depths, which can also affect the reconstructed picture of standing inner gravity waves;
- the reconstruction of the structural-tectonic section along the gravitational and magnetic fields by standard selection and analytic continuation methods, as indicated by comparisons with the traced seismic boundaries, do not reflect, with satisfactory accuracy, the relief features of the bedding surfaces described above.

We consider the gravitational field in Bouguer reduction ΔG_{B} to be a result of a superposition of gravitational responses with different wavelength. The regularity of its variation with depth is defined as deterministic only in the region of spatial (lateral) stationarity of the initial gravitational field. An independent analysis of its components with different wavelengths is algorithmically realised in the form of a reduction procedure with the transfer function $h(\tau)$, determined at intervals of different lengths. As the parameter of spatial stationarity of the profile signal, presented here ΔG_B, the autocorrelation radius is chosen. If

$$R(\tau) = \int_{-\infty}^{+\infty} \Delta G_B(x) \cdot \Delta G_B(x+\tau) \mathrm{d}x \quad (6.16)$$

is the autocorrelation function (ACF) of the initial gravitational field in the Bouguer reduction, reflecting the dominant of the non-periodic trend and the wavelength ranges of the periodic components, then the autocorrelation radius r_{ACF} is the magnitude of the shift τ along the x-axis, for which in the case of

- constant-sign autocorrelation function the value of $R(\tau) = 0.3 \cdot R(0)$;
- alternating-sign autocorrelation function the graph of $R(\tau)$ intersects the abscissa axis τ at the smallest distance from the point $\tau = 0$.

Algorithmically, in a sliding window, the smallest dimension of which is defined as three spatial steps between the measurement pickets on the profile, we calculate the ACF using the formula (6.16), estimate the presence of positive and negative values of the calculated ACF, and then, using one of the methods, determine r_{ACF}. This window, reaching the end of the profile, returns to its beginning, increasing its dimension by one spatial interval, after which the procedure is repeated again. The ultimate is the case when the dimension of the sliding window is half the length of the entire measuring profile. Then the calculations cease. We referred to these estimates as the reduction procedure above, and write the transformation of the set of gravitational field values in the Bouguer reduction to the set of the autocorrelation radius values as

$$r_{ACF} = h(\tau) * \Delta G_B(x). \quad (6.17)$$

The next step in the calculations is to translate each of the calculated autocorrelation radii into the depths of the anomaly-forming source. This translation is possible proceeding from general considerations, since the autocorrelation radius is a parameter of the lateral extension of the gravitational anomaly, and from the formal position this possibility is justified by the calculations of Strakhov (1979) and Serkerov (1991). In particular, it turns out that the functional dependency of the autocorrelation radius on the depth of the source is determined by the geometry of the latter; that is, it is algorithmically necessary to estimate the given geometry from a system of indirect features. For example, in the case of an isometric anomaly with a sufficiently extended extremum region, it is a question of the source of the spherical shape. For an isometric anomaly with a narrow, localised extremum region, we should speak of an infinite material vertical band. With pronounced asymmetry, when an anomaly forms on the one of its flanks a region of a sharp spatial gradient, a model of a semi-infinite horizontal material line is applicable. Of course, the spectrum of models we used in approbation is broader and the examples given above allow us to write here the particular cases of the relationship between the radius of autocorrelation and the depth of occurrence h:

- for a source of spherical shape

$$r_{\mathrm{ACF}} = \sqrt{8}h; \tag{6.18}$$

- for an infinite material vertical band

$$r_{\mathrm{ACF}} = \pi h_1 \frac{(m-1)^2}{4} / \ln\left(\frac{2^{m+1}m^m}{(m+1)^{m+1}}\right), m = h_2/h_1, \tag{6.19}$$

where h_1 and h_2—the depths of the upper and lower edges of the body;

- for a semi-infinite horizontal material line

$$r_{\mathrm{ACF}} = \pi h. \tag{6.20}$$

The result of such recalculations indicated in Fig. 6.34 has the form of a family of wave profiles that demonstrate the presence of syn- and antiforms in the section, as well as vertical correlation zones. Under the conditions when the above recalculations of the gravitational field ΔG_{B} into the structure of a family of standing internal gravitational waves are performed on the basis of the transfer function $h(\tau)$, given on an interval of variable length, but for the invariant field ΔG_{B}, the family of wave profiles has three main features:

- relief, the sign of features, the amplitude of gradient zones change monotonically with depth from 0 to 50 km;
- in some sections the wave profiles demonstrate the absence of any features of the geological structure, whereas the structure of the gravitational field in these areas can be characterised as strongly differentiated;
- the zones of vertical correlations, traced in the family of wave profiles, are more gently sloping than disjunctives, reconstructed from the seismic wave section, but having approximately the same angle of incidence.

All the noted features can be attributed to the shortcomings of the developed method. At the same time, at this stage of calculations, undoubted correlations of the obtained wave gravity section and the section reconstructed from the seismic survey data can be noted. In particular, the reflection is found by the North-Murmansk buried offset and the adjoining West-Kola rift induced downfold

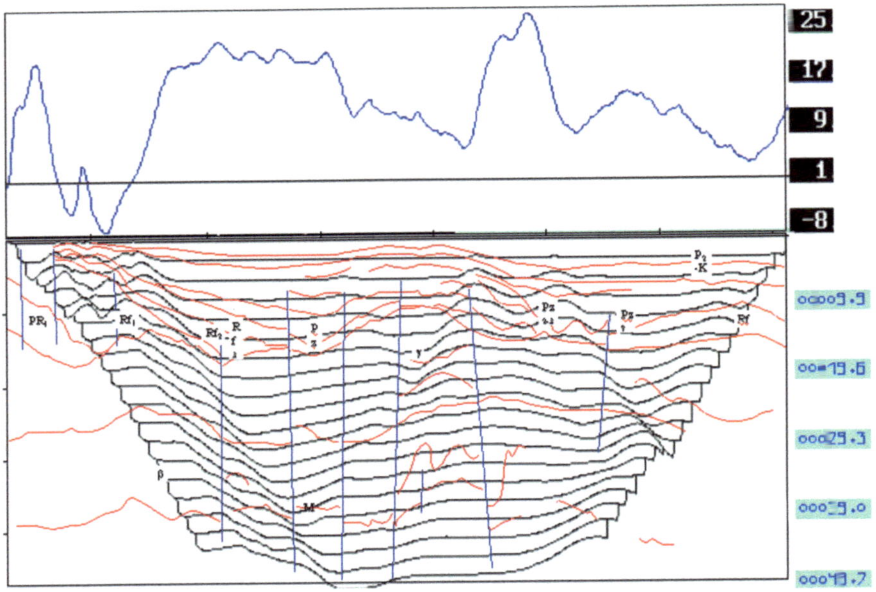

Fig. 6.34 Recalculation of the gravitational field the Bouguer reduction into the wave section, formed with profiles of standing internal gravitational waves. Calculation performed in a sliding window of variable dimensions, in each of which the same initial field ΔGB was operated (blue graph at the top). Red and blue lines, filling the plane of the section repeat Fig. 6.33, black curves —the result of recalculation of the gravitational field. According to Petrov et al. (2006)

in the south-western part of the profile. In the north-eastern part of the profile one can see the Demidov aulacogen separated by an offset from the adjacent Varanger graben. However, these correlations in the central part of the profile are valid only for the lower part of the section and are rather conditional because of the smoothed nature of the profiles of the calculated internal gravitational waves.

By improving this method of quantitative interpretation of potential fields, we assumed that a given range of the transfer function $h(\tau)$ should correspond to a narrow range of wavelengths in the structure of the initial gravitational field. In other words, now in recalculation at each of its stages not the same spatial signal ΔGB takes part, but one of its components, obtained at the stage of preliminary frequency filtering. The family of calculated profiles of multi-depth standing internal gravitational waves is indicated in Fig. 6.35.

The visual comparison of the structure of the seismic wave section with the structure of the section of standing internal gravitational waves, calculated by the modified method, demonstrates high correlations across the entire plane of the section. Suture zones separating regional structural units, including blocks of the Karelian basement, as well as Riphean, Caledonian and Hercynian rift induced downfolds, are clearly pronounced. It is possible to see a sharp, close to abrupt change in the length of standing internal gravity on the boundary between the

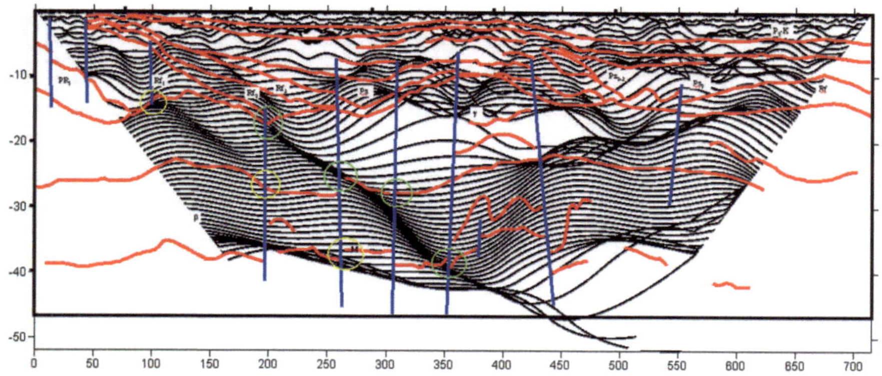

Fig. 6.35 Recalculation of the gravitational field in the Bouguer reduction to the wave section, formed with profiles of standing internal gravitational waves. Calculation performed in a sliding window of variable dimensions, in each of which one of the frequency components of the initial field ΔGB was operated. According to Petrov et al. (2006)

Paleozoic and Riphean formations, on the basis of which the roof of the consolidated foundation is confidently pronounced. The areas of violation of lateral tracing of individual bedding surfaces, combined in the seismic interpretation into subvertical disjunctivities, are distinctly distinguished as sharp gradient zones in a family of standing internal gravitational waves, while these zones, according to models of contemporary geotectonics, are smoothing out as they move from the land surface to the roof of the upper mantle.

The resulting section, as, indeed, all geophysical sections, requires additional interpretation, which determines their rejection among geologists. In particular, in our last section, in addition to a very unusual representation of geological structures, there is a discreteness—the presence of an unfilled space in the interval between individual standing internal gravity waves, while the geological section is on the average a continuous medium. To compensate for this shortcoming, we performed a recalculation of each frequency component of the gravitational field in the Bouguer reduction on the corresponding depth levels specified by a separate profile of a standing internal gravitational wave. In addition, this recalculation is supplemented by calculating the apparent density given for a prismatic body that is depth-limited above and below the underlying internal standing gravity wave profiles, and on the sides limited by the positions of the pickets. The traditional image of the distribution of apparent density in a section in the form of an isoline map (Fig. 6.36), when compared with a seismic-wave section, was unconvincing.

We propose a method of illumination with a subsequent halftone image of the final section (Fig. 6.37). Technically, this method is widely distributed and implemented by software in specialised cartographic programmes to analyse the structure of isoline maps. Physically, this method considers the distribution of density values as the distribution of the values of the absolute heights of some conditional surface, the presence in which regularly connected local forms can be

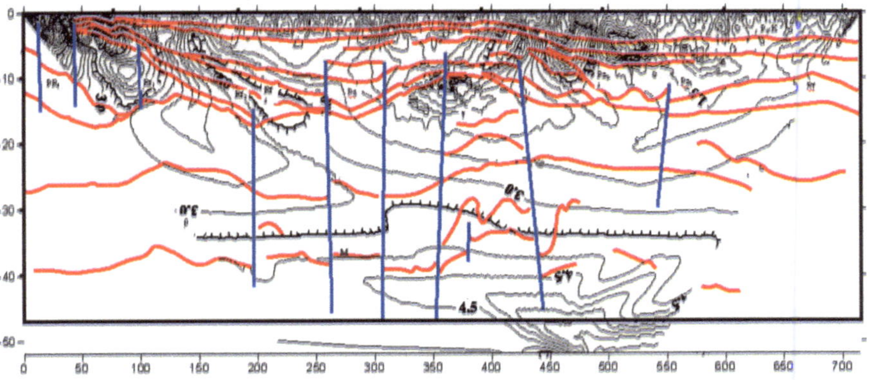

Fig. 6.36 Distribution of the apparent density (g/cm^3) in the geological section plane represented by an isoline map with an overlapping of the result of the interpretation of the seismic wave section. According to Petrov et al. (2006)

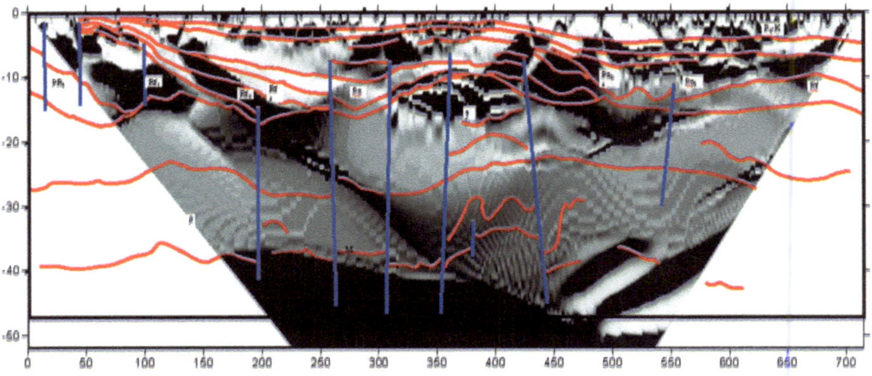

Fig. 6.37 Reconstruction of the structural tectonic section along the profile of AR-1 using a method of calculation of a system of standing internal gravitational waves and the calculation at the level of each wave of apparent density with a plutonic representation of its distribution in the plane of the section. According to Petrov et al. (2006)

revealed by illuminating under a certain azimuth. The selection of the conventional height of the light source and the suitable azimuth of the beam incidence results in the formation of a system of illuminated and shadow-closed regions joined by the eye into extended or zircoidal structures. Mathematically, the light source is assumed to be homocentric, and the scattering of rays incoming on the conditional surface satisfying the Lambert condition, then the brightness of each illuminated area of the conditional surface proves to be proportional to the sine of the angle of incidence.

A comparison of the final section, constructed on the basis of the calculation of standing internal gravity waves, with a seismic wave section, as indicated in Fig. 6.37, reveals additional correlations to those indicated in Fig. 6.35. In particular, we can talk about tracking the sole of the second structural-material complex of the terrigenous cover, represented by early Hercinides. It is clear that most local and regional syn- and antiforms are reflected in the structure of the seismic wave section both within the sedimentary cover and at the interface between the sedimentary cover of the Riphean and Proterozoic formations. We can assume the presence of a deep suture zone of the northeastern dip azimuth separating the block, designated as the North-Murmansk buried offset, and the block of the Earth's crust that corresponds to the West-Kola rift induced downfold.

We moved from a consideration of one of the properties of standing internal gravity waves, namely, the fractal dimensionality of the dissipative structures of the Earth, to the calculation and interpretation of the intrinsic internal gravity waves. In this case, if fractal dimensionality is an indirect feature on the basis of which it is possible to construct some qualitative assumptions about the spatial hierarchy and genesis of geological formations, then our proposed calculation (Petrov et al. 2006) of the family of standing internal gravity waves refers to the quantitative interpretation of geophysical potential fields and is related to the structural-tectonic reconstruction of the section. It is worth emphasising the importance of this method of reconstruction, since, firstly, its results are correlated with seismic-wave sections, which distinguishes our method from classical selection methods and analytic continuations, and secondly, the resulting structural sections reflect the multi-ordinal disjunctive and plication dislocations, which is significant in forecasting tasks.

It should be recalled that in theoretical and practical geophysics, the preference has been given for such principally new technological schemes that combine the principles of statistical, approximation and frequency approaches to the interpretation of gravitational and magnetic anomalies (Strakhov 1963, 1970, 2002; Gravity Survey. Handbook of Geophysics 1990; Petrischevsky 1990, 2004; Lebedev and Petrov 2001). There are also some very successful examples of the use of autocorrelation functions in tectonic and prognostic division (Berland and Rose 1971; Serkerov 1976; Petrishchevsky and Demura 1987; Petrishchevsky 1988, etc.).

6.3 Application of the Wave Nature of Dissipative Structuring Processes of the Geological Environment in Solving Predictive Problems

The classical theory of interpretation of geological and geophysical data in connection with the solution of predictive problems of applied geology is traditionally divided into three main approaches.

The first of these approaches refers to the methods of qualitative interpretation and is oriented towards the use of geological and geophysical data both as direct and indirect search characteristics, more or less contrastively reflecting the spatial position of geological objects, as well as their hierarchy and genesis. This approach is based on the methods of detailed study of the geology of mineral deposits and the creation on this basis of models of their formation in the geological space within the hierarchy of ore fields, nodes, as well as regional, supra-regional and planetary structures. Methodologically, this approach resembles the principle of "look for ore near ore". Noting the difficulties encountered by geologists in the implementation of this approach, one of the largest experts and specialists in the field of ore geology, Krivtsov (2006) asks: "… in understanding how the deposits emerged, do we know where to find them?!" Trying to answer this question and assuming that the wave model for the formation of geological structures also determines the discrete-periodic localisation of many endogenous ore deposits, we can obviously try to formulate new criteria for their forecasting.

Ideas on the influence of wave mechanisms on the global discretisation of geological space in the form of periodically ordered structural elements that have a decisive influence on the location of minerals, including groundwater, oil and gas, are comprised in the works of Bogatsky (1977), Vityaz (1982), Kvet and Nemets (1976), Nemets (1972), Piotrovsky (1964), Baryshev (2006), Abukov and Kartsev (1999) and others. Thus, V. V. Bogatsky considered the main provisions of the concept of the wave mechanism for the formation of ore-localising structures of magnetite deposits of the Siberian Platform. This concept assumes that energy transfer in geological processes is carried out by oscillations and can be expressed by the laws of wave mechanics. Geological structures are considered here as the areas of interference-resonant interaction of stress waves (compression-stretching) of different amplitude and frequency. The continuity and discreteness of wave stress fields also determine the spatial periodicity of geological structures for which the corresponding moiré patterns can be selected. Indeed, the flat moiré grids shown in this paper demonstrate that known features of the brecciated-cleavage ore localisation structures that determine the morphology of ore bodies can be the consequence of specific conditions for the manifestation of interference of wave fields (Bogatsky 1977).

Periodic alternation in the space of diamondiferous systems, determined by the rise of the deep material, was also indicated by the author in the works. The physical basis of this pattern is the wave nature of the advection origin. Based on the study of wavelength dependency on viscosity, a hierarchy of geological structures associated with convection cells in different layers of the Earth was compiled (Baryshev 2001a, b, 2006). According to Baryshev (2006), an advective system with upwardly changing viscosity as a whole has a fractal (tree-like) structure in which a series of small cells appears on a large one, and on each of them a series of smaller ones arise. Because of the wave nature of the uplift origin, the distances (steps) between the fractals of a certain order become commensurate. Thus, the distances between the centres of asthenospheric uplifts of the II dimensional order are 1000–2000 km; between third-order cells in the upper layers of the

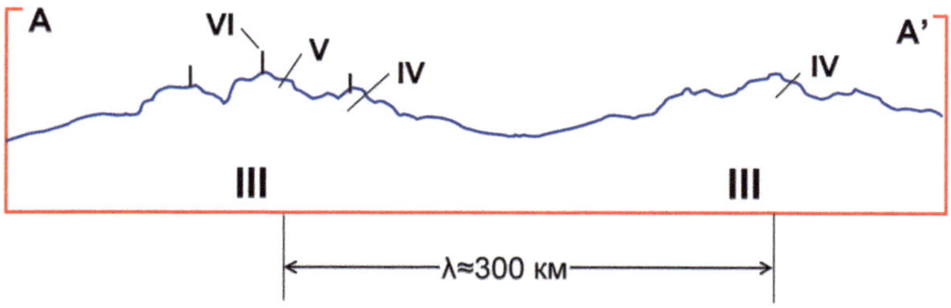

Fig. 6.38 Model of the periodic location of advective fractals of the III, IV, V, VI dimensional orders for diamond-bearing areas. Fractals: III—superregions; IV—fields; V—groups of pipes; VI—pipes (Baryshev 2006). {Legend on Fig. 6.38: $\lambda \approx 300$ km}

asthenosphere—200–350 km; between the incipient magmatic foci (IV order)—about 30 km; between chambers (V order)—1–3.5 km; between the apophyses leaving the chambers (VI order)—about 0.3 km (Fig. 6.38).

While at this stage of research with reference to diamond mineralogy there is no clearly defined criterion for determining the exact position of the centre of systems of different hierarchical levels; nevertheless, the average step between the centres of the bushes probably gives an idea of the length of the advective wave that determines the distance between the chambers in the magma focus. In Fig. 6.39, compiled according to the work of Androsov et al. (2005); Sablukov (2001) and Baryshev (2006), possible positions of centres or contours of magmatic nodes (supra-focal areas), reflecting the advective systems of the IV order, are outlined.

From the standpoint of the concept of dissipative structures, the arrangement of diamondiferous fields and regions tends towards structures of the L-hexagons type with emersion in the centre and immersion in the marginal parts of the mantle and crustal substrate. Moreover, in the marginal parts of geoblocks, which are the most promising for the detection of kimberlite pipes, the lithospheric compression environment emerges. This is evidenced by the features of kimberlite magmatism, which does not possess significant thermal energy, and the presence of the isotropic minerals of cubic syngony (diamond, garnet, pyrite), as well as readily mobile mineral components (carbonaceous matter). In contrast, according to Baryshev (2006), the central sections with a clearly expressed gneiss formation reflect the areas of metamorphogenic squeezing with the greatest difference in principal normal stresses. These facts may indicate the need to take into account the heterogeneous field of tectonic stresses for various types of dissipative structures (L-hexagons, G-hexagons, Rolls) in predictive constructions.

In this connection, of great interest is the above data on the discrete distribution of copper-nickel deposits in the Norilsk region, which were studied in detail by the author of this work. These deposits are confined to the hinge parts of isometric moulds of 65–75 km in diameter or III-order dissipative structures of the L-hexagon type with immersion in the centre and emersion in the marginal parts of the deep mantle substrate (Fig. 1.41). Geological evidence convincingly indicates that the establishment of these structures occurred at the final stage of trap

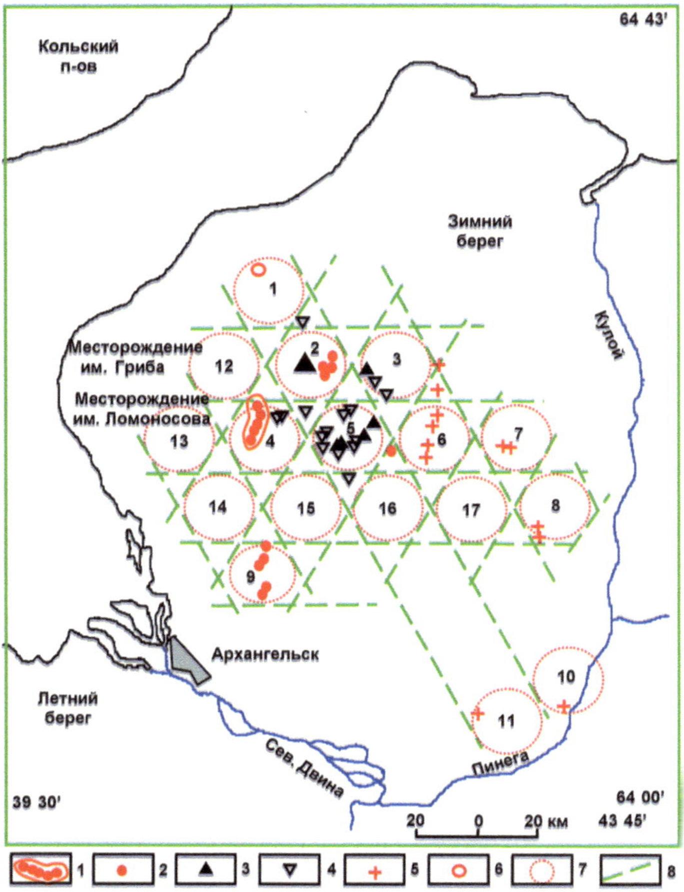

Fig. 6.39 Magmatic manifestations and their nodes in the territory of Zimny Bank (Baryshev 2006). Magmatic manifestations: aluminous series: 1—kimberlites, 2—kimmelilitites and melilitites of the iron-titanian series: 3—kimberlites, 4—kimpicrites and picrites; 5—basalts; 6—carbonatites and kimberlites; 7—assumed contours of magmatic nodes (projections of magmatic foci), including: 1—Melsky, 2—named after Grib, 4—Zolotitsky, 5—Kepinsky, 9—Izhmozhersky, 8—assumed projections of negative phases of deep waves in gravitationally unstable environment. {Legends on Fig. 6.39, t to b, l to r: Kola Peninsula, Zimny Bank, Deposit named after Grib, Kuloi, Lomonosov deposit, Arkhangelsk, Letny Bank, North Dvina, Pinega river}

magmatism, when huge masses of trap magmas were already poured onto the day surface. In the process of melting and lifting an enormous volume of lighter trap magmas, much heavier sulphide phases of the melt could accumulate in deep foci, and at the final stage, during the wave dynamics of the breakdown of the trap plateau into mould-like troughs, they were squeezed into their hinge parts in an inhomogeneous field of tectonic stresses. Such a dynamics of the copper-nickel deposit formation is in a good agreement with the huge volumes of sulphide masses, and their spatial association with sharply differentiated derivatives of trap magmatism (Petrov 1983). Strong evidence in favour of this model of the Norilsk copper-nickel deposit formation was recently obtained on the basis of the most recent isotope-geochronological and isotope-geochemical data. Thus, within the framework of the development of the isotope-geochemical model of the Norilsk copper-nickel deposit formation as well as the formation of criteria for their potential production, we investigated about 2000 of zircon grains. According to the morphology of the crystals (Pupin 1980), D-type zircons dominate, indicating the high-temperature and alkaline conditions of the mineral formation medium. The petrographic study of zircons made it possible to identify several of their varieties, characterised by a different set of melt inclusions. The concentration of thorium, uranium and rare-earth elements in the studied zircons varies within wide limits (respectively, 89–49,220, 75–18,348, 416–20,890 g/t). On the Th-U binary diagram, the zircons of the intrusive formations of the Norilsk region (Th/U = 0.8–5.5) differ from most zircons from different geological formations, but partially overlap the fields of mantle metasomatites (MARID) (Kinny and Dawson 1992). On discriminant diagrams (Sm/La) N-La and Ce/Ce*—(Sm/La) N, according to (Hoskin 2005), Norilsk zircons occupy the field of magmatic zircon compositions and also form a trend of compositions in the direction of hydrothermal zircon field.

The revealed discreteness of the distribution of U–Pb ages in the 220–260-million-year interval suggested that the crystallisation of zircons occurred in several stages. Polyphase grains also preserve the geochronological information of the formation of older zircons, which are grouped around the boundaries of 300 and 340 million years. Our new data are in agreement with the assumption of the interaction of several contrasting magmatic sources, along with the considerable duration of the processes of fractionating the components in the magmatic system. An indispensable consequence of these processes should be the achievement of a high degree of separation and concentration of ore elements and the formation of ore magmas unique in volume and concentration. To confirm these representations, the initial isotope compositions of Sr, Pb, and Nd in pyroxenes, plagioclases, and co-existing sulphides were determined by the TIMS method (Finnigan-MAT, "Triton") in industrially ore-bearing Norilsk type massifs with the largest reserves of Cu–Ni sulphide ores (about 450 isotope analyses). The initial isotope compositions of lead in plagioclases and sulphides (67 determinations) differ significantly from one another, indicating their genetic heterogeneity. Nevertheless, the most contrasting rock-forming minerals and sulphide ores differ in the initial isotope composition of strontium. In the massive Cu–Ni sulphide ores of the Kharaelakh and Talnakh intrusions with strontium content from 4 to 30 g/t, the value of ($^{87}Sr/^{86}Sr$) 0 varies from 0.7085 to 0.7111, while in the associated pyroxenes and plagioclases this value does not exceed 0.7076. It follows that the sulphide ore melt entered the magmatic chamber as an isotopically heterogeneous mechanical impurity labelled with a specific isotope composition of lead and strontium, different from that in rock-forming silicates. At the time of the introduction, the sulphides were significantly richer in the radiogenic isotope of strontium as compared with silicate minerals.

Thus, the given examples of qualitative interpretation of geological and geophysical data reveal the potential for using concepts of the Earth's dissipative structures in the tasks of predicting ore deposits. As a result, a system of criteria can be proposed for predicting endogenous deposits having a wave discrete-periodic nature of localisation and associated with three different types of thermal trans-crustal heat and mass transfer (G-hexagons, L-hexagons and Rolls type structures) and specific recommendations for optimisation of predictive and prospective works can be provided.

At the global level, this pattern was noted in the book, edited by D. V. Rundkvist "Large and super large deposits: regularities of location and conditions of formation" (2004). Based on the application of statistical methods to the analysis of the main regularities of the distribution of the largest deposits, it is established in this book that the main regions of their concentrations are located along the outline of the planetary anomalous heights of the geoid, gravitating toward the areas of inflection of the geoid's surface between positive and negative anomalies. Since, in accordance with the ideas developed here on the dissipative structures of the Earth, the anomalies of the geoid heights are shown to be of a wave dynamic nature and are associated with ascending (negative anomalies) and descending (positive anomalies) heat and substance fluxes, it may be considered quite probable that the anomalies of the geoid heights are indicators of the temperature and pressure field in the mantle, both existing and ancient, which was present during the period of mineralisation.

In essence, this work develops the idea that the voluminous deep inhomogeneities revealed by complex geophysical studies at various hierarchical levels reflect the relic systems of trans-crustal heat and mass transfer, which ultimately determine the metallogenic specialisation of large areas of the earth's crust. The task of processing and interpreting geophysical data when mapping palaeosystems of heat and mass transfer from these positions is the isolation of mass deep anomalous objects characterised by disturbance of seismic stratification of the Earth's crust from anomalous density.

Previously, we have taken a similar approach to the harmonisation of ideas about geological, mineralogical and mineral-resource components of the Earth's lithosphere on the basis of ideas about its cellular (geoblock) divisibility in the work "Harmonisation Experience …" (2001). This work was the result of the creation and analysis of the Geological and Mineralogical Map of the World at a scale of 1:15,000,000, published with explanatory notes (Parts 1–4 (2000)). The main conclusion of this work is the justification of the hierarchical stratification of the lithosphere and mantle of the Earth, which caused its cellular (geoblock) divisibility and is consequently responsible for the discrete-periodic concentration of mineral raw materials.

The second approach to the interpretation of geological and geophysical data and the solution of predictive problems in geology, the methodological and mathematical apparatus of which we have already referred to extensively in the previous section, is connected with the quantitative interpretation of geological and geophysical parameters—namely, with their recalculation in various degrees of parametrised geological, geophysical, morphometric and structural-tectonic algorithms reflecting global, regional and local discretisation of geological space. In continuing to develop the approach outlined in this paper relating to quantitative methods of analysis of the relief and the recalculation of geophysical data, we note that these methods of isolating the fractal hierarchies of the dissipative structures of the Earth, as well as the completely regular nature of the interrelationship of their various types (G-hexagons, L-hexagons and Rolls) with morphostructural elements of the relief and distribution of geostatic and geotectonic stresses in the Earth's crust, fundamentally new approaches to the prediction of groundwater, oil and gas deposits become available. The use of the wave theory in predicting problems can result in new methods of hydrogeological and oil and gas division of sedimentary basins, as well as the determination of optimal conditions for their operation, which significantly increase the water yield and the percentage of oil and gas recovery from the reservoirs. In addition, the wave nature of dissipative structures opens up new possibilities in the forecasting of earthquakes. The point is that, proceeding from the above ideas, under the lithospheric conditions, the wave processes of formation of the Earth's dissipative structures are manifested both as epeirogenetic and as seismic vibrations. The first are an example of slow super-low-frequency waves, according to Bogatsky (1977), known in fluid dynamics as standing inner gravity waves that reflect the behaviour of the geological environment as a very viscous fluid. The frequency of oscillation of these waves cannot exceed the frequency of oscillations of the geological environment. The second relatively the first ones are high-frequency oscillations characterising the geological environment as a solid body. As Bogatsky noted, "… speaking metaphorically, epeirogenesis is a slowly developing" typically" geological oscillatory process, and seismic phenomena are high-speed disturbances, that is, in a sense, "impact" geological waves …" (Bogatsky 1977).

From the standpoint of the ideas developed here in this book, it is especially important to emphasise that epeirogenesis is not only a slowly developing, typically geological oscillatory process, but, more importantly, a typically wave process. And although the frequencies of epeirogenic and seismic vibrations differ incommensurably, the wavelengths when realising the natural oscillations of the Earth as an integral object derived from the equilibrium state will completely coincide. In this sense, both of them characterise the wave properties of the Earth that are fundamental from the point of view of contemporary science as a single system, and seismicity, according to the figurative expression of some geologists (Kats et al. 1989), can be considered as an "echo" of the contemporary structural adjustments of the Earth's crust. The same authors, however, with respect only to the ring structures of the Earth had expressed a very interesting idea that "… and according to seismic data, precisely by locating the earthquake foci in space, it is possible to identify very clearly and outline different-scale ring formations …”, which, as a rule, are established based on the elements of the relief, geophysical fields and individual elements of geological structures. Since, as we can see below, ring structures in the case of their true geological nature are only particular cases of manifestation of fundamental wave processes of density instability of the interior, then these same concepts can be extended to the dissipative structures of the Earth. In connection with this, we have already mentioned earlier that the presence of a regular connection between the hierarchies of the morphometric series of surface structures and the location of foci (hypocentres) of earthquakes was first pointed out by Piotrovsky (1964). In relation to the hierarchies of the Earth's dissipative structures, this empirically-revealed regularity acquires the features of quantitative physical and mathematical definitions, since the modal hierarchies of epeirogenic oscillations from the point of view of the manifestation of the wave properties of matter in the processes of spontaneous structuring of the density unstable masses of our planet will exactly correspond to the set of modes of its own oscillations, observed by us after strong earthquakes.

The question of how the epeirogenic and seismic wave processes generate interrelated stresses in reservoir systems makes it possible to better understand the experiments of Sboev (1988). He indicated the principle possibility of transitions in a strained rock massif of elastic deformation energy into high-frequency oscillations. They noted that quasi-resonant microseismic oscillations with significant amplitudes exceeding 2–3 times the amplitudes of the provoking oscillations are provoked in “good” undisturbed blocks capable of accumulating a significant amount of potential energy from the action of geostatic pressure (Fig. 6.40). In other words, in a natural geological environment, geodynamic compression leads to an increase in the potential energy of elastic deformation, which in turn leads to the appearance of microseismic noise; the imposition of additional (any and arbitrarily small) wave epeirogenic influences on the system initiates the transfer of a part of the deformation energy to high-frequency seismic energy (Abukova and Kartsev 1999). In this case, reflecting the fractal dimensionality of the wave processes of spontaneous structuring of density unstable masses, and corresponding to the set of modes of the Earth's natural oscillations, the seismic process itself possesses hierarchy and discreteness features and thus itself possesses a complex structure of earthquake distribution in space analogous to the hierarchy of dissipative structures of the Earth. As noted earlier, Sadovsky and others (1982) have indicated that this regularity of earthquake manifestation related to the fractal properties of self-similar sets can in a sense be regarded as a spatial analogue of the law of repeatability: for each “seismic” area of a certain rank; on average, there is a certain fixed number of “seismic” areas of lower rank, independent of the magnitudes of the sites.

The interdependence of high-frequency effects of the manifestation of density instability processes entails fluid dynamic and hydrochemical phenomena of a nonlinear nature. Numerous data on hydro-geodynamic and hydrochemical effects observed during earthquakes is of a particular importance for practical applications. Thus, increased geodynamic pressure is noted in the areas of intense folding and increased seismicity. In this case, the ratio of reservoir to hydrostatic pressure often

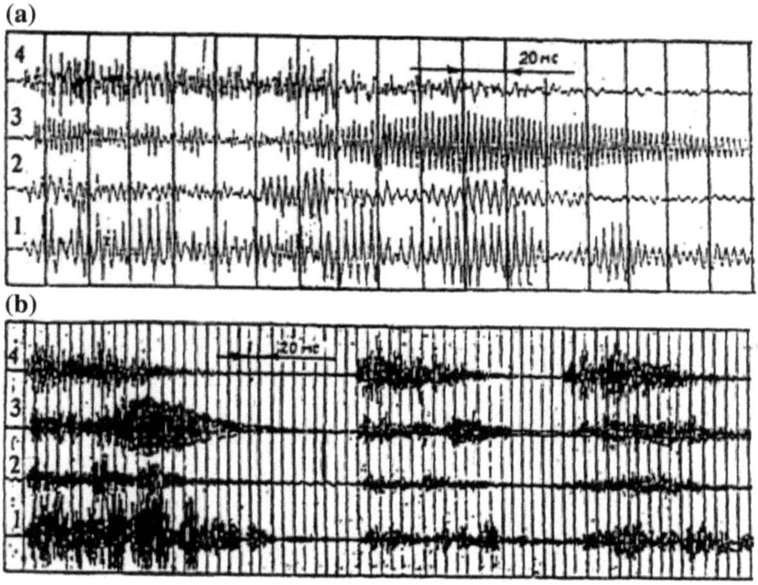

Fig. 6.40 Oscillograms of micro seismic energy excited at the blocks **a** by an initial explosion **b** by a series of four explosions. According to Sboev (1988). 1–4—numbers of the rock blocks. {Legends on Fig. 6.40: **a** 20 ms **b** 20 ms}

reaches 1.8–2.0, sometimes even more. Similar phenomena are noted, for example, in the Tajik-Afghan and Precarpathian oil-and-gas bearing basin (Abukova and Kartsev 1999).

At present, in regional geological studies, the identification of periodically ordered structural elements that have a fundamental effect on the distribution of earthquakes and the location of minerals is widely used in the interpretation of geophysical fields and remote sensing data from space. As a rule, the result of such structural interpretation is a set of maps of linear, circular and spiral landscape elements that illustrate not only the nature of regmatic fracture systems of different orders, but also the spatial relationships between discrete cellular, circular and spiral structural elements of different orders (Fig. 6.41).

Without dwelling on the historical issues, it should be noted that, beginning since the 1960s, in the national and foreign geological literature, a certain tectonic and ore-controlling role in the construction of concentric and spiral formations has been repeatedly noted (Popov 1960; Pavlovsky and Markov 1963; Wisser 1964; Popova 1966; Favorskaya 1969; Salop 1971; Thomson and Favorskaya 1973; Soloviev 1975, 1978; Khudyakov 1974; Khain 2003; Bush 1983; Glukhovskikh et al. 1983). Long before the peak of the cosmic remote sensing methods, research in this direction began in the USSR Academy of Sciences, in the PGO "Aerogeology" and in VSEGEI. Ring and spiral structural forms as independent cartographic elements were identified in the Map of Morphostructures of the

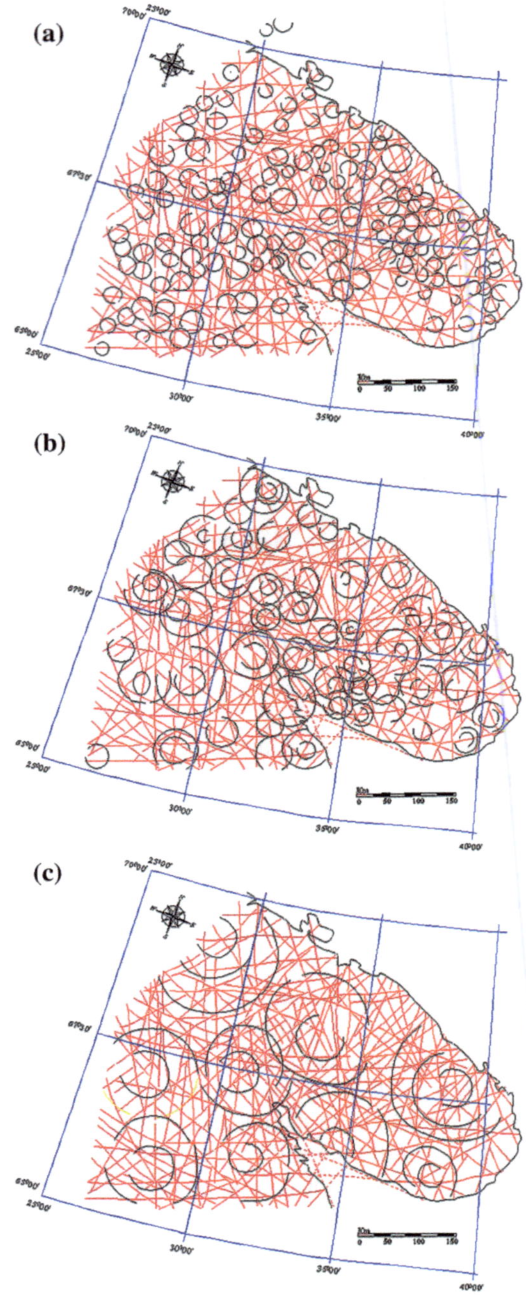

Fig. 6.41 Linear and annular topographical elements: disjunctive network of the second order and the system of the annular structures of the first—third order. The transition from the part a to part c reflects the co-ordination of the annular structures of different order (Petrov et al. 2006)

Central Type (VSEGEI) and the Satellite-Photogeological Map (PGO "Aerogeologya"), edited by Solovev (1975, 1978). On these maps these forms appear as structures of circular and spiral outlines, whose diameter varies from the first to several thousand kilometres. Thus, a discrete series of morphostructures of the central type was established, as well as a proportional relationship between the lateral dimensions and the depth of their deposition and a step-conic form of their relationship to the deep discontinuities of the crust and mantle (Soloviev 1975, 1978; Khudyakov et al. 1988). In studying the morphostructures of the central type of Siberia and the Far East, the concept of geological and geomorphologic conformity was widely used, targeting researchers to identify the spatial-genetic links between the geomorphological "form" and the geological "content" of morphostructures (Khudyakov 1974, 1978; Khudyakov et al. 1988). However, as a result of these studies, conformal rock complexes in the rank of geological facies and formations were isolated mainly for low-order morphological structures of the central type. Higher hierarchical levels are distinguished here geologically only as fragment radial-concentric structural elements, further integration of which is carried out on the basis of remote cosmogeological, morphometric and geophysical data.

Despite some uncertainty with the isolation of ring and spiral morphostructures as an independent class of geological formations, a number of important applied problems are being solved with its help in practical geology: the definition of a "translucent" structural pattern of deep cover horizons and the basement surface in areas with a multilevel tectonic structure; the zoning of territories by blocks of primary divisibility of the Earth's crust and lithosphere as a whole; quantitative determination of the amplitudes of the most recent horizontal and vertical movements; the identification of typomorphic features of the location of ore deposits in metallogenic studies. Thus, a fruitful idea was expressed by Tomson and Favorskaya (1968, 1973) from IGEM about the development of ore zoning in connection with the formation of orogenic vaults, and a group of geomorphologists (I. K. Volchanskaya, N. T. Kochneva, E. N. Sapozhnikova) developed methods for analysing the latest movements of morphostructural blocks, expressing control through zones of orogenic mineralisation.

In essence, it is these same ideas that form the basis for the concept of the metallogenic role of the processes of tectonic magmatic activation, arising under the influence of deep subcrustal motions of matter and its transformations (Shcheglov 1976; Favorskaya 1979; Krasny 1980, and others.) In full accordance with the two main types of dissipative structures of the Earth—G-hexagons and L-hexagons—the authors of the above-mentioned works singled out two main types of tectonic-magmatic activation processes, respectively. The deep activity of the first type structures proceeds under the conditions of preferential extension, of the second, "orogenic" type—under the conditions of preferential compression. The above-mentioned connections proved to be valid for geologically different regions. For example, they were successfully used by employees of VSEGEI N. V. Nikitin, D. V. Lopatin, N. R. Gorbatsevich and others for the predictive assessment of the territories of Transbaikalia, Western Siberia and the Kola-Karelian regions, and for the Central Kyzylkum region—S. S. Schultz (Jr.) and Sh. E. Ergashev. The impressive results of studies of morphological structures of the central type for predictive-metallogenic purposes were achieved by geologists and geomorphologists of Siberia and the Far East (Metallogenetic Analysis … 1977; Metallogeny of Hidden Lineaments … 1984; Morphostructure Studies … 1985; Central Type Morphotectonic Systems … 1988; Tveritinov et al. Predicting gold mineralisation in the south of Eastern Siberia and the Far East 2006).

From the point of view of using mathematical methods, the selection of the hierarchy of ring and linear structures allows us to speak about the algorithmisation of the generalisation effect. The latter is empirically recorded in geology as a natural merge of geological bodies of a smaller order and less depth into geological formations of regional significance and greater depth. In the method of automated structural deciphering of space images proposed by us (Petrov et al. 2005), the generalisation effect is formalised at the level of amplitude-frequency filters, as well as angular and dispersion criteria. Comparing maps of linear and annular landscape elements of different levels of generalisation, one can see geometric self-similarity in a family of linearised structures of smaller order with respect to a linearised structure of a larger order into which this family unites. For annular structures, this effect is often expressed through the generation of spiral formations.

Mathematical modelling and physical experiments indicate the fundamental interconnection of wave phenomena in the development of density instability and spontaneous cellular structuring of density unstable media with spiral waves. These waves are observed in many experiments. In this case, the appearance of spiral waves is in no way connected with the specific type of instability of the media under study—this is a general property of dissipative systems.

Thus, from the geometry of dissipative structuring of density unstable media, one can see that in the formation of the hierarchy of hexagonal structures, the family of streamlines circulating relative to the system of foci and converging to a system of nodal points are organised into spiral structures (Fig. 6.42). Moreover, for an individual hexagonal cell, the streamlines emanating from its geometric centre to opposite vertices of this cell, lying on the same diagonal of the hexagon, form a pair of spiral formations satisfying the dependencies of the type

$$r = \sqrt{\varphi}, r = -\sqrt{\varphi}, \quad \text{where } \varphi \in [0, \pi],$$

where φ—the polar angle, r—the radius of the spiral (Fig. 6.43). Following the described geometrical model of the development of a cellular dissipative structure, it becomes obvious that abnormal physical events, such as the formation of new unstable structures such as saddle points, are realised exclusively on one of the elements of the arc of the spiral. The vertex of the spiral formation is confined to the stationary region of the dissipative structure of a density nonequilibrium system.

Geological observations have confirmed that under the conditions of the wide development of the hierarchy of regular cellular structures of the Earth, spiral formations containing anomalous geological events on the elements of their arcs can

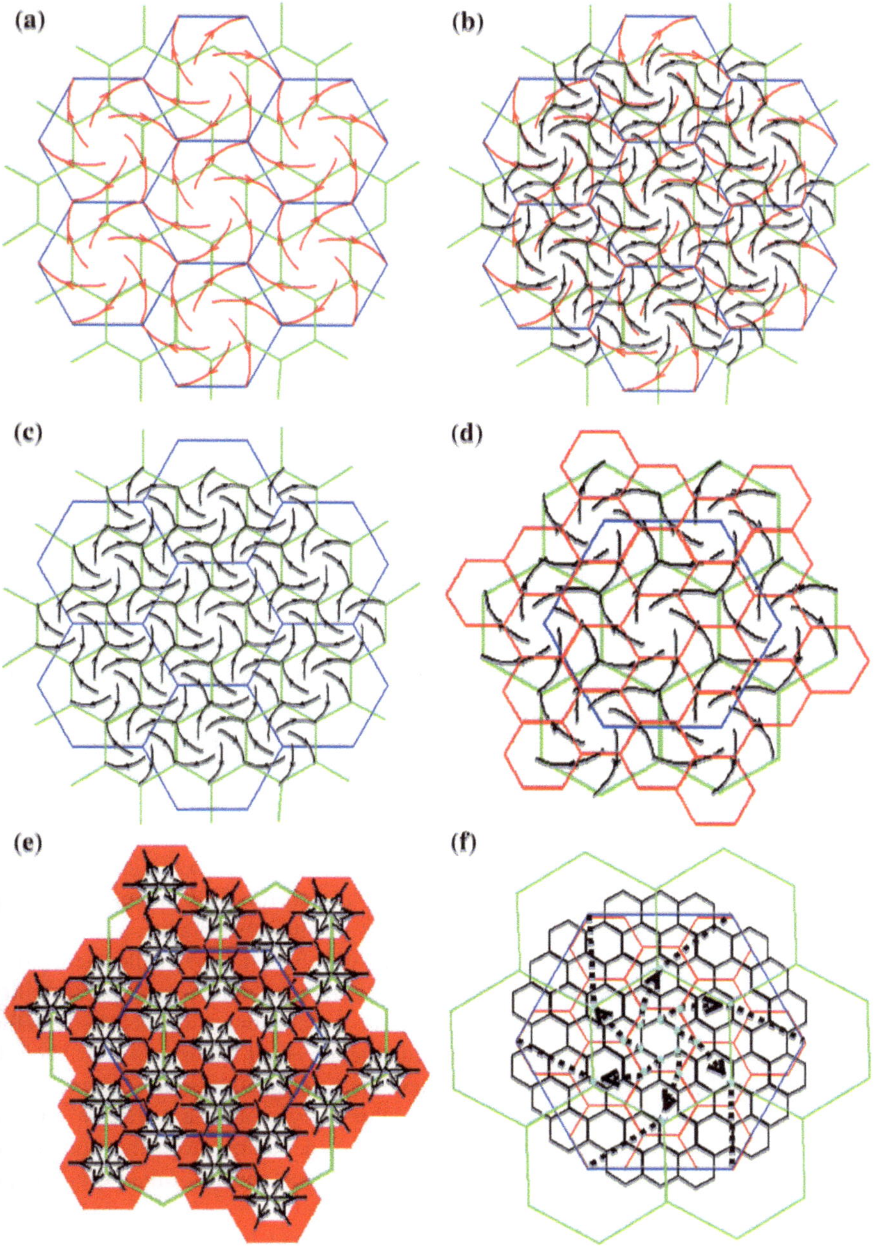

Fig. 6.42 The development of spiral flow structures with the growth of density instability **a** formation of the second-order cells in the development of spiral structures in the first-order cells; **b** development of spiral structures in second-order cells with and without superposition; **c** on first-order spiral structures; **d** formation of third-order Bénard cells; **e** development of spiral structures in third-order cells in the case of continuation of the dissipative fragmentation; **f** flow structure in the case of the limit nature of the third-order cells; **g** points of the fracture of stream lines (blue circles) in the tops of the cells of different orders—specific points

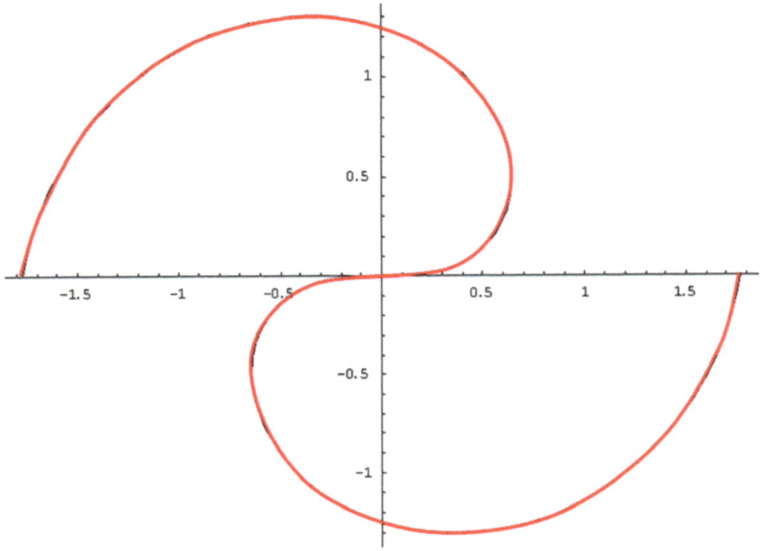

Fig. 6.43 Spiral formations approximating the course of a pair of streamlines oriented along the diagonal of a hexagonal cell from the geometric centre to its opposite vertices

also arise. The methodological justification and approbation of the approach to the isolation of similar spiral structures were previously performed by Oganezov (1997).

In Oganezov's technique, the region of maximum spatial variability of the so-called geofields—or, in other words, a system of geological and geophysical features specified on the area of a certain polygon in digital form—is adopted as a sign of an anomalous geological event. Taking into account the development of geoinformation systems and, therefore, a lot of digital representations of cartographic materials, we note that in this case the "digital form" is understood as the assignment of a geophysical field in the form of its amplitude values to nodes of a regular two-dimensional grid, each position of which is uniquely allocated either rectangular, or geographical coordinates. The parametrisation of field variability assumes dividing the territory of the polygon into equal squares, in each of them the alteration rate of each available geological and geophysical features is calculated. It is argued that the nature and quantity of geological and geophysical attributes used are not limited in any way, and any algorithmically available variation can be chosen as a parameter of their spatial variability (deviation from the norm), for example, the variation of the number of isolines in each square of the polygon, the length of lineaments, statistical moments.

Based on the results of recalculation of the geofield into the parameter of its spatial variability, an isoline map is constructed in which regions with the smallest and greatest variability—or, in other words, anomalous regions—are identified. This contour map should be converted to a vector diagram. Here each vector is drawn from a square with a smaller value of the spatial variability parameter of the geofield to the nearest square with a larger value of this parameter. The vertices of the vector family that are plotted from the same point are united by a certain nonlinear envelope. The latter, according to Oganezov's statement, displays the arc element of the so-called logspiral, which has a variable radius of curvature. The region on the cartographic plane, where an anomaly with an increased value of the parameter of the geofield spatial variability falls on the element of the arc of logspiral, is referred to by Oganesov as a singular point. The region promising for prospecting mineral deposits is connected with this singular point. The additional value of this technique is attached to the fact that, in addition to the absence of the need for a reference sample, a forecast of the ore component of the deposit is allowed. The main criterion here is the radius of the lognormal spiral. For example, using a radius of 3–4 km, we can refer to the presence of groundwater; with a radius of 150–300 km—the presence of oil and gas manifestations. It is recommended that the above recalculations be performed several times on different-scale cartographic materials pertaining to the same area. The appearance of additional geological and geophysical features should be considered as specifying material on its spatial variability.

In general, Oganezov's technique is quite complete both in terms of ease of execution as compared to other developments, both domestic and foreign, as well as in terms of design (copyrights are protected by the Russian Federation patent No. 2097794). However, the main drawback of the methodology should be attributed to its qualitative nature: no analytical calculations are used, whether in the description of the calculation of the spatial variability parameter, the definition of the character of the spiral structure (with the exception of the not commonly accepted term "lognormal") or in the representation of the vector diagram. At the same time, the methodology does not employ measurements, but only recalculations. In this sense, the effect of the interconnection of standing internal cellular gravitational waves with spiral structures, reflected in our wave model of the development of dissipative structuring, makes it possible to formalise and optimise a methodology for forecasting anomalous geological events, qualitatively stated by Oganezov.

The third approach to the interpretation of geological and geophysical data in connection with the solution of the predicting problems of applied geology, which stands out from the standpoint of using the mathematical apparatus, but is closely connected through geological data with the two previous ones, refers to the problem of image recognition and the localisation of promising ore-bearing areas. Turning to the most popular methods underlying this approach, we note that traditionally they are based on the recognition of images with learning. The essence of the method consists, firstly, in the employing the reference sample—a system of numerically given geological and geophysical features describing the reference object, represented, for example, by some ore deposit of a certain genesis. Secondly, mathematically n geological and geophysical characteristics form n-dimensional conditional feature space, in which the reference sample occupies a local area, usually of a complex geometry. Thirdly, based on the analysis of average variances and covariances of the reference and arbitrary samples, a linear discriminant function is constructed in the system of measured geological and geophysical features (Davis 1977), geometrically reflected in the n-dimensional conditional characteristic space as a hyperplane separating the reference and close to it (by statistical sampling criteria) samples from all other samples. Finally, measuring pickets, in which the values of geological and geophysical features are defined as close to the reference ones, are marked as promising for the desired object, and all other pickets are labelled as "empty" areas. The resulting map or section may be referred to as predictive, although they are subject to further sorting, for example, by qualitative indicators of structural control. For all its determinism, the described method has two main disadvantages. The first lies in the linear nature of the statistical estimates in the construction of the discriminant function, whereas the relationship in the system of measured geological and geophysical features can be essentially nonlinear. Traditionally in geology, this feature is attributed to the errors of the method and is not specifically considered as a separate problem requiring the improvement of measuring and/or computational procedures. The second drawback is the need for a reference sample, which is the basis for an expert opinion on the presence or absence of perspective territories.

Without going into further analysis of the shortcomings or advantages of this or that approach to the interpretation of geological and geophysical data with a view to solving the predicting problems of applied geology, we note that in recent decades, quite unexpectedly, one of the results of their use turned out to be the identification of a number of general laws. Along with the empirically discovered commonality in describing the diverse phenomena of nature in other sections of natural science, they convincingly testify to the fundamental role of the wave properties of matter in the development of geological processes. The ideas about the influence of wave mechanisms on the global discretisation of the geological space in the form of the periodically ordered structural elements that have a determining effect on the location of mineral resources, including groundwater, oil and gas, are found in the works of many Russian and foreign researchers. However, our new data on the wave nature of processes of the spontaneous structuring of density unstable masses allows all these isolated examples to be considered as a general reflection of the fundamental laws of natural development that underlie the new paradigm of theoretical and practical geology.

References

Khain V.E. Regional geotectonics. M.: Nedra; 1971. (In Russ.).

Anokhin V.M. Global disjunctive network of the Earth: structure, origin and geological significance. SPb.: Nedra; 2006. (In Russ.).

Mandelbrot B.B. Fractals: Form, Chance and Dimension. San Francisco: W.H. Freeman; 1977.

Mirlin E.G. Fractal discreteness of the lithosphere and geodynamics. Dokl. RAN. 2001; 379 (2):231–234. (In Russ.).
Vadkovskii V.N., Sokolov S.D., Zakharov V.S. et. al. Accretionary tectonics and fractal dimensionality. The abstracts of the conference "Geofizicheskie chteniya im. V.I. Fedynskogo". M.: GEOS; 2002:57–58. (In Russ.).
Malinverno A. Fractals and ocean floor topography. A review and a model. Fractals in the Earth Sciences. Ed. by C.C. Barton and P.R. La Pointe. NV, Plenum Press; 1995.
Dubois J., Cheminee J.L. Fractal analysis of eruptive activity of some basaltic volcanoes. J. Volcanology and Geothermal Res. 1991; 45(3–4):197–208.
Sherman S.I., Gladkov A.S. Fractals in studies of faulting and seismicity in the Baikal rift zone. Tectonophysics. 1999; 308:133–142.
Gzovskii M.V. Mathematics in geotectonics. M.: Nedra; 1971:97–99. (In Russ.).
Nikolaev N.I. Neotectonics of USSR. Proceedings of the commission for the study of the Quaternary period. M. L.: AN SSSR. 1949; 8. (In Russ.).
Gerasimov N.P. Experience in the geomorphological interpretation of the general scheme of the USSR geological structure. Problemy fizicheskoi geografii. AN SSSR. 1946; 12:33-46. (In Russ.).
Meshcheryakov Yu. A. The main elements of the Earth's morphological structure and the problem of its origination. Izv. AN SSSR. Ser. geogr. 1957; 4:3–15. (In Russ.).
Filosofov V.P. Remarks on the theory and practice of the morphometric method. Morphometric method in geological studies. Saratov: Sarat. SU; 1963:249–261. (In Russ.).
Piotrovskii V.V. The use of morphometry to study the relief and structure of the Earth. Earth in the Universe M.: Mysl'; 1964: 278–297. (In Russ.).
Nikol'skii Yu. I. Map of the anomalous gravitational field of Russia and the adjacent offshore areas. Geological Atlas of Russia. Section 2. Geological structure and geophysical characteristics of subsoil. M.-SPb.; 1996:35–51. (In Russ.).
Odesskii I.A. Rotational-pulsation regime of the Earth and its geological consequences. SPb.: Politekhn. un-t; 2005. (In Russ.).
Petrov O.V., Movchan I.B. Self-similarity and dimension in the dissipative structuring of the Earth. Regional geology and metallogeny. 2003; 19:33–47. (In Russ.).
Landau L.D., Lifshits E.M. Theoretical physics. Fluid dynamics. M.: Nauka; 1988. (In Russ.).
Strakhov V.N. Basic ideas and methods of extracting information from the data of gravitational and magnetic observations. Collection of articles "Theory and technique of interpretation of gravitational and magnetic anomalies" M.: IFZ; 1979:146–264. (In Russ.).
Serkerov S.A. Spectral analysis in gravity and magnetic prospecting. M.: Nedra; 1991. (In Russ.).
Petrov O.V., Movchan I.B., Yakovleva A.A. Interpretation of the gravitational field along the AR-1 profile on the basis of wave analogies with the density unstable media. Regional geology and metallogeny. 2006;28:174–180. (In Russ.).
Strakhov V.N. On some estimates of the depth of occurrence of perturbing masses. Bulletin of the Academy of Sciences of the USSR. Geophysics Series. 1963; 1:90–109. (In Russ.).
Strakhov V.N. On the state and problems of the mathematical theory of the interpretation of gravitational and magnetic anomalies. Izvestiya of the Academy of Sciences of the USSR. Physics of the Solid Earth. 1970; 5:112–119. (In Russ.).
Strakhov V.N. The formation of a new paradigm is the destruction of the prevailing stereotype of thinking (on the example of gravimetry and magnetometry). Izvestiya, Physics of the Solid Earth. 2002; 3:3–20. (In Russ.).
Petrishchevskii A.M. Statistical gravitational models of the lithosphere of the Russian Far East. Published summary of the Doctor's dissertation. M.: 1990. (In Russ.).
Petrishchevskii A.M. Gravitational autocorrelation indicators of deep geological structures. Russian Journal of Pacific Geology. 2004; 23(4):13–24. (In Russ.).
Lebedev A.N., Petrov A.V. Statistical sounding of geofields. Proceedings of Higher Schools. Geology and Exploration. 2001; 3:106–110. (In Russ.).
Berlyand N.G., Roze E.N. The appliance of correlation analysis for classification of potential fields. Geomagnetism and Aeronomy. 1971; 2:43–48. (In Russ.).

Serkerov S.A. The application of gravity data and magnetic prospecting for tectonic zoning. Use of geological and geophysical data for studying regional tectonics of oil and gas bearing areas. M.: Nedra; 1976:36–61. (In Russ.).

Petrishchevskii A.M., Demura G.V. Gravitational sounding of the Western Pacific continent-ocean transition zones. Tez. dokl. III Tikhookean. shkoly. Ch. 2. DVO AN SSSR. Vladivostok; 1987:16–18. (In Russ.).

Petrishchevskii A.M. Statistical gravitational models of the lithosphere of the Russian Far East. Vladivostok: Dal'nevost. un-t; 1988. (In Russ.).

Krivtsov A.N. While understanding how the deposits originated, do we know where to search for them? Otech. geologiya. 2006; 5:117–119. (In Russ.).

Bogatskii V.V. Wave mechanism of formation of ore-localising structures of magnetite deposits of the Siberian platform. Geology of Ore Deposits. 1977; 19(3):3–18. (In Russ.).

Vityaz' V.N. Periodicity of location of geological structures of Siberian platform areas. M.: Nauka; 1982. (In Russ.).

Kvet R., Nemets V. Symmetry of equidistant planetary discontinuous systems. Structural symmetry of geological bodies. M.: 1976; 2:8–9. (In Russ.).

Nemets V. Regularity of geological structures - mathematical geology - geodynamics. Internal geodynamics. General questions of the internal geodynamics. L.: 1972; 1:32–33. (In Russ.).

Baryshev A.N. Periodic location of diamondiferous systems and related problems of diamond geology. Otech. geologiya. 2006; 6:20–35. (In Russ.).

Abukova L.A., Kartsev A.A. Fluid systems of sedimentary oil-and-gas bearing basins. Otech. geologiya. 1999; 2:11–16. (In Russ.).

Androsov E.A., Verzhak V.V., Larchenko V.A., Minchenko G.V. About the structural control of the location of kimberlite bodies (on the example of the Arkhangelsk diamond province). The geology of diamonds is the present and future of geology: to the 50th anniversary of Mirny and the diamond mining industry in Russia. Collection of articles. Voronezh; 2005:31–43. (In Russ.).

Sablukov S.M. Diamond depositaries of the Russian North. Science in Russia. 2001; 1:82–88. (In Russ.).

Petrov O.V. Crystallization of natural sulphide melts and stages of copper-nickel ore formation. Scientific work collection of PGO "Sevmorgeologiya" "Forecasting and evaluation of the nickel content of new ore areas in the north of the Siberian Platform". L.; 1983:95–99. (In Russ.).

Pupin J.P. Contrib. Mineral. Petrol. 1980; 73:207–220.

Kinny P.D., Dawson J.B. Nature. 1992; 360:726–728.

Hoskin P.W.O. Trace element composition of hydrotermal zircon and the alteration of Hadean zircon from the Jack Hills, Australia. Geohimica et Cosmohimica Acta. 2005; 69:637–648.

Kats Ya.G., Poletaev A.I., Sulidi-Kondrat'ev E.D. Ring structures of the face of the planet. (New in life, science and technics. Ser. "Nauki o Zemle", № 5). M.: Znanie, K 62; 1989. (In Russ.).

Sboev V.M. Investigation of microseismic processes occurring in an rock massifs of underground mines. (Preprint № 25). Novosibirsk: IGD SO AN SSSR; 1988. (In Russ.).

Popov V.I. Nuclear theory of the development of the Earth's crust. Ed. B.F. Vasil'evskii. Tashkent: SSU; 1960. (In Russ.).

Pavlovskii E.V., Markov M.S. Some general questions of geotectonics (About the irreversibility of the development of the Earth's crust). M.: Tr. GIN. 1963; 93:9–53. (In Russ.).

Wisser E. Mineralisation connection with dome structures in the North American Cordilleras. Problems of endogenous deposits. Issue 2. M.; 1964. (In Russ.).

Popova G.Z. Ring and linear morphostructures of the Kazakh folded country. Alma-Ata: Nauka; 1966. (In Russ.).

Favorskaya M.A. The issues of bonding mineralisation with magmatism. Itogi nauki. Geokhimiya. Mineralogiya. Petrografiya. 1969:55–83. (In Russ.).

Salop L.I. The main features of the stratigraphy and tectonics of the Precambrian of the Baltic Shield. Collection of VSEGEI articles. L.; 1971; 175:3–87. (In Russ.).

Tomson I.N., Favorskaya M.A. On the types of focal structures and the associated mineralisation. Regularities in the location of mineral resources. V. 10. M.: Nauka; 1973:49–63. (In Russ.).
Solov'ev V.V. The mapping of central type morphostructures in forecasting and searching for deposits of endogenous mineral resources. Problems of geomorphological mapping. L.: ONTI VSEGEI; 1975:62–64. (In Russ.).
Solov'ev V.V. The structures of the central type of the USSR territory according to geological and morphological analysis: app. to a map of morphostructures of the central type of the USSR territory. Scale 1:10 000 000. L.: VSEGEI; 1978. (In Russ.).
Khudyakov G.I. Principal foundations of morphotectonic studies (on the example of studying the southern mainland part of the Soviet Far East). Published summary of the Doctor's dissertation. Novosibirsk; 1974. (In Russ.).
Khain V.E. The main issues of modern geology. 2nd edition. M.: Nauchnyi mir; 2003. (In Russ.).
Bush V.A. Systems of transcontinental lineaments of Eurasia. Geotectonics. 1983; 3:15–31. (In Russ.).
Glukhovskii M.Z., Kats Ya., Moralev V.M. About nuclears of the world continents. Proceedings of Higher Schools. Geology and Exploration. 1983; 8:14–19. (In Russ.).
Khudyakov G.N., Kulakov A.P., Ezhov B.V. et. al. Morphotectonic systems of the central type of Siberia and the Russian Far East. M.: Nauka; 1988. (In Russ.).
Khudyakov G.I. New ways – new problems in geomorphology. Geomorphostructure of the Far East. Vladivostok; 1978:5–11. (In Russ.).
Tomson I.N., Favorskaya M.A. Ore-controlling structures and principles of local forecasting of endogenic mineralisation. Sov. geologiya. 1968; 10:6–21. (In Russ.).
Shcheglov A.D. Fundamentals of metallogenic analysis. M.: Nedra; 1976. (In Russ.).
Favorskaya M.A. On some new trends in tectonic-magmatic and prognostic-metallogenic studies. Analysis of space images in tectonic magmatic and metallogenic studies. M.: Nedra; 1979:17–24. (In Russ.).
Krasnyi L.I. The geology of the Baikal-Amur Mainline Region. M.: Nedra; 1980. (In Russ.).
Petrov O.V., Movchan I.B., Yakovleva A.A., A.A. The method of automated structural interpretation of space images. Regional geology and metallogeny. 2005; 23:105–114. (In Russ.).
Davis J. Statistics and analysis of geological data. M.: Mir; 1977. (In Russ.).
Egorov A.S. Deep structure and geodynamics of the lithosphere of Northern Eurasia. Published summary of the Doctor's dissertation. SPb.; 2000. (In Russ.).
Turner J. Buoyancy effects in fluids. M.: Mir; 1977. (In Russ.).
Sherman S.I., Sorokin A.P., Cheremnykh A.V. A new approach to the tectonic zoning of the Amur region according to the fractal dimensionality of the crustal faults. Dokl. RAN. 2001; 381(3):1–5. (In Russ.).
Baryshev A.M. The position of the largest ore regions and nodes in the system of advective structures of the Earth. Otech. geologiya. 2001; 2:6–11. (In Russ.).
Baryshev A.M. Tectonic and tectonophysical conditions for the formation of small intrusions. Rudy i metally. 2001; 6:19–26. (In Russ.).
Tveritinov Yu.I., Tveritinova T.Yu., Brant S.B. et al. Forecasting gold mineralisation in the south of Eastern Siberia and the Far East: geological and isotope-geochemical aspects. Irkutsk: In-t zemnoi kory SO RAN; 2006. (In Russ.).
Oganezov A.V. Method of Determining Spatial Coordinates of Geological Formations. Pat. RF: RU 2097794 C1. 11/27/97 (In Russ.).
Khain V.E., Koronovskiy N.V. Planet Earth. From the Nucleus to the Ionosphere: Coursebook. M.: KDU; 2007. (In Russ.).
Kirsanov A. A. [et al.]. Landsat 7 multispectral data: modern multi-purpose remote basis for geological maps and geological exploration / Regional Geology and Metallogeny. 2005; 24:60–66. (In Russ.).
Petrov O.V., Movchan I. B., Yakovleva A. A. Interpretation of the gravity field along the AP-1 profile based on wave analogies with density-unstable media. Regional Geology and Metallogeny. 2006; 28:174–180. (In Russ.).

Gravimetric Map шт 1:2.5 M USSR/ Edited by Stepanov P.P., Yanushevich M.A. Moscow: VNIIgeofizika; 1999. (In Russ.).
Central Type Morphotectonic Systems in Siberia and the Far East. Ed.-in-chief V.V. Soloviev. – M .: Nauka; 1988. (In Russ.).
Morphostructural Studies: Theory and Practice. Ed.-in-chief N.V. Vasilkovsky. M.: Nauka; 1985. (In Russ.).
Metallogeny of Hidden Lineaments and Convergent Structures / I.N. Thomson [et al.]. M.: Nedra; 1984. (In Russ.).
Metallogenic Analysis in Activation Areas (Case study: Transbaikalia). Ed.-in-chief Shatalov E.T.. M .: Nauka; 1977. (In Russ.).
Geological Metallogenic Map of the World 1:15M: Explanatory Note in 4 Parts. Part 1: Geology and metallogeny of continents, transitals and the World Ocean / Krasny L.I. [et al.]. St. Petersburg: VSEGEI Cartographic Factory Publishing House; 2000. (In Russ.).
Geological Metallogenic Map of the World 1:15M: Explanatory Note in 4 Parts. Part 2: Mineral resources of continents and active transitals. Edited by Krasny L.I. St. Petersburg: VSEGEI Cartographic Factory Publishing House; 2000. (In Russ.).
Geological Metallogenic Map of the World 1:15M. Expl. Note in 4 parts. Part 3: Oil resources of continents and transitals. Geological and economic assessment / Nazarov V.I. [et al.]. St. Petersburg: VSEGEI Cartographic Factory Publishing House; 2000. (In Russ.).
Geological Metallogenic Map of the World 1:15M. Expl. Note in 4 parts. Part 4: Mineral resources of the oceans / Datsenko V.A. [et al.]. S.-Petersburg: VSEGEI Cartographic Factory Publishing House; 2000.(In Russ.).
Tsuboi S. Gravity. George Allen and Urwin (Publishers)Ltd. (London), 1983. 254 p.

Conclusion

Geology, remaining to this day a descriptive science, has accumulated a huge number of facts that characterise the Earth as an unstable thermogravitational natural system. Contemporary developments in natural science attest to the fact that such systems imply a new conception of dynamic states of matter, which reflect the wave nature of their interaction with the environment. All this, as Ilya Prigozhin writes, "… leads to a new view of matter, according to which it is no longer perceived as a passive substance described from the standpoint of mechanical views, but is associated with spontaneous activity." In our understanding, this spontaneous activity is associated with the manifestation of the wave properties of matter in the processes of structuring and shaping of density-unstable masses. When taking such an approach towards understanding density (or convective) instability phenomena, they can be characterised using the concepts of amplitude, phase, wavelength, period and frequency. The change of state of such systems acquires a collective character, in which questions concerning individual characteristics of distinct elements are set to one side. In the description of such systems, the wave properties of matter acquire a decisive importance.

Thus, from the standpoint of contemporary natural science, in common with fractal theory and fractal geometry in general, we consider fractal dimensions found in density-unstable geological media as an abstract mathematical reflection of the manifestation of one of the fundamental wave properties of matter in nature, namely, the expansion of the wave oscillatory process into modes. Like wave processes, fractal objects are characterised by self-similarity, i.e., in terms of the similarity of the parts to the whole. In this connection, the wave properties of matter acquire a paradigmatic significance in which the question concerning the relationship between the continuous and discrete (or the whole and its parts) in nature, that exercised philosophers, sculptors, architects and artists from Antiquity to the Renaissance, once again becomes the cornerstone of contemporary natural science. These fundamental wave properties of matter, which impart such striking geometric and dynamic properties to natural objects and characterise the common developmental processes of heterogeneous objects, from atoms up to galaxies and residents

O. V. Petrov, *The Earth's Dissipative Structures*, Springer Geophysics,
https://doi.org/10.1007/978-3-319-93614-7

of all worlds in between, have repeatedly revealed themselves in the history of the development of science and art.

The Fractal Dimension as the Highest Manifestation of the Structural and Functional Perfection of the Relations of the Whole and Its Parts in Nature

Long before the beginning of the twentieth century, when Georg Cantor, one of the founders of modern mathematics, constructed the first example of his famous set having the characteristic fractal dimension $D \approx 0.632$, the outstanding Italian mathematician Leonardo Fibonacci (1180–1240) discovered the numerical sequence known today as the Fibonacci series: 0, 1, 1, 2, 3, 5, 8, 13, 21, 34, 55… In this sequence, each subsequent term is derived through summation of the two previous terms. The unique property of this series is that the ratio of two neighbouring terms tends to a constant irrational value, approximately equal to 1.6180339887—inversely proportional to the value 0.618. Just like a wave process, the Fibonacci series represents a power-law algorithm for expansion into modes.

Subsequent scientific discoveries showed that both living and inanimate nature is subject to this relationship throughout the universe. It can be seen in the spirals of sea shells, in the form of flowers and beetles, as well as in the beautiful proportions of the human body (Figs. A.1 and A.2). Due to this universal property, it has been referred to in such terms as divine proportion, harmonic proportion and golden section.

This ratio of the whole and its parts to each other in the proportion of 1.618 (0.618) was already being applied in the Ancient World: by Euclid, to create geometry, and by Phidias to his legendary sculptures. Plato already believed that these principles applied throughout the entire universe. This harmonic correlation, co-occurring in all living and nonliving nature, has been used in architecture since even earlier times: the golden section was discovered in the Egyptian pyramids and later appeared in the construction of the Parthenon and the earliest Russian churches. The golden section was meticulously applied in the work of Leonardo da Vinci and Michelangelo. According to them, the ratio of the distance from the navel to the foot to the distance from the crown to the navel in an ideal human body is 1.618 (inverse ratio 0.618).

This is illustrated by the famous drawing by Leonardo da Vinci, from around 1490, which was accompanied by explanatory notes in one of his journals (Fig. A.2). According to da Vinci's notes, the drawing consisted of a study of the proportions of the human body as described in the treatises of the ancient Roman architect Vitruvius. The rediscovery of the strict mathematical proportions of the human body in the 15th century, made by da Vinci and other figures of the Renaissance art world, was one of the fundamental achievements leading up to the Italian Renaissance. The sketch itself is often referred to as a potent symbol of latent symmetry—of the human body in particular, as well as that applying throughout the

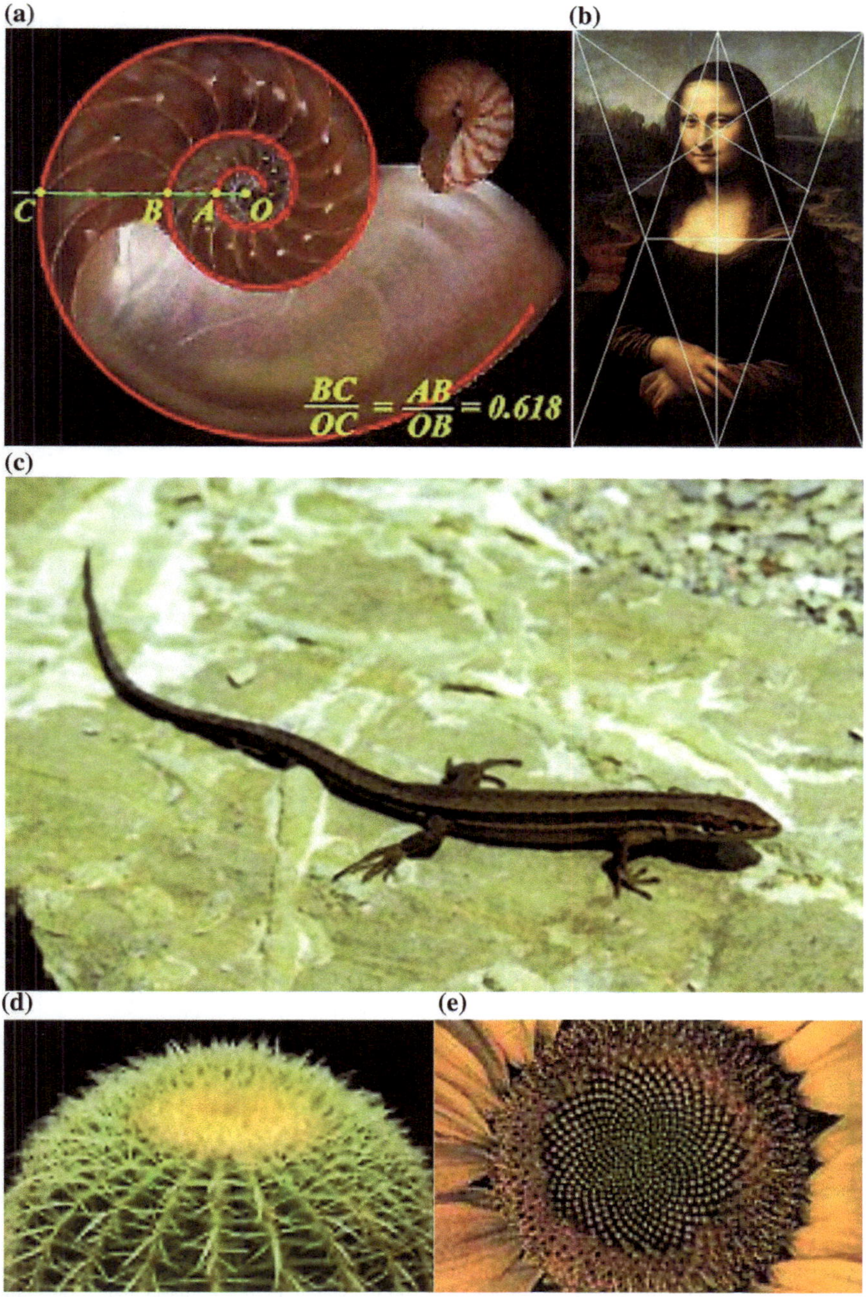

Fig. A.1 Principle of the golden section in nature and art **a** the whorls of the Nautilus cochlea, which are in the ratio 0.618; **b** location of the elements on the master's canvas is in the ratio 0.618; **c** proportion of the length of a lizard's tail to the rest of its body (62/38 $\approx$ 1.6); **d** formation of spirals in plants, often conforming to the Fibonacci sequence; **e** ordering of sunflower seeds in two rows of spirals: the number of seeds in each spiral is an individual member of the Fibonacci sequence

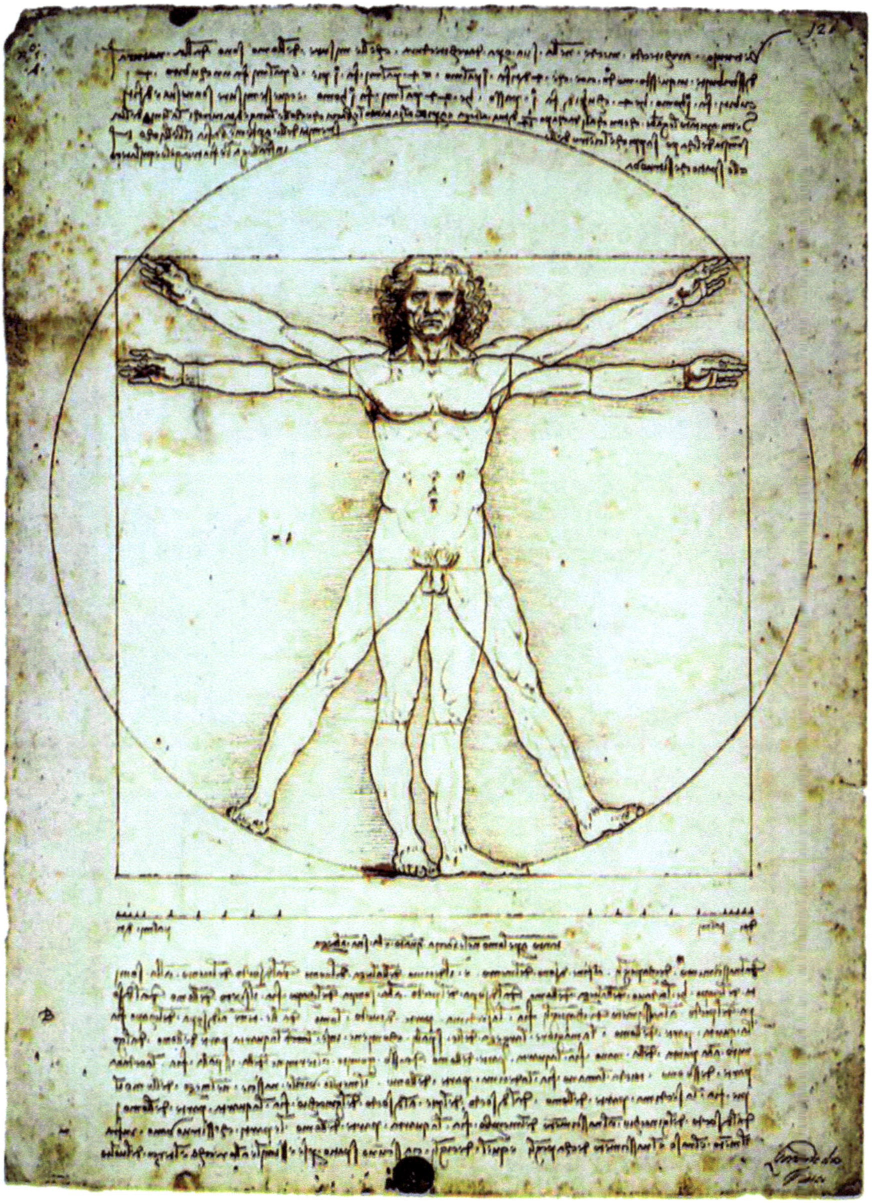

Fig. A.2 The Vitruvian man, a famous image, accompanied by explanatory signatures, produced by Leonardo da Vinci around 1490. The picture is signed with a pen, ink and watercolour using a metal pencil; its size is 34.3 by 24.5 cm. Currently located in the Gallerie dell 'Accademia collection in Venice entire universe. In the process of studying the drawing, it can be seen that the combination of hands and feet forms four distinct poses. One of them, composed by hands outstretched to the sides and legs brought together, is inscribed in a square. The figure of the same person, stretching his arms up and with legs to the sides, fits into a circle. When changing the poses, a visual sensation of the movement of the centre of the whole figure is created due to the displacement of the geometrical centre of the square and the centre of the curvature of the circle relative to each other, although the true centre of the human figure that coincides with its navel remains immobile. Many centuries later, the Swiss-French architect Le Corbusier was to compile his own positioning scale based on this intuitive image, which would influence the aesthetics of the architecture of the 20th century.

In terms of facial anatomy, the line of the golden section runs along the tip of the nose. A contemporary adherent of this theory, the Californian surgeon Stephen Marquardt, went further than the ancient Greeks who calculated the golden facial cross section across one parameter and developed a virtual mask (Fig. A.3) in which all the parts of the face correlate in a ratio of 1.618 (0.618).

Thus, it turns out that the established fractal dimensions in density-unstable geological media are a reflection of the fundamental wave properties of matter—that is to say, of the highest manifestation of the structural and functional perfection of the relation of the whole to its parts in art, science, technology and nature.

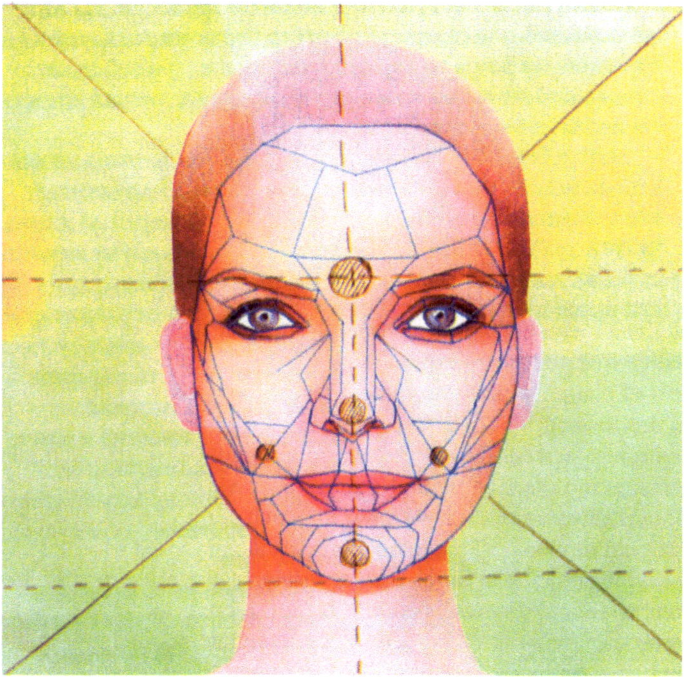

Fig. A.3 Virtual mask of Stephen Marquardt, in which all parts of the face are correlated in a proportion of 1.618 (0.618). Gala magazine, 2007, No. 6

The perfect relations of the whole and its parts were thus able to be empirically demonstrated; however, in the middle of the 18th century, science also faced another fundamental problem of the manifestation of the wave properties of matter, since it turned out that the fractal dimension is concealed in the dual nature of the wave process itself. This dual character of the waves—travelling and standing—can often be observed in nature. Employing a musical analogy, we can imagine the waves that arise when we pluck a violin string. The plucking creates a musical tone consisting of the main tone and the overtones, or the higher harmonics of the string. Such tones consists of nothing else but sound waves formed by groups of waves in a string. At first glance, it seems obvious that the movements of the string can easily be explained either by waves traveling back and forth along the string after being reflected at its ends, or in the form of standing vibrations of the string perpendicular to its length. However, the equal validity of these two interpretations, as well as the solution of their corresponding equations, is far from obvious. This question of the dual nature of waves became the basis for a scientific discussion that unfolded in the middle of the 18th century, known as the "vibrating string controversy". It was attended by eminent mathematicians and mechanics Leonard Euler, Jean Léron D'Alembert and Joseph Louis Lagrange. Later, Jules Henri Poincaré, Erwin Schrodinger, Max Planck, Max Born, Albert Einstein, Paul Dirac, John von Neumann, Georg Kantor, William Rowan Hamilton, Alexander Mikhailovich Lyapunov, Andrei Nikolayevich Kolmogorov, Vladimir Igorevich Arnold, Jurgen Kurt Moser, Leonid Isaakovich Mandelstam, Alexander Andronov, Ilya Prigogine and many other outstanding physicists and mathematicians, who would turn their attention to the fundamental principles of this discussion with respect to continuous media and subtle physical continua. Now that the development of various areas of natural science has made it possible to detect the manifestation of the wave properties of matter and fractal dimensions in most objects of animate and inanimate nature, we understand that this discussion has remained incomplete. After all, even the huge number of empirically obtained partial results characterising the fractal nature of the ratio of the whole to its parts cannot substitute for the generalising ideas of transition to a new, deeper level of understanding of the physical essence of the natural phenomena under consideration. From our point of view, it is precisely in this duality that the new formulations of the fundamental wave laws of the development of nature, which everywhere reveal themselves in the form of a fractal dimension, are located.

This emerging perception of the world around us is the natural result of developments taking place in natural science as a whole. The classical Newtonian mechanics, familiar from our school days, in which convection is described in terms of the vibrational motions of individual particles along closed trajectories, derives a given number of particles in an elementary volume (or phase point). In this case, their motion in a continuous medium is considered to be isolated from the remaining particles, without taking into account the interaction between the selected particles and the continuous medium. In fact, since the works of Henri Poincaré, contemporary natural science inevitably leads to the conclusion that such an approach is excessively idealistic for continuous media comprising large numbers of particles. New regularities of convective oscillatory processes are manifested from the more holistic approach taken when considering systems consisting of a large number of interacting particles as a whole. Vibrational processes in a system comprising a large number of elements are referred to in terms of wave properties. The states of such systems can be principally explained by the wave nature of the propagating energy, i.e. it is the wave properties of matter manifesting themselves at all levels of its organisation that acquire the determining significance in describing the fundamental laws of the development of such systems. Just as the revelations at the beginning of the 20th century led the scientific community to a cognisance of the dual wave-particle nature of light, today, at the beginning of the 21st century, we are standing on the verge of a new major scientific paradigm shift in which conclusions about the discrete-wave nature of the structure of any dissipative systems for which the emergence of a special kind of structural order and qualitatively different principles of spatial and temporal organisation are characteristic.

Fractal Hierarchies of the Earth's Dissipative Structures—A New Theoretical and Practical Geological Paradigm

The determination of the topic of our research by the wave nature of the spontaneous structuring processes of density-unstable masses allowed us to reach the conclusions drawn above concerning the causes of the appearance of fractal geometry in nature. They consist in the fact that since the spontaneous structuredness of nonequilibrium systems in a density relation is of a wave nature, the size of the structures varies according to the law of wave dispersion and the appearance (relief) of the structures allows the movement of matter within the system to be predicted. In applying these conclusions with respect to the Earth, which, in the first approximation on a geological timescale can be regarded as an unstable system in terms of density, we are breaking new geological ground, in which wave-processes are considered to be the root cause of the structuredness of the Earth's crust and mantle. Moreover, as we noted earlier, the question of the primary source of internal standing gravitational waves, be it convection, plume-tectonics or another hypothetical process, loses its practical significance. In order to establish the connection between the heterogeneous density structures and determine the nature of the motion of matter within the planet's interior according to the external topographical structure of the Earth's surface, it is sufficient to invoke wave laws. On this basis, a methodology is developed that allows the surface forms of the geoid and its topographical relief, along with remote geophysical measurements, to be used to reconstruct the distribution pattern of the deep regions of thermogravitational instability and describe the motion dynamics of matter in the interior of our planet. Under these circumstances, the three-dimensional *dissipative structures of the Earth become the elementary object of study.*

The practical application of the theory of standing internal gravitational waves in describing the mechanisms of the motion of matter in the interior of our planet is determined by the fact that these waves are essentially three dimensional. In hydrodynamic systems, they reveal a feedback between the internal structure of the motion of matter and the topography of the exposed surface. For convective cells, this connection was first revealed and described in detail in 1901 by Bénard, who observed that the central part of each isolated section of the circulation is formed by the concave surface; i.e., the centre of each internal ascent corresponds to the centre of the lowering of the exposed surface, and vice versa, with the dividing crest corresponding to the subduction of the fluid. The existence of such a connection between the bulk motion of matter and the relief of the free surface was used in this paper to identify the cellular hierarchies of the Earth's thermogravitationally-unstable dissipative structures of the L-hexagon, G-hexagon and Rolls types on the basis of an investigation of the relief of its exposed surface. A dissipative structure is considered here as part of the geological space organised by the wave processes of the Earth's thermogravitational instability with its external typomorphic lithospheric relief also orthomorphic to its internal geological and geophysical boundaries. With this approach, the dissipative structures of thermogravitational instability (or cellular standing inner gravity waves) of the Earth are defined as a new type of geodynamic (geotectonic) objects reflecting the general orientation of contemporary geological mapping towards a comprehensive study of large three-dimensional parts of the Earth's crust. This orientation is taken up in the works of many geologists, geophysicists and geomorphologists of the twentieth century (N.P. Gerasimov, Yu.A. Meshcheryakova, G.I. Khudyakova, V.P. Filosofova, V.E. Khain, V.V. Piotrovsky), who believed that no consideration of large volumetric elements of the Earth's crust can be sufficiently complete without an analysis of their external orographic expression.

The discovery of fractal hierarchies of dissipative structures of the Earth is a consequence of the objective manifestation of the fundamental wave properties of matter in processes of spontaneous structuring of density-unstable masses, associated both with internal sources of energy and matter, as well as with external heat sinks. Fractal hierarchies can be considered as a previously undiscovered yet objectively existing stable connection between otherwise irregular and chaotic phenomena in the material world, thus introducing a radical change in the level of understanding, that is, in terms of the discovery of a new underlying pattern in the structure and development of the Earth. The essence of the breakthrough is encompassed in the colossal global structures of the Earth's gravitational instability of G-hexagon type, having immersion in the centre and emersion in the marginal parts of the deep mantle substrate in the eastern hemisphere, and the L-hexagon type, having emersion in the centre and immersion in the marginal parts of the deep mantle substrate in the western hemisphere. The indicated global structures reflect the predominant heterogeneity of the Earth—a liquid core, having a radius comprising almost half of the planet, overlapped with a crystalline mantle, having unusual geometric and dynamic properties, i.e. a fractal (or fractional) dimension. On the one hand, due to their consisting in interconnected and interpenetrating formations, in which their constituent parts appear to be common to both structures, such that any given point within the structure can be referred to as the limiting one (that is, in any neighbourhood there are points of the set), it is not generally possible to draw the boundary of the global structure of G-hexagon or L-hexagon types in the form of a dividing line. On the other hand, like any wave process, these global structures repeat themselves at different scales. In this way, they determine the predominant confinement of multi-ordinal and heterochronous gravitationally unstable structures of the L-hexagon type to the continents of the western hemisphere, and structures of the G-hexagon type to the Eurasian continents of the eastern hemisphere. The oscillation frequency of these standing waves (or structures) cannot exceed the buoyancy frequency of the geological medium; in this sense, it is comparable with the duration of geological periods in the Earth's evolution. Here, from the perspectives of wave theory, planes between cellular standing waves and dynamic structures unstable in terms of density ratio—or, more precisely, three-dimensional morphostructures—are effaced.

Since the fractal dimension inherent in dissipative wave structures is characteristic of irregular chaotic regimes, objective difficulties are inevitably faced in terms of recognising these structures as an independent and distinct group of geological objects. This is explained by the fact that, despite the inheritance and inversion dynamics of their development, we often observe the uncoordinated location of dissipative structures characterising the new wave dynamics of the density instability of the interior of the Earth with respect to mappable geobodies and ancient tectonic deformation zones that are reflective of the general integral time picture, portrayed in terms of historical categories.

Paleogeodynamic reconstructions have also discovered that periodic "crises" and "geodynamic catastrophes" associated with the emergence of new generations of dissipative structures (or standing waves) against the background of smooth evolutionary structural and temperature changes over hundreds of millions of years have been observed in geological history. The periodicity of such "thermogravitational crises" covers different time intervals, from 850 million years in the Early and Late Proterozoic to 62 million years in the Mesozoic. The model of probabilistic selection of the processes of structuration as a result of the selection of the dominant wavelengths, commensurate to the thickness of the overlying nonequilibrium cooled, gravitationally unstable interfaces under conditions of decreasing heat flux, leads to the generation of dissipative structures of increasingly large sizes at each succeeding historical stage. This fact allows the entire variety of tectonic regimes, successively replacing each other over the course of geological history, to be accounted for on a single physical basis. The geodynamical features of Laurasia and Gondwana are explained in the presence of various types of thermogravitationally-unstable structures. In this sense, the dissipative structures of the Earth, reflecting its thermodynamic evolution, also appear as historical categories. Therefore, it is evident that many of the associated endogenous tectonic regimes are manifested only in the past and in principle cannot be detected in the contemporary state of the lithosphere. The emergence of the generation of dissipative structures of increasingly large dimensions at each subsequent historical stage, as evidenced by paleogeodynamic reconstructions, increases the significance of horizontal tectonic movements of the lithosphere. With an increase in the size of the structures, the tectonic regimes associated therewith, ranging from fine-grained greenstone and granulite, protoaulacogenic and protogeosynclinal to global riftogenic and marginal-plate types (Table 5.1), also change. In this connection, the tectonics of lithospheric plates (or plate tectonics), at least as it is represented nowadays, began to manifest itself only in the late Mesozoic era. Thereby, the discovery of the dissipative structures of the Earth provides a theoretical basis for the elaboration of a wide variety of paleogeodynamic circumstances that determined the tectonic development of the Earth, as well as pointing to the limited applicability and ultimate impossibility of understanding Earth's tectonic history solely from the standpoint of contemporary plate tectonics.

The dissipative structures of the Earth, representing the fractal hierarchy of the various structural types of its density instability (L-hexagons, G-hexagons and Rolls), reflect fundamental laws according to which the wave properties of matter are manifested in the spontaneous structuring processes of density-unstable masses. Thus, it is dissipative (spontaneous) structuring that appears in our proposed interpretation as the primary factor of planetary tectogenesis. Thus, this new theoretical and practical geological paradigm, which reveals spatio-temporal wave-based connections of externally disconnected and mutually remote geological objects and processes, becomes the key to solving a wide range of geological problems. From the standpoint of the possibilities of a wide spectrum of remote-sensing methods for identifying the dissipative structures of the Earth, the most pressing practical task for geology involves the field of prospecting for deep-lying mineral deposits, the use of new principles for hydrogeological and oil and gas zoning of territories and the determination of optimal locations for production wells with a view to significantly increasing water, oil and gas recovery, as well as forecasting potentially dangerous geological processes and phenomena. In this sense, the dissipative structures of the Earth as discrete manifestations of wave processes of the structuring of deep matter at the global, supra-regional, regional and local levels seem to be the most practical basis for the creation of new technologies for three-dimensional geological mapping, metallogenic zoning and forecasting of mineral deposits.

Bibliography

Anderson D. Scale of the mantle convection.Tectonophys. 1998; 284(1–4):1–17.
Artyushkov E.V. Density differentiation of the Earth's matter and the phenomena associated with it. Izv. AN SSSR, Fizika Zemli; 1970:5. (In Russ.).
Atlas of geological and mineralogical maps of the Late Jurassic - Quaternary sedimentary cover in the oceans and continents. Ed. I.S. Gramberg et al. 1992. (In Russ.).
Barenblatt G.N., Zhivago A.V., Neprochnov Yu.P., Ostrovskii A.A. Fractal dimension: quantitative characteristics of the ocean bottom relief. Oceanology. 1984; 24(6):924–928. (In Russ.).
Baryshev A.M. The position of the largest ore regions and nodes in the system of advective structures of the Earth. Otech. geologiya. 2001; 2:6–11. (In Russ.).
Baryshev A.M. Tectonic and tectonophysical conditions for the formation of small intrusions. Rudy i metally. 2001; 6:19–26. (In Russ.).
Belousov B.P. Periodically acting reaction and its mechanisms. The collection of reference papers on radiation medicine for 1958 year. M.; 1959:145–146. (In Russ.).
Belousov V.V. Geotectonics. M.: Izd-vo MSU; 1976. (In Russ.).
Bogolepov K.V. Mesozoic tectonics of Siberia. M.; 1987. (In Russ.).
Botev E.A., Burmakov J.A., Treskov A.V. Crust and uppermantle inhomogeneties beneath the central part of the Balkan region. Earth and Planet. Inter. 1982; 51(1–3):133–152.
Bozhko N.A. The Late Precambrian of Gondwana. M.: Nedra; 1984. (In Russ.).
Bryukhanov V.N., Bush V.A., Glukhovskii M.Z. et. al. Ring structures of the Earth's continents. M.: Nedra; 1987. (In Russ.).
Bush V.A., Yakhimovich N.N., Terekhov V. About the possible origin of ring structures of closed regions. Earth investigations from space. 1983; 2:5–13. (In Russ.).
Busse F.H. Patterns of convection in spherical shells. J. Fluid Mech. 1975; 72(1):67–85.
Carslaw H.S., Jaeger J.C. Operational method in applied mathematics. London: Oxford University Press. 1953.
Chemekov Yu.F., Ganeshin G.S., Solov'ev V.V. et. al. Methodological guide to geomorphological research. L.: Nedra; 1972. (In Russ.).
Chernyshov N.M., Egorkin A.V. et al. The deep structure of the north of the West Siberian plate based on seismic data. Sov. geologiya. 1978; 9:46–58. (In Russ.).
Chumakov N.M. Precambrian tillites and tillitoids (the problems of Precambrian glaciation). M.: Nauka; 1988. (In Russ.).
Davies G.F. Dynamics Earth. Plates, plumes and mantle convection. Cambridge. 1999.
Davies G.F., Richards M.A. Mantle Convection. J. Of Geol. 1992; 100(2):151–206.

O. V. Petrov, *The Earth's Dissipative Structures*, Springer Geophysics,
https://doi.org/10.1007/978-3-319-93614-7

Dobretsov N.L. Mantle superplumes as the cause of the main geological periodicity and global rearrangements. Dokl. RAN. 1997; 357(6):797–800. (In Russ.).

Precambrian continents. Basic features of their tectonics. Novosibirsk: Nauka; 1977. (In Russ.).

Dziewonski A.M., Anderson D.L. Seismic tomography of the Earth's interior. Am. Scientist. 1984; 483–494.

Egorov D.G. Information measures for the analysis of geological self-organising systems. Ed. P.M. Goryainov. RAN. Kol. NTs, Geol. in-t. SPb.: Nauka; 1997:41–51. (In Russ.).

Egorov D.G. Introduction to nonlinear Geology. Apatity; 1997. (In Russ.).

Ez V.V. Formation and destruction of inhomogeneities during the development of metamorphic complexes. Heterogeneity of the tectonosphere and development of the Earth's crust. M.: Nedra; 1986. (In Russ.).

Ez V.V Folding in the Earth's crust. M.: Nedra; 1985. (In Russ.).

Fedorov A.E. Hexagonal grids of linear inhomogeneities of the Earth. M.: Nedra; 1991. (In Russ.).

Filosofov V.P. Essays on the general geomorphology of plains. Scientific annual collection for 1954. Saratov: Sarat. SU; 1955:409–411. (In Russ.).

Filosofov V.P. Fundamentals of the morphometric method of searching for tectonic structures. Saratov: Sarat. SU; 1975. (In Russ.).

Fukuyama E., Madariaga R. Integralequation method for plane crack with arbitrary shape in 3D elastic medium. Bull. Seismol. Soc. Amer. 1995; 85(2):614–628.

Fukuyama E., Mikumo T. Dynamic rupture analysisinversion for the source process of the 1990 Izu-Oshima, Japan, earthquake (m = 6.5). J. Geophys. Res.Solid earth. 1993; 98(4):6529–6542.

Geodynamic reconstructions (Guidance manual for the regional geological research). Ed. V.A. Unksov L.: LO Nedra; 1989. (In Russ.).

Geology of the World Ocean bottom: Atlantic. Biostratigraphy and tectonics (Results of drilling of the 38th voyage "Glomar Challenger"). Collection of articles of AN SSSR. Ed. G.B. Udintsev. M.: Nauka; 1982. (In Russ.).

Goncharov M.A. The western and northern components of the continent drift as a result of forced convection in the mantle according to the Ampere's rule. Tektonika i geofizika litosfery. M.: GEOS. 2002; 1:128–131. (In Russ.).

Goncharov M.A., Tapitskii V.G., Frolova N.S. Introduction to tectonophysics. M.: Knizhnyi dom Universitet; 2005. (In Russ.).

Grachev A.F. Map of the neotectonics of Northern Eurasia. Razvedka i okhrana nedr. 1996; 10:2–7. (In Russ.).

Grachev A.F. The main issues of the neotectonics and geodynamics of Northern Eurasia. Izvestiya, Physics of the Solid Earth. 1996; 12:5–36. (In Russ.).

Gramberg I.S., Kos'ko M.K., Lazurkin V.M., Pogrebitskii Yu.E. The main stages and boundaries of development of the Arctic continental margin of the USSR in the Neogene. Sov. geologiya. 1984; 7:32–41. (In Russ.).

Gramberg I.S., Ronkina Z.Z. Late Cenozoic - Mesozoic stage in history paleogeographic development of the Soviet Arctic. The Mesozoic Soviet Arctics. Novosibirsk; 1983:12–19. (In Russ.).

Gravity prospecting. Reference book of geophysics. M.: Nedra; 1990. (In Russ.).

Haken H. Synergetics. Hierarchies of instabilities in self-organising systems and devices. M.: Mir; 1985. (In Russ.).

Hart J.E., Glatzmaier G.A., Toomre J. Simulations of thermal convection in a rotating shell with radial gravity. J. Fluid Mech. 1986; 173:519–544.

Heezen B.C., Fornari D.J. Pacific Ocean. Atlas geologique du monde - Geological world atlas 1/10 000 000. Coordonnateurs generaux, G. Choubert et A. Faure Muret; Travaux cartographiques, P. Chanteux. Paris: Unesco. 1976–1983; Sheet 20; 54 × 55 cm.

Holmes A., Cahen L. African geochronology - results available to I September I954. Colon. Geol. Min. Res. 1955; 5(1):3–39.

Initial reports of the Deep Sea Drilling Project: A project planned by and carried out with the advice of the Joint Oceanographic Institutions for Deep Earth Sampling. Scripps Institution of Oceanography. [Washington: Nat. Science Foundation]. 1969–1983; 96 vols.

Ivanov M.K. Tectonic-magmatic and prognostic classification of the fields of nickel-bearing Siberian traps. Petrologic features and predictive zoning of nickel-bearing trap fields in the north of the Krasnoyarsk Territory. Proceedings of the All-Union Scientific Research Institute of Oceanology. L.: LO Nedra; 1983:191. (In Russ.).

Kanla W.M. Global gravity and tectonics. The Nature of Solid Earth. N.Y. 1972; 385–405.

Kats Ya.G., Poletaev A.I., Sulidi-Kondrat'ev E.D. Ring structures of the Earth: myth or reality. Ed. V.E. Khain. M.: Nauka; 1989. (In Russ.).

Keodzhan V.P. Model of chemical-density convection in the Earth's mantle. Izv. AN SSSR. Fizika Zemli. 1980; 8:3–21. (In Russ.).

Khain V.E. On the origin of ancient platforms (cratons). Moscow University Geology Bulletin. 1984;2:32–37. (In Russ.).

Khain V.E. Regional geotectonics. Oceans. Synthesis. M.: Nedra; 1985. (In Russ.).

Khain V.E., Koronovskii N.V. Planet Earth. From the core to the ionosphere. Tutorial. M.: KDU; 2007. (In Russ.).

Khain V.E., Seslavinskii K.B. Global changes in the endogenous activity of the Earth in the Paleozoic. Moscow University Geology Bulletin. 1990; 6:3–26. (In Russ.).

Khain V.E., Seslavinskii K.B. . Historical geotectonics. Paleozoic. M.: Nedra; 1991. (In Russ.).

Khudyakov G.I. Geomorphological component in geomorphostructural studies. Morphostructures of the Far East. Vladivostok: DVNTs AN SSSR; 1982:5–13. (In Russ.).

Kirsanov A.A., Lipiyainen K.L., Kiseleva E.A., Kunina E.L. Landsat 7 multispectral data as a contemporary multi-purpose remote basis for geological maps and geological exploration. Regional Geology and Metallogeny. 2005; 24:60–66. (In Russ.).

Computers and nonlinear phenomena. Informatics and modern natural science. M.: Nauka; 1988. (In Russ.).

Koronovskii N.V. Geodynamic environments of the late Cenozoic volcanism of the Aegean, Anatolian and Caucasus regions: the central part of the Alpine fold belt. Moscow University Geology Bulletin. 1994; 1. (In Russ.).

Kosygin Yu.A. Salt tectonics of platform areas. M., L.: Gostekhizdat; 1950. (In Russ.).

Krasnyi L.I. The problem of tectonic systematics - a new look. Otech. geologiya. 1992; 2:13–26. (In Russ.).

Krasnyi L.I., Gramberg I.S., Petrov O.V., Blyuman B.A. et. al. Experience of harmonisation of ideas about geological, mineralogical and mineral-resource components of the Earth's lithosphere. SPb.: Izd-vo VSEGEI; 2001. (In Russ.).

Large and super large deposits: patterns of location and conditions of formation. Ed. D.V. Rundkvist M.: IGEM RAN; 2004. (In Russ.).

Kulikov G.V., Vartyanyan G.S. et. al. 27th International Geological Congress. Hydrogeology. Section No. 16. Reports. M. 1984; 16. (In Russ.).

La Fond E.C. Internal wave motion and its geological significance. Mahadevan volume. A collection of geological papers in commemoration of the sixtyfirst birthday of Prof. C. Mahadevan. Hyderabad (India). 1961; 61–77.

Lavrova L.D., Pechnikov V.A., Pleshakov A.M. et.al. New genetic type of diamond deposits. M.: Nauchnyi mir; 1999. (In Russ.).

Lebedeva N.B. About the formation mechanism of clay diapirs. Sov. geologiya. 1958; 11:76–87. (In Russ.).

Leonov Yu.G. Continental rifting: contemporary ideas, problems and solutions. Fundamental problems of general tectonics. M.: Nauchnyi mir. 2001:155–173. (In Russ.).

Lisitsyn A.P. Flows of matter and energy in the outer and inner spheres of the Earth. Global changes of the natural evironment. M.: SO RAN, GEO affiliation. 2001:163–248. (In Russ.).

Liu Sh., Dixon J.M. Centrifuge modelling of thrust faulting; structural variation along strike in foldthrust belts. Tectonophysics. 1991; 188(1–2):39–62.

Lobkovskii L.I., Nikishin A.M., Khain V.E. Contemporary issues of geotectonics and geodynamics. M.: Nauchnyi mir; 2004. (In Russ.).

Malich K.N., Petrov O.V., Badanina I.Yu., Presnyakov S.L. Zircons from ultramaficmafic intrusions at Noril'sk area (Russia): a compositional and U—Pb study. Geochim. et Cosmochim. Acta. 2007; 71(15):616.

Mandelbrot B.B. The fractal geometry of nature. San Francisco: W.H.Freman; 1982.

Milanovskii E.E. USSR gelolgy in three parts. M.: Izd-vo MGU; 1987. (In Russ.).

Milanovskii E.E. Rifting in the Earth's history: rifting in mobile belts. M.: Nedra; 1987. (In Russ.).

Monin A.S. Sorokhtin O.G. Thermal evolution of the Earth with a volumetric mechanism differentiation of its subsoil. DAN SSSR. 1982; 266(1):63–67. (In Russ.).

Morphostructural studies: theory and practice. DVNTs AN SSSR. M.: Nauka; 1985. (In Russ.).

Mossakovskii A.A., Pushcharovskii Yu.M., Ruzhentsev S.V. Tectonic segments of the Earth and features of their development in the Neogaea. Dokl. RAN. 1992; 324(5):1069–1075. (In Russ.).

Nikolis G., Prigogine I. Self-organisation in nonequilibrium systems. From dissipative structures to ordering through fluctuations. M.: Mir; 1979. (In Russ.).

Neotectonics of Northern Eurasia. Explanatory note to the map of the neotectonics of Northern Eurasia. Scale 1: 5 000 000. Ed. A.F. Grachev. M.: GEOS; 1998. (In Russ.).

Oganezov A.V. The method for determining the spatial coordinates of geological formations. Patent RF: RU 2097794 C1. 27.11.97. (In Russ.).

Olsow P., Silver P.G., Carlson R.W. The largescale structure of convection in the Earth's mantle. Nature. 1990; 344:209–215.

Pavlovskii E.V. The problem of the "greenstone belts" of the Early Precambrian. Izv. AN SSSR. Ser. geol. 1979; 4:5–18. (In Russ.).

Petrov O.V., Abramovich I.I. Mantle plumes and mineralization. Materials of the All-Russian Conference, dedicated to the 90th anniversary of Academician N.A. Shilo (12th Annual Meeting of the North-Eastern Branch of VMO) "Geodynamics, magmatism and mineralogy of the continental margins of the North Pacific". 2003; 3:24–25. (In Russ.).

Petrov O.V., Daragan-Sushcheva L.A., Sobolev N.N., Daragan-Sushchev Yu.I. Construction of the pre-Jurassic basement of the northern part of the West Siberian plate. Regional geology and metallogeny. 2005; 26:153–168. (In Russ.).

Petrov O.V., Kostyuchenko S.L. Deep sedimentary depressions of Siberia as a consequence of the impact of small mantle plumes on the Earth's lithosphere. Regional geology and metallogeny. 2002; 15:58–74. (In Russ.).

Petrov O.V., Kostyuchenko S.L. Plume tectonics in the evolution of the sedimentary depressions of the East European platform. Regional geology and metallogeny. 2002; 16:5–22. (In Russ.).

Petrov O.V., Krasnyi L.I., Gramberg I.S., Belonin M.D. et al. Geological and mineralogical features of the Earth. Regional geology and metallogeny. 2000; 11:95–105. (In Russ.).

Petrov O.V., Movchan I.B. Dissipative structuring of the Earth's crust and mantle as a reflection of wave processes. Regional geology and metallogeny. 2002; 17:53–65. (In Russ.).

Pijush K. Kundu Fluid Mechanics. London: Academic Press; 1990.

Piper J.D.A. Palaeomagnetic evidence for a Proterozoic supercontinent. Phil trans. Roy. Soc. Lond. 1976; 280:469–490.

Prigogine I.R., Stengers I. Order out of chaos: Man's new dialogue with nature. Ed. V.I. Arshinov, Yu.L. Klimontovich and Yu.V. Sachkov. M.: Progress; 1986. (In Russ.).

Problems of global correlation of geological phenomena. Ed. A.V. Peive, Yu.G. Leonov. M.: Nauka; 1980. (In Russ.).

Problems of tectonic movements and the newest structures of the Earth's crust. Ed. N.I. Nikolaev. M.: Nauka; 1968. (In Russ.).

Pushcharovskii Yu.M., Peive A.A. Real inhomogeneities in the oceanic lithosphere and geodynamic consequences. Geotectonics. 1993; 4:15–26. (In Russ.).

Ramberg H. Ramberg, X. Modelling deformations of the Earth's crust with the use of a centrifuge. Ed. V.V. Belousov. M.: Mir; 1970. (In Russ.).

Faults and horizontal movements of the platform areas of the USSR. M.: Nauka; 1977. (In Russ.).
The results of deep-sea drilling in the Atlantic Ocean in the 38th voyage of the "Glomar Challenger": Lithology and petrography: collection of articles. Ed. G.B. Udintsev. Institute of Oceanology of RAS named after P.P. Shershov. M.: Nauka; 1979. (In Russ.).
Romanenko P.N. Heat and mass transfer and friction in the gradient flow of liquids M.; 1971. (In Russ.).
Sadovskii M.A., Nersesov I.L., Pisarenko V.F Hierarchical discrete structure of the lithosphere and seismic process Modern tectonic activity of the Earth and seismicity. M.: Nauka; 1987:182–191. (In Russ.).
Schuster H.G. Deterministic Chaos. M.: Mir; 1988. (In Russ.).
Semenov V.S., Solov'ev I.A., Platunova A.P., Turchenko S.I. Features of the evolution of the Earth's crust of the supercontinent Gondwana in the Early Precambrian. M.; 1987. (In Russ.).
Semikhatov M.A. The chronometric scale of the Archean (proposed by the International Subcommission on Precambrian Stratigraphy). Izv. RAN. Ser. geol. 1992;9:153–157. (In Russ.).
Seyfert C.K. Convection and motion of plates. Structural geology and plate tectonics. T. 1. Mir; 1990. (In Russ.).
Shcheglov A.D. Fundamentals of metallogenic analysis. 2nd edition. M.: Nedra; 1980. (In Russ.).
Shtengelov E.S. Short-period Earth pulsation and geological cataclysm in 1999–2002. Odessa: Astroprint; 1999. (In Russ.).
Shtengelov E.S., Shtengelov A.R. Forecast of global activation of geological and geophysical processes in 1997–2003. Dokl. RAN. 1999; 364(2):231–234. (In Russ.).
Shults S.S. Geological studies of continental deposits of part of the Iliiskaya valley. Izv. GGRU. 1931;50(46). (In Russ.).
Shults S.S. Report on the work of the Jarkent Geological Party in the summer of 1930 in the Dzungarian Alatau. L.: Geolgiz; 1932. (In Russ.).
Shults S.S. Tectonics of the Earth's crust (based on the analysis of the latest movements). L.: LO Nedra; 1979. (In Russ.).
Transparent ore-concentrating structures. Collection of articles. Ed. M.A. Favorskaya and I.N. Tomson. M.: Nauka; 1989:52–58. (In Russ.).
Solov'ev V.V. Rational complex of geomorphological research methods for forecasting mineral deposits. L.: LO Nedra; 1976. (In Russ.).
Solov'ev V.V., Ganeshin G.S. The development of the relief and formation of Quaternary deposits of Sakhalin. M.: Nedra; 1971. (In Russ.).
Solov'ev V.V., Ryzhkova V.M. Morphostructural map of the central type of the USSR territory. Scale 1:10 000 000. L.: VSEGEI; 1978. (In Russ.).
Sorokhtin O.G. Global evolution of the Earth. M.: Nauka; 1974. (In Russ.).
Sorokhtin O.G., Ushakov S.A. The nature of tectonic activity of the Earth. Physics of the Earth. V. 12. M.; 1993. (In Russ.).
Stille H. Selected works. M.: Mir; 1964. (In Russ.).
Stokes G.G. On the theories of the internal friction of fluids in motion, and of the equilibrium and motion of elastic solids. Philos. Mag. 1846; 29:60–62.
Strange attractors. Collection of articles. Ed. A.N. Kolmogorov, S.P. Novikov. M.: Mir; 1981 (In Russ.).
Structural geology and plate tectonics. Ed. C.K. Seyfert. M.: Mir; 1990. (In Russ.).
Su W.J., Dziewonski A.M. Shear velocity throughout the mantle. Proposal "A science facility for studying the dynamics of the solid Earth". Harvard univ.; 1995.
Suvorov A.I. Horizontal movements and the formation of the continental crust of the Earth. Faults and horizontal movements of the USSR platform areas. M.: Nauka; 1977:9–19. (In Russ.).
The connection between magmatism and endogenous minerageny with block tectonics. Ed. M.A. Favorskaya, I.N. Tomson. M.: Nedra; 1969. (In Russ.).
Talbot C.J. A plate tectonic model for the Archaean crust. Phil. Trans. Roy. Soc. London A. 1973; 273:413-428.

Tectonics of continents and oceans. Explanatory note to the international tectonic map of the World. Scale 1:15 000 000. M.: Nauka; 1988. (In Russ.).

Turing A.M. The chemical basis of morphogenesis. Philos. Trans. Royal Soc. London. Ser. B. Biological Sci. 1952; 237(641):37-72.

Tveritinov Yu.I., Tveritinova T.Yu., Brant S.B. et al. Forecasting gold mineralisation in the south of Eastern Siberia and the Far East: geological and isotope-geochemical aspects. Irkutsk: In-t zemnoi kory SO RAN; 2006. (In Russ.).

Vikhert A.V. About the folded structure emerging from the decomposition of layered strata (according to experimental data). Geotectonics. 1980; 3:27–34. (In Russ.).

Vyayuryunen Kh. Crystalline basement of Finland. M.: IL; 1959. (In Russ.).

Zonenshain L.P., Kuz'min M.I. Paleogeodynamics. M.: Nauka; 1992. (In Russ.).

Zonenshain L.P., Kuz'min M.I., Natapov L.M. Tectonics of lithospheric plates of the USSR territory. Book 1. M.: Nedra; 1990. (In Russ.).

Zverev A.T. A multi-epoch model of the lithospheric dynamics. DAN SSSR. 1989; 307( ):168–170. (In Russ.).

MIX
Papier aus verantwortungsvollen Quellen
Paper from responsible sources
FSC® C105338

If you have any concerns about our products,
you can contact us on
ProductSafety@springernature.com

In case Publisher is established outside the EU,
the EU authorized representative is:
Springer Nature Customer Service Center GmbH
Europaplatz 3, 69115 Heidelberg, Germany

Printed by Libri Plureos GmbH
in Hamburg, Germany